CRIMINAL JUSTICE ILLUMINATED

Prisons: Today and Tomorrow

Second Edition

Joycelyn M. Pollock, Ph.D., J.D.
Editor
Texas State University-San Marcos

JONES AND BARTLETT PUBLISHERS
Sudbury, Massachusetts
BOSTON TORONTO LONDON SINGAPORE

World Headquarters
Jones and Bartlett Publishers
40 Tall Pine Drive
Sudbury, MA 01776
978-443-5000
info@jbpub.com
www.jbpub.com

Jones and Bartlett Publishers Canada
6339 Ormindale Way
Mississauga, Ontario L5V 1J2
Canada

Jones and Bartlett Publishers International
Barb House, Barb Mews
London W6 7PA
United Kingdom

Jones and Bartlett's books and products are available through most bookstores and online booksellers. To contact Jones and Bartlett Publishers directly, call 800-832-0034, fax 978-443-8000, or visit our website www.jbpub.com.

Substantial discounts on bulk quantities of Jones and Bartlett's publications are available to corporations, professional associations, and other qualified organizations. For details and specific discount information, contact the special sales department at Jones and Bartlett via the above contact information or send an email to specialsales@jbpub.com.

Production Credits
Chief Executive Officer: Clayton E. Jones
Chief Operating Officer: Donald W. Jones, Jr.
President, Higher Education and Professional Publishing: Robert W. Holland, Jr.
V.P., Sales and Marketing: William J. Kane
V.P., Production and Design: Anne Spencer
V.P., Manufacturing and Inventory Control: Therese Connell
Publisher, Public Safety Group: Kimberly Brophy
Acquisitions Editor: Stefanie Boucher
Associate Editor: Janet Morris
Production Editor: Jenny L. McIsaac
Director of Marketing: Alisha Weisman
Interior Design: Anne Spencer
Cover Design: Kristin E. Ohlin
Composition: Auburn Associates, Inc.
Cover Images: photo © Corbis/PictureQuest; column © Ron Chapple/Thinkstock/Alamy Images
Chapter Opener Image: © Masterfile
Text Printing and Binding: Malloy, Inc.
Cover Printing: Malloy, Inc.

ISBN-13: 978-0-7637-2904-2
ISBN-10: 0-7637-2904-3

Library of Congress Cataloging-in-Publication Data

Prisons : today and tomorrow / [edited by] Joycelyn M. Pollock.— 2nd ed.
p. cm.
Includes bibliographical references and index.
ISBN 0-7637-2904-3 (hardcover)
1. Prisons—United States. 2. Corrections—United States. 3. Imprisonment—United States. I. Pollock, Joycelyn M., 1956–
HV9471.P73 2006
365—dc22

2005024820

6048

Printed in the United States of America
10 09 08 07 06 10 9 8 7 6 5 4 3 2

Dedication

To Gregory and Eric, as always.

Contents

Acknowledgments

I would like to thank the contributors of this second edition for sharing their perspectives of the prison world. I would especially like to thank Robert Johnson who so graciously volunteered his poems to open the sections. Dr. Johnson is gaining a well-deserved reputation as an accomplished poet in addition to his already established reputation as scholar in the criminal justice field. His witty and insightful poetry cuts straight to the heart of the prison experience. I heartily recommend his book of poetry: *Poetic Justice* (2004) and a book of short stories he published with Victor Hassine and Ania Dobrzanska: *The Crying Wall* (2005). The graduate students who assisted me in checking sources were Jessica Contreras and Brooke Miller and I owe them my gratitude. The editors and staff at Jones and Bartlett have been dedicated and gracious and I thank them for their input and assistance in developing this second edition and bringing it to a completed form—special thanks to Stefanie Boucher and Jenny McIsaac. Their efforts have improved the finished product. Finally, as always, I thank my family—Gregory and Eric, for putting up with me.

Desperately Seeking Freedom

O'er the cell a mark still lingers
Of where a convict's bloodied fingers
Could make stone speak of life's hard ends
With words that shine like darkling gems
I was here
I am a man
I bleed, therefore I am . . .
Alive, in a manner of speaking
It's raw, sweet freedom I'm desperately seeking
A prison cell's a coffin reeking
Of dreams gone sour
Of life died by the hour
Of death by decree
Until you're set free
In this life or the next

Robert Johnson

Preface

Crime and the criminal justice system permeate every political race, every evening newscast, and every newspaper. Even though crime rates have been falling, "what to do about crime" is a stock element in politicians' platforms—from mayoral races to presidential elections. The answer to crime always seems to be "more"—more punishment, more police, and, of course, more prisons.

This text's underlying theme is that prisons do very little to solve the crime problems and that they sometimes do a great deal of damage to those who end up serving time. Prisons have many purposes: punishment, deterrence, rehabilitation, and incapacitation. It will become clear by the end of this book that the only goal that prisons accomplish quite well is incapacitation. Prisons now often serve as "human warehouses" where individuals are kept in captivity for short or long periods of time and then let loose again to pick up where they left off. While we incarcerate many more people that we have in the past, prison time isn't usually as painful. Physical punishment and severe conditions are no longer present, except in isolated situations, as prisons have become less oppressive and inhumane due to the recognition of legal rights of prisoners. Of course, the deprivations suffered in prison (of family, autonomy, and freedom) are painful, and, for some inmates, prison causes severe emotional trauma, even mental breakdowns. But for others, prison actually may be more comfortable and safer than living on the outside. It may provide a "home away from home" or just another place to "do business." For these types of inmates—some of whom consider prison a rite of passage—prison is hardly a deterrent. In fact, prison sentences have become as common as poverty for large segments of our society.

As to rehabilitation, we hardly even pay lip service to this goal anymore. There are attempts to provide education and vocational training, but no official rhetoric places much emphasis on the reformative goals of prison. Recidivism is high and hopes are low that anything we do "to" an offender will change his or her behavior once released. Actually, the only answer that seems to be offered by most politicians, policy makers, and the public at large is "more time." If a prisoner recidivates after a short sentence, let's give them a longer sentence; thus we have habitual sentence laws ("three strikes" laws) that have created the scenario where 24-year-old men are facing life sentences without parole. Whether we've actually thought through the effects of putting young men in prison for the next 60 years is another matter. Obviously some individuals are dangerous and need to be removed from society, and for those who break the law, there must be some consequence. The disagreement comes in answering the questions of who should be locked up and for how long. The chapter authors of this text do not necessarily agree with each other on these questions, but it becomes clear that at least on one thing there is agreement: Prisons have been a negative force in the lives of most of those who are touched by them, and if essential, they are essential evils in the society of humans.

In this text, the first three chapters introduce us to prisons. We look at the ideology that formed the rationale for their creation in Chapter 1, and explore the history of prisons in Chapter 2. We see cycles of great optimism and then neglect, corruption and overcrowding; and these cycles seem to be with us still today. Present-day events are echoes of the past. In Chapter 3, the sentencing patterns that led to current overcrowding problems are examined with information also presented on the extent and effects of prison overcrowding. In the next section of the book, we explore features of inmate life. Chapter 4 looks at the prison subculture and the theories that attempt to explain its origin and existence. Chapters 5 and 6 examine some of the formal activities that prisoners are either offered or forced into by prison staff. In these chapters, special attention has been given to the experiences of minorities and women. In the next section of the book, the focus is shifted to formal authorities. Correctional officers and managers are examined in Chapter 7. The legal holdings regarding the relative rights and responsibilities of correctional authorities and inmates are examined in Chapter 8. The final section of the book turns to release and community corrections. Jails are a type of community facility and Chapter 9 re-examines all the major topics explored in the text with a special focus on jails. Finally, the concept and reality of release and reintegration is explored in the last chapter.

The book may be used as a supplementary text in a Corrections class and presents more detail than a survey text's chapters on prisons. Many curriculums now have upper division courses in penology or prison, for which this text would be ideal. We also suggest using smaller supplemental texts—many of which can be found in each chapter's source list—for highlighting specific elements of the prison experience. Finally, no course on prisons is complete without the daily newspaper. The instructor must provide what we cannot—up-to-date, current events in the reader's own state. During any semester, one may read news announcing the opening of new prisons, erupting scandals involving corruption, escapes, private prisons contracting with the state to open new facilities, new programs, or exposed brutality or sexual harassment by correctional officers. More recently news items have been appearing announcing that a state is scaling down plans for new prisons, or passing legislation that reduces mandatory minimums or increases the use of parole in an effort to reduce correctional budgets that now are described as bloated. And so the cycle turns.

Joycelyn Pollock, October 2005

Contributors

Biographical Sketches

Kelly Cheeseman is a doctoral candidate at Sam Houston State University with an anticipated graduation date of May 2006. Ms. Cheeseman received a Bachelor of Science in Criminal Justice and Psychology from Youngstown State University in 1997 and a Master of Arts in Criminology and Criminal Justice from Sam Houston State University in 1999. She has published articles in revered journals such as *Deviant Behavior*, *Corrections Management Quarterly* and *Criminal Law Bulletin*. Current research interests include institutional corrections, deviance, and women and crime. She is currently employed as the Unit Culture Profile Coordinator for the Texas Department of Criminal Justice—Correctional Institutions Division.

Ania Dobrzanska is a research associate at The Moss Group in Washington, D.C. Formerly with the American Correctional Association, she holds a B.A. in Psychology and Administration of Justice from Rutgers University and an M.S. in Justice, Law & Society at American University. Ms. Dobrzanska's current interests include the problem of rape in prisons, leadership in corrections, and the design of correctional intervention programs. She has published several articles and is co-editor and contributor of *The Crying Wall and Other Prisons Stories: Fiction True to Life*.

Dennis Giever is Professor and Chair of the Department of Criminology at Indiana University of Pennsylvania (IUP). He holds a Ph.D. from Indiana University of Pennsylvania in Criminology and a master's degree in Criminal Justice with a graduate minor in Experimental Statistics from New Mexico State University. Dr. Giever has worked on a number of large research projects including two National Institute of Justice grants: Evaluating a Metropolitan Area—Driving While Intoxicated (DWI) Drug Court and a National Evaluation of the Gang Resistance Education and Training (G.R.E.A.T.) program. He was also the principle investigator on a $2 million grant funded by the Department of Justice to develop a Security Technology Program. In addition, he has published 19 research articles or book chapters in diverse areas such as juvenile transfers, jails, fear of crime, parental management and self-control, and police pursuits.

Robert Johnson is a Professor of Justice, Law and Society at American University in Washington, D.C. He is the author or editor of eight books, including *Hard Time: Understanding and Reforming the Prison* and *Death Work: A Study of the Modern Execution Process*, as well as over 30 articles published in journals and an-

thologies. Dr. Johnson is a Distinguished Alumnus of the Nelson A. Rockefeller College of Public Affairs and Policy, University at Albany, State University of New York.

John McLaren is an Associate Professor of Criminal Justice at Texas State University at San Marcos and has been associated with the university for 24 years. His primary teaching interests are in Criminal Procedure and the law of Civil Rights, especially as it pertains to persons in custody of, or under supervision of, criminal justice agencies. He is an attorney licensed to practice in both state and federal court. He has taught a wide range of courses, graduate and undergraduate, and served three years (1998-2001) as Acting Chair of the Department of Criminal Justice. He holds degrees from Texas Tech University and the University of Texas School of Law. He has presented and published numerous works including the *Texas Intermediate Sanctions Bench Manual* (2003) a manual for the judiciary concerning community corrections sanctioning options and resources throughout Texas.

Alida V. Merlo is Professor of Criminology at Indiana University of Pennsylvania. She holds a Ph.D. from Fordham University. She has conducted research and published in the areas of corrections, juvenile justice, and women and the law. She is the Past-President of the Academy of Criminal Justice Sciences (1999–2000), and a co-recipient (with Peter J. Benekos, William J. Cook, and Kate Bagley) of the Academy of Criminal Justice Sciences, Donal MacNamara Outstanding Publication Award for 2002. Dr. Merlo was also the recipient of the Academy of Criminal Justice Sciences Fellow Award in 2004 and the Founder's Award in 1997. Her published works include *Women, Law, and Social Control*, 2nd Edition, (with Joycelyn Pollock), *Controversies in Juvenile Justice and Delinquency* (with Peter J. Benekos), *What's Wrong with the Criminal Justice System: Ideology, Media and Politics* (with Peter J. Benekos), and *Dilemmas and Directions in Corrections* (with Peter J. Benekos).

Seri Palla is presently a Ph.D. candidate in Justice, Law, and Society at American University. She holds a BS in Psychology from the University of Michigan, Ann Arbor, and an MS in Justice, Law and Society from American University. Ms. Palla is presently interning as a Statistician for the U.S. Department of Justice, Bureau of Justice Statistics. Her primary research interest is the analysis of prisoner adjustment.

Joycelyn M. Pollock is a Professor of Criminal Justice at Texas State University-San Marcos. She holds a Ph.D. from State University of New York at Albany and a J.D. from the University of Houston. She has published numerous books, including *Women, Law and Social Control*, 2nd Edition (co-editor with Alida Merlo), *Dilemmas and Decisions: Ethics in Crime and Justice*, 4th ed., *Women, Prison and Crime*, 2nd Edition, *Prisons and Prison Life: Costs and Consequences*, *Counseling Women in Prison*, *Sex and Supervision*, *Morality Stories* (co-author with Michael Braswell and Scott Braswell), and *Exploring Corrections in America* (co-editor with John Whitehead and Michael Braswell). Her published chapters and articles are in the area of ethics, female criminality and corrections.

David Spencer is presently a Lecturer in the Criminal Justice Department at Texas State University-San Marcos. He holds a J.D. from the University of Texas School of Law (1972) and a Ph.D in Educational Psychology from the University of Texas at Austin (2000). He was previously a member of the Criminal Justice faculty from 1991 to 1998. From 1998 to 2000 he was a prosecutor with the International Criminal Tribunal for Rwanda, operating both in Kigali, Rwanda and Arusha, Tanzania. He has 32 years of experience as a lawyer, during which time he has been a prosecutor, a defense lawyer and a judge.

William Stone is a Professor of Criminal Justice at Texas State University. He has over 25 years of teaching experience and teaches a broad range of courses in corrections and research methods. In addition to his faculty experience, Dr. Stone has worked professionally in both corrections and law enforcement. During his tenure at the Texas Department of Corrections, he authored numerous research reports and a detailed history of the Texas correctional system.

Chad R. Trulson is an Assistant Professor in the Department of Criminal Justice at the University of North Texas. He has published in numerous professional journals and recently co-authored *Juvenile Justice: System, Process, and Law* (Wadsworth) with Rolando v. Del Carmen. His current research interests involve recidivism among institutionalized delinquents, and racial desegregation and violence in prisons.

I

The Philosophy and History of Prisons

Poetic Justice

Build prisons
Not day-care
Lock 'em up
What do we care?

Hire cops, not counselors
Staff courts, not clinics
Wage warfare
Not welfare

Invest in felons
Ripen 'em like melons
Eat 'em raw, then
Ask for more

More poverty
More crime

More men in prison
More fear in the street

More ex-cons among us
Poetic justice

Robert Johnson

The Rationale for Imprisonment

1

Joycelyn M. Pollock
Texas State University–San Marcos

Chapter Objectives

- Understand the definition of punishment.
- Be able to articulate the retributive and utilitarian rationales for punishment.
- Understand the social contract and how it supports the right of society to punish.
- Distinguish between incapacitation and punishment.
- Understand the restorative justice philosophy.

Why do we punish? Why do we use prison instead of other types of punishment? In this chapter, we are concerned with the fundamental rationale for the existence of prisons. What do we want them to be? Punishment is a natural response to fear and injury, and prison seems to be our favorite punishment.

Philosophy of Punishment

Most people would agree that hurting someone or subjecting them to pain is wrong. However, **punishment**, by definition, involves the infliction of pain. Does this make punishment wrong? Philosophers are divided on this issue. One group believes that inflicting pain as punishment is fundamentally different from inflicting pain on innocents, and therefore is not inherently wrong. Another group believes that punishment is a wrong that can be justified only if it results in a "greater good" (Murphy 1995).

punishment a pain or unpleasant experience inflicted upon an individual in response to a violation of a rule or law by a person or persons who have lawful authority to do so.

Those who hold the first view do not feel it necessary to justify punishment beyond the fact that the individual deserves it. This would be considered a *retributive* approach. The second view justifies punishment through the secondary rationales of deterrence, incapacitation, or rehabilitation. This will be called the *utilitarian* approach (Durham 1994).

Retributive Rationale

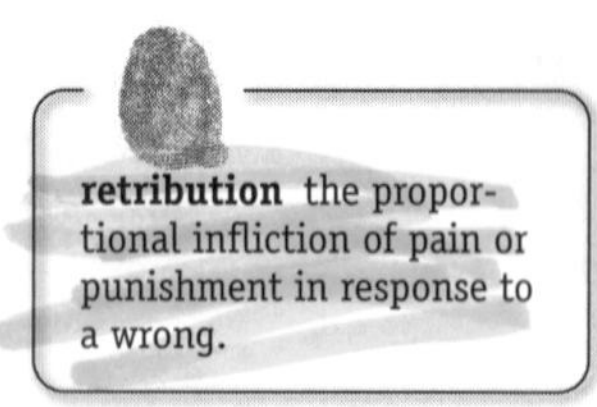

retribution the proportional infliction of pain or punishment in response to a wrong.

The first philosophical approach (or rationale) is that punishment, strictly defined, is not evil. **Retribution** is a term that means balancing a wrong through punishment. While revenge is personal and not necessarily proportional to the victim's injury, retribution is impersonal and balanced. Newman, although recognizing the difficulty of defining punishment, defines it in this way: "Punishment is a pain or other unpleasant consequence that results from an offense against a rule and that is administered by others, who represent legal authority, to the offender who broke the rule" (Newman 1978, 6–7). The supposition is that by strictly limiting what can be done, to whom and by whom, the evilness of the action is negated. There are two equally important elements to this view: first, that society has a right to punish, and second, that the criminal has the right to be punished.

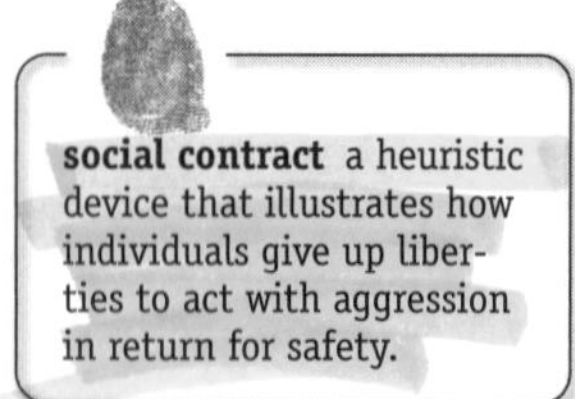

social contract a heuristic device that illustrates how individuals give up liberties to act with aggression in return for safety.

The right of society to punish is said to lie in the **social contract**. Although this idea dates back to the ancient Greeks, it gained its greatest currency during the Age of Enlightenment in the 17th and 18th centuries and is associated with Thomas Hobbes (*Leviathan* 1651), John Locke (*Two Treatises on Government* 1690), and Jean-Jacques Rousseau (*Du contrat social* 1762). Basically, the concept proposes that all people freely and willingly enter into an agreement to form society by giving up a portion of their individual freedom for the return benefit of protection. If one transgresses against the rights of others, one has broken the social contract, and society has the right to punish (Mickunas 1990).

One problematic element to the social-contract theory of punishment is the fiction that everyone willingly plays a part or had a part in the agreement to abide by society's laws. Many authors have suggested that certain groups in society are, in effect, disenfranchised from the legal system and play no part in its creation. To assume that such groups break a "contract" they had no part in creating (nor benefit from) weakens the legitimacy of this theory. If we believe that our political process and even our justice system is operated for the benefit of only certain groups of citizens, we would also believe that the social contract is a weak rationale for punishment.

The second element of the retributive rationale is that the criminal *deserves* the punishment and, indeed, has a *right* to be punished. Only by forcing the individual to suffer the consequences of his actions does one accord them the rights of an equal citizen. Herbert Morris explains this view:

> *[F]irst, . . . we have a right to punishment; second . . . this right derives from a fundamental human right to be treated as a person; third . . . this fundamental right is a natural, inalienable, and absolute right; and, fourth, . . . the denial of this right implies the denial of all moral rights and duties (Morris, in Murphy 1995, 75).*

To do anything other than to punish is to treat the person as less than equal, perhaps even less than human. Under this view, correctional treatment is infinitely more intrusive than punishment because it doesn't respect the individual's ability and right to make choices. It regards their behavior as "controlled" by factors that can be influenced by the intervention (Morris, in Murphy 1995, 83).

It is a primitive, almost instinctual, response of humankind to punish wrongdoers, as noted by French sociologist Émile Durkheim and cited in Durham (1994, 22). Punishment is believed to be an essential feature of civilization. The state takes over the act of revenge and elevates it to something noble rather than base, something proportional rather than unlimited. Immanuel Kant (1724–1804) supported a retributive rationale:

> *Juridical punishment . . . can be inflicted on a criminal, never just as instrumental to the achievement of some other good for the criminal himself or for the civil society, but only because he has committed a crime: for a man may never be used just as a means to the end of another person. . . . Penal law is a categorical imperative. . . . Thus, whatever undeserved evil you inflict on another person, you inflict on yourself (Kant, cited in Borchert and Stewart 1986, 322).*

In conclusion, the **retributive rationale** for punishment holds that because of natural law and the social contract, society has the right to punish, and the criminal has the right to be punished. It is not an evil to be justified, but rather, represents the natural order of things. According to Newman (1978, 287), "There is little grace in punishment. Only justice."

retributive rationale the justification for punishment that proposes that society has a right to punish, as long as it is done lawfully and proportionally to the wrong committed by the offender.

utilitarian rationale the justification for punishment that proposes that society has a right to punish, as long as it results in a greater good for the majority of the population.

utilitarianism the ethical system whereby good is defined as that which results in the greatest good for the greatest number.

hedonistic calculus Jeremy Bentham's concept that the potential profit or pleasure from a criminal act can be counterbalanced with the risk of slightly more pain or punishment. If this is done then rational people will choose not to commit the act.

Utilitarian Rationale

The **utilitarian rationale** defines punishment as essentially evil, and seeks to justify it by the greater benefits that result. Under a utilitarian philosophical system, or utilitarianism, what is good is that which benefits "the many." Thus, even if it were painful to the individual, if the majority benefit from a certain act, then **utilitarianism** would define that act as good. In our discussion, if punishment did *deter* or *incapacitate* or facilitate *rehabilitation*, then "the many" (all of society) would benefit, and punishment, by definition, would be good.

This rationale for punishment is ancient. Plato argued that punishment is a benefit to the person because it improves their souls or characters (cited in Murphy 1995, 17). Jeremy Bentham (1748–1832), the classical advocate of utilitarian punishment, believed that punishment could be calibrated to deter crime. His idea of a **hedonistic calculus** involved two concepts: first, that mankind was essentially rational and hedonistic (pleasure-seeking), and would seek to maximize pleasure and reduce pain in all behavior decisions; and second, that a legal system could accurately determine exactly what measure of punishment was necessary to slightly outweigh the potential pleasure or profit from any criminal act. Thus, if done correctly, the potential pain of punishment would be sufficient to outweigh the potential pleasure or profit from crime, and all people would rationally choose to be law-abiding. (See **Box 1-1**.)

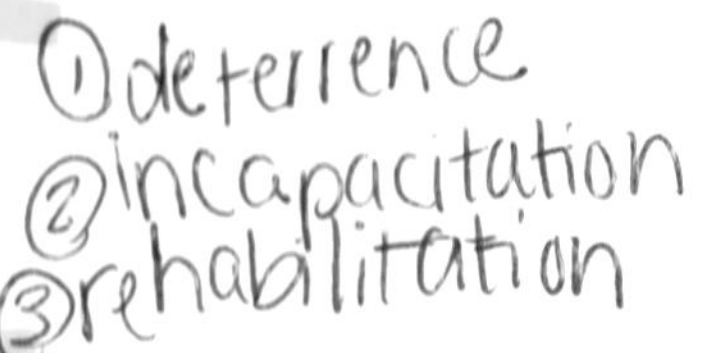

Under the utilitarian rationale, punishment is evil, but it is justified when punishment accomplishes more good than the evil it represents. Cesare Beccaria (1738–1794), another utilitarian thinker, suggested that in some instances the benefits of punishment *do not* outweigh the evil, as illustrated by the following quote.

Box 1-1
Philosophers of Punishment and Penology

Cesare Beccaria (1738–1794). Beccaria was an Italian writer during the age of the Enlightenment, a historical era marked by great advances in political and social thought. He wrote a treatise on criminal law that was highly critical of the practices of the day, and advocated major reforms that included ideas that were widely adopted, such as the right to defend oneself against one's accusers. The philosophical rationale for these reforms was utilitarianism. He believed that the objective of punishment should be deterrence, and that the effectiveness of punishment was based on certainty, not severity. He was largely responsible for major criminal-law reforms in Europe and America.

Jeremy Bentham (1748–1832). Bentham was an English philosopher, economist, and theoretician. Among his many works was *The Rationale of Punishment* (1830), in which he proposed a utilitarian rationale for punishment. Mankind, according to Bentham, was governed by the pursuit of pleasure and the avoidance of pain. These two masters affected all behavior decisions and could be utilized to deter criminal behavior through a careful application of criminal law. He is also known for his design of the "Panoptican Prison."

"Free will" decisions

Immanuel Kant (1724–1804). Kant was a German philosopher who wrote in the areas of metaphysics, ethics, and knowledge. He is the founder of "Kantianism," a philosophical tradition that explores the limits of human reason and establishes a philosophy of morality based on duty. His views on punishment would be considered purely retributive. He believed that the criminal deserved to be punished, but that to punish for other purposes, such as deterrence, was to violate the "categorical imperative," specifically, that one should not use others for one's own end.

> *But all punishment is mischief: all punishment in itself is evil. Upon the principle of utility, if it ought at all to be admitted, it ought only to be admitted in as far as it promises to exclude some greater evil. . . . It is plain, therefore, that in the following cases punishment ought not to be inflicted.*
>
> 1. *Where it is groundless: where there is no mischief for it to prevent; the act not being mischievous upon the whole.*
> 2. *Where it must be inefficacious: where it cannot act so as to prevent the mischief.*
> 3. *Where it is unprofitable, or too expensive: where the mischief it would produce would be greater than what it prevented.*
> 4. *Where it is needless: where the mischief may be prevented, or cease of itself, without it: that is, at a cheaper rate* (Beccaria, cited in Murphy 1995, 24).

ex post facto laws laws that make an act criminal "after the fact," so that individuals would not have received due notice that the behavior would be punished. Our Constitution prohibits these laws.

Situations in which punishment does not deter include ex post facto laws (because people cannot be deterred from some action they do not know to be illegal when they decide to do it), and infancy or insanity (because people cannot be deterred if they cannot control their behavior). This approach views prevention of future harm as the only justifiable purpose of punishment, with retribution having no place because "what is done can never be undone" (Hirsch 1987, 361).

The *social contract* is also the basis for a utilitarian rationale for punishment. In this case, the social contract gives society the right to punish—not because of the offender's violation, but rather, to protect all members of society against future harms. The right of society to punish comes from the responsibility of society to protect. The utilitarian approach of punishment sees it as a means to an end—the end being deterrence (general or specific), incapacitation, or rehabilitation (reform).

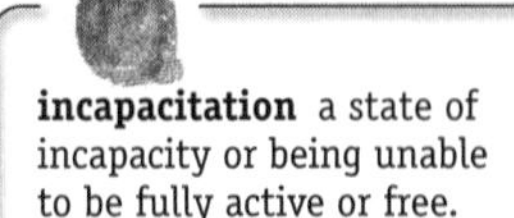

incapacitation a state of incapacity or being unable to be fully active or free.

Incapacitation and rehabilitation are not really related to punishment at all. **Incapacitation** prevents an individual from inflicting further harm for at least as long as the individual is under control. Strictly speaking, it is not punishment because it does not necessarily imply pain. To put all criminals under a drug that induced sleep would be to incapacitate them, not necessarily to punish them. If one takes away the ability of the criminal to commit crime, this also would be incapacitation; for instance, chemical castration has been discussed and, in some cases, inflicted on sexual offenders. Note that there is no physical pain involved, only the incapacitating nature of the chemical. This is obviously a punishment, but it could also be termed incapacitation because it takes away the ability to commit the particular crime. House arrest, electronic bracelets, or other means of monitoring the movements of criminals have all been suggested as less expensive alternatives to incapacitating criminals in prisons. Prison, of course, has become synonymous with incapacitation because as long as the person is incarcerated, they cannot commit crimes against the rest of us. Of course, prisoners continue to commit crimes in prison against other inmates, and there is at least some limited ability to continue to commit some crimes, for instance, credit-card abuse over prison phones or computer fraud using computers provided in vocational programs.

selective incapacitation the concept that we can predict who is going to be highly recidivistic or violent and incarcerate these individuals longer than others.

One issue of incapacitation is how long to hold the individual. **Selective incapacitation** is the policy of holding some offenders longer because of their likelihood of recidivism. Unfortunately, there is little confidence in our ability to predict how long someone may be dangerous or who may continue to commit crime. Auerhahn (1999) found that even the best predictions had an error rate of 48–55 percent, meaning that the prediction would be wrong about half the time—in effect, no different from chance. Thus, until our ability to predict future risk improves, it does not seem to be a legitimate argument to use incapacitation to prevent future harm.

Because incapacitation is forward-looking, it is assumed that the incapacitative period should last as long as the risk exists. This may be inconsistent with principles of justice, even assuming we could predict risk accurately. For instance, forgers have high recidivism rates but are not especially dangerous; should we hold them longer than murderers who have lower recidivism rates? Should the period of incapacitation be tied somehow to the seriousness of the risk (severity of the crime), as well as to the extent of the risk itself (likelihood of recidivism)? Again, this discussion assumes that we can accurately predict risk, an extremely problematic assumption. Although, strictly speaking, incapacitation is not punishment, it usually does involve some deprivation of liberty, and therefore is painful to those who value liberty and autonomy.

rehabilitation the process of internal change brought about by external agents.

Rehabilitation is not punishment either, although punishment may be used as a tool of reform. **Rehabilitation** is defined as internal change that results in a cessation of the targeted negative behavior. It may be achieved by inflicting pain as a learning tool (behavior modification) or by other interventions that are not painful at all (for example, self-esteem groups, education, or religion). Under the retributive philosophy described earlier, rehabilitation and treatment are considered more intrusive and less respectful of the individuality of each person than pure punishment because they attack the internal psyche of the individual. They seek to change offenders, perhaps against their will. This is probably more sophistry than reality, as anyone who has worked with offenders can attest. Very few people enjoy the experience of being a drug addict or sex offender, and most prison programs have limited capacity to change individuals against their will anyway. In a later chapter, we will explore the concept of rehabilitation and the various modes of individual change.

To conclude, the utilitarian rationale for punishment must determine that the good coming from punishment outweighs the inherent evil of the punishment itself. The beneficial aspects of punishment include deterrence, incapacitation, and rehabilitation or reform.

Methods of Punishment

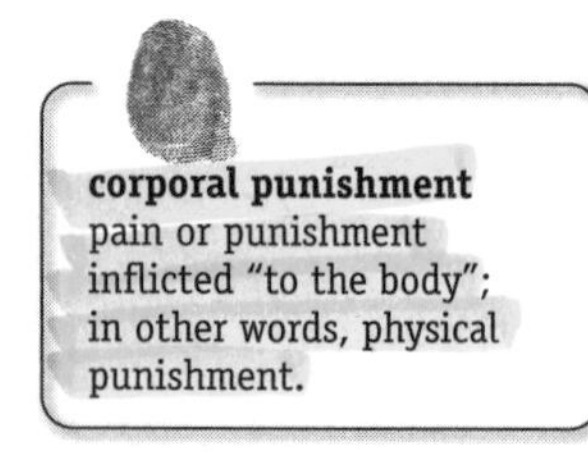

corporal punishment pain or punishment inflicted "to the body"; in other words, physical punishment.

Targets of punishment include one's possessions, one's body, or one's psyche. Punishments throughout the ages have attacked the body. **Corporal punishment** (meaning "to the body") included drawing and quartering, flaying, whipping, beheading, dismembering, and numerous other means of torture or death (Newman 1978). Fines and dispossession of property also have been common throughout history. Conley (1992) writes that fines were more common than physical torture during many time periods. Execution was an economic as well as a corporal punishment because the person's estate was forfeited to the monarch.

bridewells early English institutions that held the itinerant poor, many of whom probably had committed petty crimes. The name derived from the location of the first such institution.

Economic and physical sanctions gradually have given way to imprisonment or lesser deprivations of liberty (probation or parole). We have reached the point today (at least in this country) where punishment is almost synonymous with imprisonment. As early as the end of the 14th century, the purpose of imprisonment changed from custody until physical punishment was inflicted, to custody as punishment itself. An increasing number of laws emerged with precisely defined prison sentences. The church also used imprisonment as a punishment for clerics (Conley 1992). Gradually, imprisonment for crime became almost indistinguishable from the other institutions that developed for vagrants and idlers—the **bridewells**, workhouses, and **gaols** all were responses to the same class of citizens. They held the itinerant poor: individuals who often were forced into petty crime because of their poverty. Chapter 2 explains this history in greater detail.

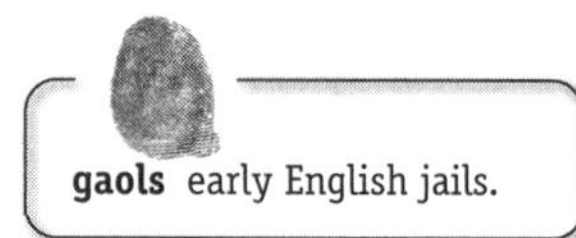

gaols early English jails.

Philosophy of Imprisonment

Of all the punishments described above, prison is perhaps the most complex. It affects the prisoners' material possessions because they can earn little or no income while incarcerated, they may lose their job or livelihood, spend their life savings, and have their total lifetime earning capacity affected. It affects the prisoner's body because he or she is under the control of others and very little freedom exists. Imprisonment may result in actual physical harm, from attacks by correctional officers or other inmates or from illnesses or injuries left untreated. Prison also attacks the psyche by attempts at reformation and through the mental deterioration that occurs because of the negative environment of the prison. Many describe prison as a "psychological punishment" (Mickunas 1990, 78).

According to some, prison in its most severe form attacks "the soul"; it acts on the "heart, the thoughts, the will, the inclinations of the prisoner" (Howe 1994, 87). Prison critics allege that the most detrimental effects are not physical deterioration, but mental and moral deterioration. "You are nothing!" is a theme that prison inmates live with during the course of their imprisonment, and the mental toll that prison takes on its population is very difficult to measure.

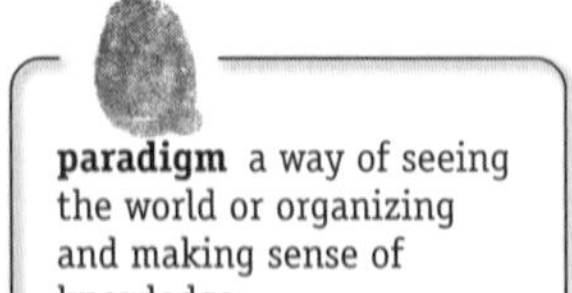

paradigm a way of seeing the world or organizing and making sense of knowledge.

Paradigms and Prison

A paradigm is a way of seeing the world or of organizing and making sense of knowledge. We can use the well-worn paradigms of conservatism and liberalism to illustrate the philosophy of imprisonment. The conservative ideology operates under the assumption that human beings have free will, can make rational choices, and deserve the logical outcomes of their choices. The liberal view of human behavior holds that behavior is influenced by upbringing, by affluence or poverty, by education, and by life experiences in general. The radical paradigm calls into question the very existence of the social order; radicals reject private ownership of property and are in favor of restructuring socioeconomic relations (Durham 1994, 17–20).

Figure 1-1 A penitentiary cell with modest appointments. ***Source:*** *Reprinted with permission of the American Correctional Association, Lanham, Maryland.*

With these elements in mind, it is clear that the conservative approach to imprisonment is one of deterrence and incapacitation. Prison life should be uncomfortable—even painful—so that rational people, will be deterred from committing crime (**Figure 1-1**). If a short prison term doesn't work, the next sentence should be longer. The liberal approach embraces rehabilitation and reform. The purpose of prison should be to change the individual. Rehabilitative programs and reintegrative assistance, such as job-placement assistance, will help the person avoid future imprisonment by addressing problems of drug addiction, poor self-esteem, and no job skills. The radical approach would abolish prisons because it views them as tools of the powerful to enslave the powerless. The only solution to recidivism and crime, according to a radical perspective, is to reform law and society's resources so that everyone gets a "fair share" (Durham 1994, 28). At least two of these three perspectives can be roughly represented by different eras of prison history, each with a predominant philosophy of penology.

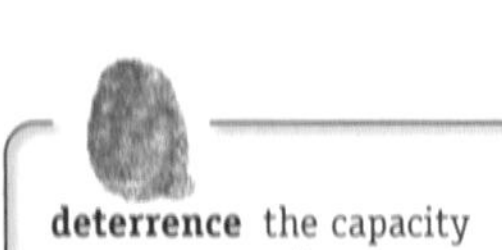

deterrence the capacity to prevent or discourage an individual or individuals from committing an act.

Conservatism: Deterrence and Incapacitation

The conservative approach characterized by views of deterrence and incapacitation was strong throughout pre-Jacksonian America and Europe. The philosophy of punishment in general, and of prison specifically, was to deter and punish.

> *[C]learly the colonists relied on societal retribution as the basis for punishment and viewed the execution of punishment as a right of the society to protect itself and to wage war against individual sin. Deviance was the fault of the offender, not the breakdown of society or the community. . . . (Conley 1992, 42).*

The use of prison was seen as a more humane form of punishment than earlier corporal punishments, but it was not necessarily viewed as reformative. The individual was seen as evil or weak, someone that society needed to protect itself against. Prison became a type of banishment. Earlier societies had banished wrongdoers to the wilderness; prisons (which were isolated far away from urban areas) became the "new wilderness." If individuals were not deterred by the thought of that punishment (**general deterrence**), then they might be after experiencing incarceration (**specific deterrence**). At the least, society was protected as long as the offender was away (incapacitation).

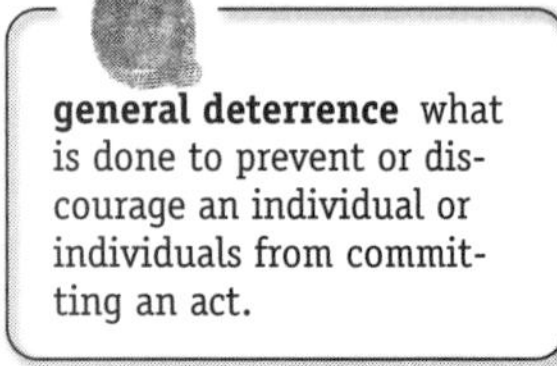

general deterrence what is done to prevent or discourage an individual or individuals from committing an act.

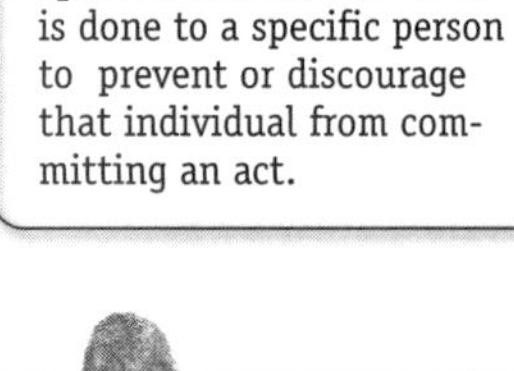

specific deterrence what is done to a specific person to prevent or discourage that individual from committing an act.

Liberalism: Reformation and Rehabilitation

At some point during the 19th century, the philosophy behind imprisonment changed. Prison became viewed as more than an alternative to brutal corporal punishments. It was seen as redemptive and capable of changing the individuals within to become better people (Conley 1992).

David Rothman (1971), one of the definitive authorities on the reformative origins of the prison, proposes that the idea of reforming the individual criminal was at odds with the Calvinist doctrine of original sin. Before the 1800s, punishment remained retributive and was associated with **expiation** (a religious term meaning personal redemption through suffering). People were viewed as not capable of reform. Once the possibility of individual change was born, the idea of prison developed as the site of the "reform" (Hirsch 1987).

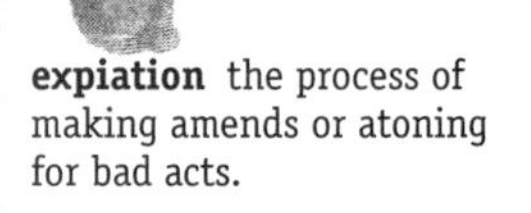

expiation the process of making amends or atoning for bad acts.

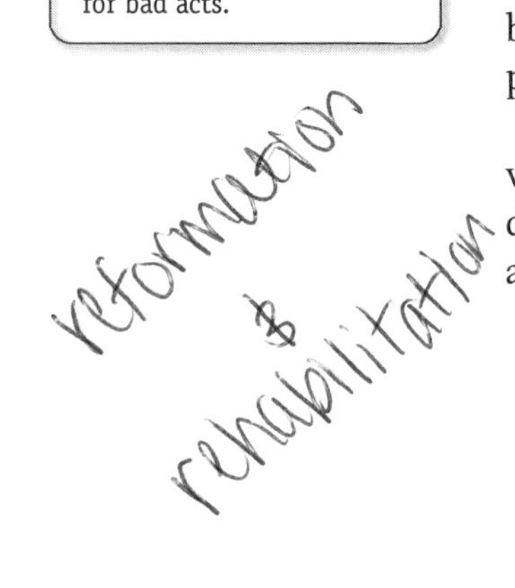

Although the penitentiary might have been an idea born in Europe, its development was purely American. Hirsch (1987) and others (McKelvey 1987) describe a shift in penal philosophy as the concept developed in the United States and Europeans began to look to American models of penal institutions.

> *The deluge of European delegations [to American prisons] in the 1830s masked a subtle shift in the intellectual center of penal reform. Before 1800, European theorists dominated the field of criminology, supplying the basic concepts and programs on which American facilities were built (Hirsch 1987, 429).*

Separation, obedience, and labor became the trinity around which officials managed the penitentiary (Crosley 1986). Convicts were "men of idle habits, vicious propensities, and depraved passions," who had to be taught obedience as part of their reformation (Rothman 1971, 579). By teaching convicts these virtues, prison officials reinforced their value for all of society. The penitentiary would reawaken the public to these "virtues," and "promote a new respect for order and authority" (Rothman, 585).

The early reformative ideals, although corrupted by greed and co-opted by practicality, evolved into the rehabilitative era of the 1960s and early 1970s. Reformation was the dominant theme of the 1870 Prison Congress, which laid out the "Principles of Corrections," and these were endorsed again, almost without change, in the 1970 Prison Congress. The 1870 and 1970 Prison Congresses endorsed such philosophical principles as:

- *"corrections must demonstrate integrity, respect, dignity, fairness . . . ,"*
- *"sanctions imposed by the court shall be commensurate with the seriousness of the offense," and*
- *"offenders . . . shall be afforded the opportunity to engage in productive work, participate in programs . . . and other activities that will enhance self worth, community integration, and economic status" (American Correctional Association Statement of Principles, 1970/2002).*

Progressive Era refers to the early 1900s when there was an explosive growth of the "sciences" and the optimism that humans could, through science, understand and control the world.

The **Progressive Era** (early 1900s) was the time period during which educated professionals entered penology believing that science would solve individual prisoners' problems. Indeterminate sentences and individualized treatment were the tools to accomplish this task. Scientific objectivity and professionalism replaced the missionary zeal of earlier penologists. The prison was no longer viewed as a utopia for society to emulate. It was viewed instead as a laboratory in which social work and psychiatry would work to help change people's behavior.

Liberalism, however, is more clearly represented by the "rehabilitative era" of the 1970s. For a brief period of time, the general philosophy and mission of prisons changed to one of reformation and rehabilitation. Even the name of the prison changed to "correctional institution" and correctional programs proliferated. However, the rehabilitative era was over by the 1980s as prison systems struggled to house rising numbers of prisoners. Penal institutions once again settled back into a less ambitious mission of providing punishment and incapacitation.

Radicalism: Prison and Economics

Rothman (1971) accepted the rhetoric of penal philosophy at face value. That is, the writings of the time indicated that the motivation and purpose of prison was to reform offenders, and this goal was accepted as fact. Others see the rhetoric of early prison reformers as masking a more subtle and insidious philosophy of imprisonment, one based on economy rather than reformation, and on power rather than benevolence. Rusche and Kirchheimer (1939) suggested that imprisonment emerged as the dominant method of punishment because of a desire to exploit and train captive labor. A scarcity of labor served as the impetus for the modern prison because of its role in training and exploiting labor reserves.

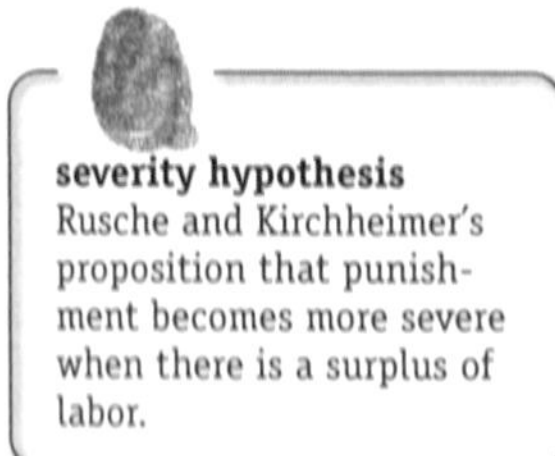

severity hypothesis Rusche and Kirchheimer's proposition that punishment becomes more severe when there is a surplus of labor.

The so-called **severity hypothesis** of Rusche and Kirchheimer proposes that punishment becomes more severe when there is a surplus of labor, and more lenient when labor is scarce and convicts are more valuable in the labor force. Some authors have supported this theory by using case histories of prison systems and comparing the treatment of prisoners to economic conditions. Other authors have not found any support for the theory, at least not measurable by standard methods (Gardner 1987). For instance, in his study of New York prison history, Gardner found that harsher punishment often resulted from attempts to maintain and increase the production of essential commodities in overcrowded, tumultuous prisons.

However, even critics of the Rusche-Kirchheimer view mention economic elements in their explanations of motivations for the development of prison. Gardner (1987) proposes the idea that prisons would have developed much earlier in England and Europe if the Board of Trade in England had not been so vigorously opposed to their creation and the competition that would develop from prison labor. He also points out that in American prison history, the promises of prison officials that prisons could be self-supporting were subverted by the low fees contractors were allowed to pay for leased labor, not to mention the economic boon of the prison itself to a local economy. In fact, he goes so far as to point to the economic benefits of the prison to certain interest groups as the reason for "the persistence and expansion of an otherwise politically and economically anachronistic form of punishment" (Gardner, 106).

The earliest origins of the prison are tied to economics because prisons targeted the "idle poor" and were first cousins to the *bridewells* and *workhouses*, institutions that absorbed the vagrant classes of Europe and early American cities. Authors disagree, however, as to the meaning of labor within the prison. Some see labor as a reformative element, helping the inmate take on the industriousness and good habits of a perfect citizen. Others describe prison labor as more purely punishment where legislators instructed prison administrators to institute labor "of the hardest and most servile kind, in which the work is least liable to be spoiled by ignorance, neglect or obstinacy" (cited in Rothman 1971, 570). There was no question that prison labor in southern prison systems was purely exploitive: "[P]enal slaves were herded about the camps by armed guards, and at night they were shackled in 'cribs.' The lease-holders were interested in making as large a return as possible for the least outlay of money" (Crosley 1986, 21).

"factory prison" model derived from the Auburn Prison and was more common in the northeast. These prisons utilized prison labor in factory settings.

In the North, economics favored the "**factory prison" model**. Inmates were housed and worked together and were better utilized in factorylike labor conditions (Melossi and Pavarini 1981). Chapter 5 explores current issues of prison labor.

The radical view sees economics as the central issue in all social relations. Those who have economic power also have legal and social power. The legal system, including the sanction of imprisonment, is viewed as a tool of those in power. The purpose is variously described as to capture and exploit the labor pool, to hold a portion of the labor class inactive to keep down labor costs, or to serve as a dumping ground for those who are expendable in a capitalist system. Theorists who advocate this philosophy of imprisonment point to the continued existence of an institution that seems to have failed miserably in its original goal of reformation. For instance, Reiman (1995, 4) says, "On the whole, most of the system's practices make more sense if we look at them as ingredients in an attempt to maintain rather than to reduce crime," and "[the criminal justice system] projects a distorted image that crime is primarily the work of the poor." It keeps the public fearful and unsympathetic toward the disenfranchised, and keeps attention away from economic power holders who are the real perpetrators of most of the injury and loss in society.

Whether the system focuses on individual responsibility for crime (and therefore punishment) or on individual deviance (and therefore treatment), the result is the same: the existing social order is excused from any charge of injustice. The radical theorists point out that this is the true reason why prisons fail to cure or deter. Even theorists who are not necessarily radical point out that imprisonment is futile without addressing social problems, such as unemployment, homelessness, poverty, discrimination, inadequate health care, and unequal education (Selke 1993).

Foucoult (1973) presented a slightly different view of prison—one based less on economics, but still premised on the need of those in power to discipline and control the populace. In his history of the emergence of the prison, he sees the prison as one part of the institutionalization of society—the prison housed the poor and criminal, the mental institutions housed those who couldn't take care of themselves, and poorhouses housed those without economic means. All controlled and contained the class of people who were considered expendable. All normalized the idea of containment and deprivation of liberty as a natural right of society.

Although the radical view has had some support for many decades, it has never been the dominant viewpoint. There is continuing evidence, however, of the potency of the idea that the underlying philosophy of imprisonment has always been economic. Today, we see that private prisons have emerged as a profit-generating industry (see Pollock 2004b). In Chapters 3 and 10, we will discuss their growth and the profits that are generated in both state and federal systems. Powerful companies such as CCA (doing business now as Prison Realty Trust) and Wackenhut are public companies, and their stock is traded on the New York Stock Exchange. Further, the biggest growth in private-prison construction has been in small towns that have seen manufacturing jobs disappear. It is not an exaggeration to say that the towns' very economic livelihoods depend on a continuing stream of prisoners to fill the prisons to provide the jobs for the townspeople. The radical view would hold that the continued existence and, perhaps, growth of prisons is assured when they generate profit for someone.

The New Conservatism: Justice and "Just Desserts"

Since the late 1970s and early 1980s, there has been disillusionment and dissatisfaction with the idea of prison as a reformative tool. Some believe that the only purpose of prison should be punishment. The first and most vocal critics of the rehabilitation ethic were Von Hirsch (1976) and Fogel and Hudson (1981). Although different in tone, both critique the idea that prison should be anything more than a measure of punishment. Their approach blends a curious mixture of utilitarianism and retributivism to form a **new retributivism**. This philosophy is actually quite old, and more similar to pre-Jacksonian deterrence and incapacitation than anything seen for the past 100 years. Von Hirsch justifies and limits the role of punishment by retributive proportionality:

new retributivism a term used to describe those in the late 1970s and early 1980s who proposed abandoning the "rehabilitative ideal" and returning to a system based on retribution.

1. The liberty of each individual is to be protected so long as it is consistent with the liberty of others.
2. The state is obligated to observe strict parsimony in intervening in criminals' lives.

imprisonment is seen as fully deserved & proper consequence of criminal behavior

3. The state must justify each intrusion.
4. The requirements of justice ought to constrain the pursuit of crime prevention (that is, deterrence and rehabilitation) (Von Hirsch 1976, 5).

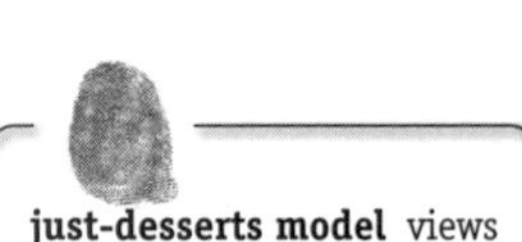

just-desserts model views retribution as the sole rationale for punishment. What is done to the individual criminal should be based solely on the wrong that was committed and measured accordingly.

The so-called **just-desserts model** also views punishment as being justified solely by retributive ends rather than utilitarian ones. This view utilizes the "social contract" again to justify punishment for those who break the law. It promotes the idea that the only goal of the justice system should be justice, not reform of the individual (Fogel and Hudson 1981). This view advocates using determinate forms of sentencing rather than indeterminate, separating treatment options from release decisions, and circumscribing the goals of custody to retribution rather than reformation. It has found popular and political favor probably because it sounds punitive, although advocates of this philosophy were most likely reacting to the abuses of power engendered by a utilitarian treatment ethic that allowed a great deal more control over the individual offender's body and mind "for their own good." There is a distinct difference between a penal philosophy that holds that we should do no more to the individual than she deserves (that is, not keep an offender imprisoned longer for treatment), and a penal philosophy that holds that the only thing an offender deserves is punishment. Despite their differences, both of these approaches have contributed to penal policies today.

The Effect of Retributivism and a New Era

We are currently in an era where neither liberalism nor rehabilitation is the dominant correctional philosophy, despite the continuing theme of the popular press and politicians who propose "getting tougher" as the answer to the crime problem. There are two fallacies to this rhetoric. The first is that we are not "tough" enough. To the contrary, some perceive the United States as a "gulag nation," and the U.S. incarceration rate exceeds that of any other Western nation. The other fallacy is that we have a crime problem. Crime is at its lowest point in 30 years, yet today the dominant penal philosophy, which developed in the late 1970s and early 1980s, continues to be conservative and punitive.

penal-harm movement a term coined to describe the punitive approach that has characterized the justice system and corrections since the 1980s when rehabilitation and reform were, to a great extent, abandoned.

Clear (1994) refers to this era as the "**penal-harm movement**." The phrase encapsulates the idea that, far from rehabilitation, the objective and goal of reformers in the late 1970s was to make prisons painful and increase the measure of punishment inflicted. This approach has been with us for at least 20 years, and has its roots in traditional retributivism. This philosophy has been pervasive in the politics and rhetoric surrounding corrections, and no doubt has contributed to the phenomenal growth of the incarceration rate and the proliferation of prisons. There is some small evidence, however, that the ever-increasing use of prison as punishment may be slowing down, and that other responses to crime might be gaining favor. It is possible that the "new retributivism" or "penal-harm movement" might be on the wane.

Restorative Justice: An Alternative Philosophy?

A relatively new philosophy, restorative justice, has emerged in criminal justice that is quite contrary to the penal-harm movement. The roots of such a philosophy might be found in the Ethics of Care (Pollock 2004a) and, to some extent, in utilitarianism. Adherents also find support in religion, arguing that in the Bible, the "eye for an eye" reference is to reparation, not restitution (Schweigert 2002, 29).

Much of the concept has been borrowed from aboriginal peoples, including the Inuit, Maori, and Navajo (Perry 2002, 5). Basically, the idea of restorative justice is that the objective is not to inflict punishment on the offender, but rather, to restore all parties to a prior state of "wholeness." This philosophy can also be called reparative justice or peacemaking justice.

One basic tenet of restorative justice is the involvement of victims in a search for a resolution that meets the needs of all parties (Van Ness and Strong 1997). The offender must meet the victim (either literally or figuratively) and take responsibility for his or her actions by in some way repairing the damage done to the victim. For instance, in one restorative-justice case, a DWI offender who had killed his best friend in a traffic accident offered to pay child support to the victim's wife and young child until the child reached college (Perry 2002, 10). This is a much different outcome from the typical system response to a DWI fatality. In this case, some attempt has been made to meet the needs of the victims; in a typical prison sentence, needs are ignored in favor of vengeance.

The idea of restorative justice is that victims must be made whole; however, part of the solution might be meeting the needs of the offender as well. An important component of this philosophy is that the offender is not to be condemned, but rather, is helped to see how he or she can repair the damage. The idea that the offender continues to be a part of the community is very important. Far from being banished or stigmatized by the experience, the offender should feel more fully integrated into his or her community (Braithwaite 1989).

Mediations and conferences between the victim and offender are often a part of restorative-justice efforts (Braithwaite 1989; Bazemore and Maloney 1994). Restitution is also consistent with the ideals of restorative justice, but only if it is tied with the specific needs of a victim and is meaningful to both (Schweigert 2002, 21). Adherents of this approach see it as a return to older forms of justice rather than as a new philosophy of justice. They note that the oldest forms of justice were concerned with restoring loss and repairing injury rather than with punishment. Further, justice was administered by and kept within the community, not abdicated to a higher state authority (Schweigert 2002, 25).

So how do prisons fit with restorative justice? Actually, they don't. In the majority of cases where prison is used as the response to an offense, a restorative-justice rationale would argue that community service, restitution, or some type of mediation would be a better alternative. Only in cases of serious violent crime would mediation and restitution not be appropriate. Prison is banishment. Individuals who are banished and feel pain via imprisonment are not likely to feel close to the community that banished them; thus, the "circle" of society has been broken. Prison not only injures the individual, but also injures the community because of the loss of the individual from his or her community. Thus, prison is basically inconsistent with a restorative-justice philosophy; however, some argue that prison might become restorative if it were to fundamentally shift its emphasis and objective to reparations to specific victims and to safeguarding the dignity and humanity of the offenders (Perry 2002, 14).

Utilitarian Caring: The Reintegrative Movement

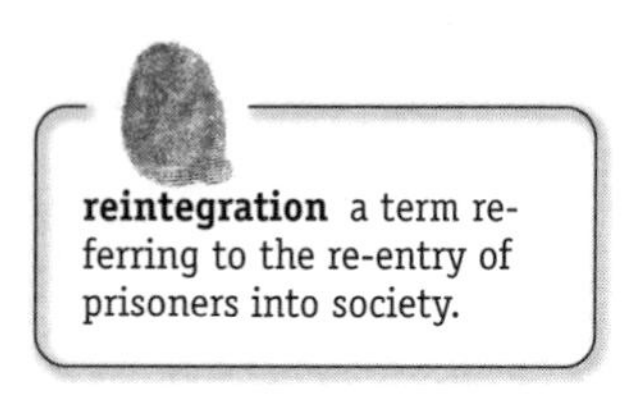

reintegration a term referring to the re-entry of prisoners into society.

Recently we have seen a renewed attention to the needs of prisoners reentering society (Mauer, Chesney-Lind, and Clear 2002). Interestingly, in the 1980s this problem was addressed and the term **reintegration** was coined. The federal government has recently budgeted some monies toward studying what can be done to aid the offender in reentry efforts (Murphy 2002b). While some might think this trend is part of the restorative-justice movement, a more likely philosophy underpinning the effort is utilitarianism. It cannot go unnoticed that more than 600,000 prisoners are reentering our society every year (King and Mauer 2002, 3). Further, at least one study indicates that the recidivism rate is worse today than 20 years ago (Murphy 2002a). Most citizens would prefer that released offenders have some means to support themselves, and it benefits us all if offenders have access to programs that may help them withstand the temptations of alcohol, drugs, and/or committing future crimes. This has translated into more funds for parole, job-placement services, and other assistance to newly released inmates (Ward 2004).

It makes sense for the public to be in favor of community alternatives to prison because such alternatives are less costly. If prison costs $45 a day compared to $2 for probation (Ward 2004), and if there is no substantial risk in choosing the cheaper alternative, it would seem that even a punitive utilitarian should prefer that the offender be placed in the cheaper correctional alternative—it is best for the offender, but more importantly, it is best for us all.

Conclusions

In this chapter, we have surveyed some historical and current philosophies of punishment and prison. An implicit assumption of this chapter is that what we do has some relationship to what we believe. Is it important to review the motivations and purposes behind the prison? One benefit of this exercise is that we become more clear about what we expect from prison. For instance, many people, including many inmates, believe that the prison's main function is to rehabilitate. In reality, this has not been a major element in the "mission" of prisons for more than two decades. Although we will discuss education and vocational training in Chapter 6, and other rehabilitation programs in Chapter 5, much of the philosophical rationale for these interventions has been discarded by penologists and the politicians who fund the prison enterprise.

Another issue to consider is whether there is any evidence in support of the rationales for punishment discussed in this chapter. If one believes in a penal philosophy based on utilitarian deterrence, is it not important to have evidence that prison deters? How does one know whether prison has deterred someone from committing crime? Some argue that the declining crime rates point to prison's effectiveness in deterrence. However, as we shall see in Chapter 3, others point out that crime rates and prison rates (both of which vary among the states) bear no relationship to each other, thus undercutting the assumption that it is imprisonment that has led to the decline of crime.

In Chapter 6, correctional treatment programs are discussed, along with the mixed findings of the evaluations that have been conducted on correctional programs. While some indicate that there is little or no evidence that treatment programs reduce crime, others argue that there are positive effects. The same mixed findings are found in evaluations of education and vocational training.

The penal enterprise has always had more than one philosophy or rationale. It is a slippery fish: if we criticize it for not rehabilitating, we are told it deters; if we ask for evidence of deterrence, we are told it is retributive. If the public is at all squeamish about locking their brethren up in cages, we are taken on tours of educational buildings and carpentry-apprentice programs to show that it is "for their own good." If the public rails against prison as the "Holiday Inn for criminals," one can show them prison chain gangs. One prevailing aspect of penal philosophy may be its amorphous content. Prisons can be all things to all people. The radical theorists may be right that the prison has been successful in diverting public attention away from the transgressions of the economically powerful by defining and reviling a "criminal" class, but they are less successful in any attempt to envision a society without prison.

KEY TERMS

bridewells—early English institutions that held the itinerant poor, many of whom probably had committed petty crimes. The name derived from the location of the first such institution.

corporal punishment—pain or punishment inflicted "to the body"; in other words, physical punishment.

deterrence—the capacity to prevent or discourage an individual or individuals from committing an act.

expiation—the process of making amends or atoning for bad acts.

ex post facto laws—laws that make an act criminal "after the fact," so that individuals would not have received due notice that the behavior would be punished. Our Constitution prohibits these laws.

"factory prison" model—derived from the Auburn Prison and was more common in the northeast. These prisons utilized prison labor in factory settings.

gaols—early English jails.

general deterrence—what is done to prevent or discourage an individual or individuals from committing an act.

hedonistic calculus—Jeremy Bentham's concept that the potential profit or pleasure from a criminal act can be counterbalanced with the risk of slightly more pain or punishment. If this is done then rational people will choose not to commit the act.

incapacitation—a state of incapacity or being unable to be fully active or free.

just-desserts model—views retribution as the sole rationale for punishment. What is done to the individual criminal should be based solely on the wrong that was committed and measured accordingly.

new retributivism—a term used to describe those in the late 1970s and early 1980s who proposed abandoning the "rehabilitative ideal" and returning to a system based on retribution.

paradigm—a way of seeing the world or organizing and making sense of knowledge.

penal-harm movement—a term coined to describe the punitive approach that has characterized the justice system and corrections since the 1980s when rehabilitation and reform were, to a great extent, abandoned.

Progressive Era—refers to the early 1900s when there was an explosive growth of the "sciences" and the optimism that humans could, through science, understand and control the world.

punishment—a pain or unpleasant experience inflicted upon an individual in response to a violation of a rule or law by a person or persons who have lawful authority to do so.

rehabilitation—the process of internal change brought about by external agents.

reintegration—a term referring to the re-entry of prisoners into society.

retribution—the proportional infliction of pain or punishment in response to a wrong.

retributive rationale—the justification for punishment that proposes that society has a right to punish, as long as it is done lawfully and proportionally to the wrong committed by the offender.

selective incapacitation—the concept that we can predict who is going to be highly recidivistic or violent and incarcerate these individuals longer than others.

severity hypothesis—Rusche and Kirchheimer's proposition that punishment becomes more severe when there is a surplus of labor.

social contract—a heuristic device that illustrates how individuals give up individual liberties to act with aggression in return for safety.

specific deterrence—what is done to a specific person to prevent or discourage that individual from committing an act.

utilitarianism—the ethical system whereby good is defined as that which results in the greatest good for the greatest number.

utilitarian rationale—the justification for punishment that proposes that society has a right to punish, as long as it results in a greater good for the majority of the population.

REVIEW QUESTIONS

1. Explain the difference between the retributive rationale for punishment and the utilitarian rationale.
2. What is the social contract?
3. Discuss the three benefits of prison under the utilitarian rationale of punishment.
4. Discuss the differences between the conservative, liberal, and radical approaches to penal philosophy. What time period in history is associated with the conservative approach? The liberal approach?
5. What is the importance of "separation, obedience, and order"?
6. Explain the severity hypothesis and the economic theories of penal philosophy.
7. Discuss the elements of the new retributivism. What is the "just-desserts" model?
8. Discuss "restorative justice" and how this approach is or is not consistent with imprisonment.
9. Discuss the philosophical rationale for reintegration efforts.
10. To sum up, describe the two most common rationales for prison.

Chapter Resources

FURTHER READING

Braithwaite, J. 1989. *Crime, Shame, and Reintegration.* Cambridge: Cambridge University Press.

Foucoult, M. 1973. *Discipline and Punish: The Birth of the Prison.* New York: Vintage.

Murphy, J. 1995. *Punishment and Rehabilitation,* 3d ed. Belmont, CA: Wadsworth.

Rothman, D. 1971. *The Discovery of the Asylum: Social Order and Disorder in the New Republic.* Boston: Little, Brown.

Rusche, G., and O. Kirchheimer. 1939. *Punishment and Social Structure.* New York: Russell and Russell.

REFERENCES

American Correctional Association. 1970/2002. Retrieved January 2, 2004, from www.aca.org/pastpresentfuture/principles.asp.

Auerhahn, K. 1999. "Selective Incapacitation and the Problem of Prediction." *Criminology* 37, 4: 705–734.

Bazemore, G., and D. Maloney. 1994. "Rehabilitating Community Service Toward Restorative Service Sanctions in a Balanced Justice System." *Federal Probation* 58, 1: 24–35.

Borchert, D., and D. Stewart. 1986. *Exploring Ethics.* New York: Macmillan.

Braithwaite, J. 1989. *Crime, Shame, and Reintegration.* Cambridge: Cambridge University Press.

Clear, T. 1994. *Harm in American Penology: Offenders, Victims, and Their Communities.* Albany, NY: S.U.N.Y.–Albany Press.

Conley, J. 1992. "The Historical Relationship Among Punishment, Incarceration, and Corrections." In S. Stojkovic and R. Lovell (Eds), *Corrections: An Introduction,* pp. 33–65. Cincinnati, OH: Anderson.

Crosley, C. 1986. *Unfolding Misconceptions: The Arkansas State Penitentiary, 1836–1986.* Arlington, TX: Liberal Arts Press.

Durham, A. 1994. *Crisis and Reform: Current Issues in American Punishment.* Boston: Little, Brown.

Fogel, D., and J. Hudson. 1981. *Justice as Fairness.* Cincinnati, OH: Anderson.

Foucoult, M. 1973. *Discipline and Punish: The Birth of the Prison.* New York: Vintage.

Gardner, G. 1987. "The Emergence of the New York State Prison System: A Critique of the Rusche-Kirchheimer Model." *Crime and Social Justice* 29: 88–109.

Hirsch, A. 1987. "From Pillory to Penitentiary: The Rise of Criminal Incarceration in Early Massachusetts." In K. Hall (Ed.), *Police, Prison, and Punishment: Major Historical Interpretations,* pp. 344–434. New York: Garland.

Howe, A. 1994. *Punish and Critique: Towards a Feminist Analysis of Penality.* London: Routledge.

King, R., and M. Mauer. 2002. *State Sentencing and Corrections Policy in an Era of Fiscal Restraint.* Washington, DC: Sentencing Project. Retrieved August 12, 2002, from www.SentencingProject.org.

Mauer, M., M. Chesney-Lind, and T. Clear. 2002. *Invisible Punishment: The Collateral Consequences of Mass Imprisonment.* New York: Free Press.

McKelvey, B. 1987. "Penology in the Westward Movement." In K. Hall (Ed.), *Police, Prison, and Punishment: Major Historical Interpretations*, pp. 457–479. New York: Garland.

Melossi, D., and M. Pavarini. 1981. *The Prison and the Factory: Origins of the Penitentiary System*. London: Macmillan.

Mickunas, A. 1990. "Philosophical Issues Related to Prison Reform." In J. Murphy and J. Dison (Eds.), *Are Prisons Any Better? Twenty Years of Correctional Reform*, pp. 77–93. Newbury Park, CA: Sage Publications.

Murphy, J. 1995. *Punishment and Rehabilitation*, 3d ed. Belmont, CA: Wadsworth.

Murphy, K. 2002a. "State Prisoners Often Return, Report Shows." Stateline.org. Retrieved June 10, 2002, from www.stateline.org/story.do?storyID=240988.

Murphy K. 2002b. "States Get Grants to Help Ex-Offenders" Stateline.org. Retrieved July 17, 2002, from www.stateline.org/story.do?storyID=248888.

Newman, G. 1978. *The Punishment Response*. New York: Lippincott.

Perry, J. 2002. "Challenging the Assumptions." In J. Perry (Ed.), *Restorative Justice: Repairing Communities Through Restorative Justice*, pp. 1–18. Lanham, MD: American Correctional Association.

Pollock, J. 2004a. *Dilemmas and Decisions: Ethics in Crime and Justice*. Belmont, CA: Wadsworth.

Pollock, J. 2004b. *Prisons and Prison Life: Costs and Consequences*. Los Angeles: Roxbury.

Reiman, J. 1995. *The Rich Get Richer and the Poor Get Prison*. Boston: Allyn and Bacon.

Rothman, D. 1971. *The Discovery of the Asylum: Social Order and Disorder in the New Republic*. Boston: Little, Brown.

Rusche, G., and O. Kirchheimer. 1939. *Punishment and Social Structure*. New York: Russell and Russell.

Schweigert, F. 2002. "Moral and Philosophical Foundations of Restorative Justice." In J. Perry (Ed.), *Restorative Justice: Repairing Communities Through Restorative Justice*, pp. 19–39. Lanham, MD: American Correctional Association.

Selke, W. 1993. *Prisons in Crisis*. Bloomington: Indiana University Press.

Van Ness, D., and K. Heetderks Strong. 1997. *Restoring Justice*. Cincinnati, OH: Anderson.

Von Hirsch, A. 1976. *Doing Justice*. New York: Hill and Wang.

Ward, M. 2004. "Rehab Trend for Convicts Gaining Favor." *Austin American Statesman* (December 7): A1, A12.

2 The American Prison in Historical Perspective: Race, Gender, and Adjustment

Robert Johnson, Ania Dobrzanska, and Seri Palla
American University

Chapter Objectives

- Be able to distinguish the separate system and the congregate system.
- Understand the experiences of women and minorities in each era of prison history.
- Be able to distinguish the reformatory from the penitentiary.
- Be able to discuss each of the eras of prison history.
- Understand the legacy of early prisons.

The prison is an institution marked by great staying power but modest achievement. We have had prisons of one sort or another since at least biblical times (Johnson 2002). Though prisons have varied in their internal regimes and in their stated aims, the main achievement of the prison has been its most basic mandate—to contain and restrain offenders (Garland 1990). Rehabilitation has been a recurring aim of prisons, and at times this goal could be described as a grand dream, but rehabilitation is a dream of reformers, not of the criminals who were to be its beneficiaries. The use of prison as a sanction has grown steadily since the advent of the penitentiary at the turn of the 19th century (Cahalan 1979, 37), and indeed has come to dominate criminal justice.

The growth in the use of prisons has been particularly pronounced for blacks and, more recently, women. It is thus telling that comparatively little attention has been paid to the prison experiences of minorities and women. Women, to be sure, have always been drastically underrepresented in our prisons, and this partly explains why limited attention has been paid to their prison experience. Even with the current accelerated growth in rates of confinement for women, only 6 or, at most, 7 percent of the overall prison population are women (Merlo and Pollock 1995; Pollock 2002).

Minorities, by contrast, have always formed a sizable portion of the prison population. In fact, ethnic and, after the Civil War, racial minorities have almost certainly been overrepresented in American prisons (Sellin 1976). Black women have been confined in disproportionate numbers in prisons for women; this trend is particularly evident in high-custody institutions, which traditionally are reserved for those female offenders seen by largely white officials as tough, manlike felons beyond the reach of care or correction (Rafter 1990; Dodge 2002; Johnson 2003). Similarly, black males, and especially young black males, have been overrepresented in our nation's more secure prisons; once again, settings reserved for those deemed least amenable to rehabilitation. These racial disparities are long-standing and must be understood in historical context.

The first minorities in our prisons were European immigrants, with but a sprinkling of offenders of African descent. This is apparent in the writing of a penitentiary inmate named Coffey, (1823, 105) who observed:

> *Emerging from my sequestered room, I was introduced into a spacious hall, where four-fifths of the convicts, eat their daily meals. Here were to be seen, people from almost every clime and country: Spaniards, Frenchmen, Italians, Portuguese, Germans, Englishmen, Scotchmen, Irishmen, Swedes, Danes, Africans, West-Indians, Brazilians, several Northern Indians, and many claiming to be citizens, born in the United States (1823, 105).*

Our current minorities are predominantly African-Americans, together with a small but growing contingent of Latin Americans and, in some areas of the country, Native Americans. Notes Cahalan (1979, 39): "Since 1850, when the first [prison statistics] reports were published, the combined percentage of foreign-born persons, blacks and other minority groups incarcerated by the criminal justice system has ranged between 40 and 50 percent of all inmates present." It is almost as if the prison treats minorities as interchangeable commodities. "As the percentage of foreign-born in our jails and prisons has declined, the proportion of blacks and Spanish-speaking inmates has increased" (Cahalan). We have had, if you will, a steady overrepresentation of one minority or another since the advent of modern prison statistics. If anything, this trend is worsening for African-Americans as we approach the close of the 20th century. In what follows, we will review the main lines of the history of modern prisons, with attention to the plight of minorities and women.

Penitentiaries

The penitentiary was the first truly modern prison. In a sense, it was the template or model from which most, if not all, subsequent prisons were cast. Some authorities claim that the penitentiary was a uniquely American institution. There is some truth to this claim—America adopted the penitentiary with a more thoroughgoing passion than did other countries—but it is important to note that penitentiaries did not exist in the original colonies. The first American penitentiary,

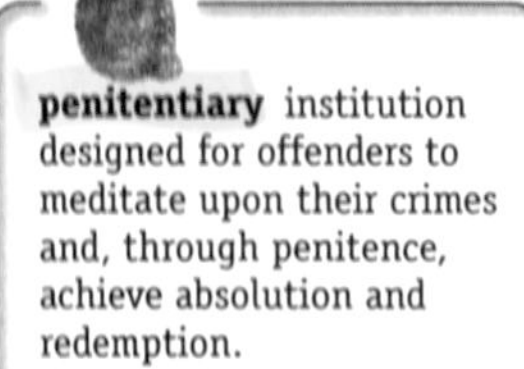

penitentiary institution designed for offenders to meditate upon their crimes and, through penitence, achieve absolution and redemption.

Walnut Street Jail the first institution that followed penitentiary ideals; i.e., single cells, individual handcrafts, isolation from temptation, classification of prisoners, a mission of reform rather than simple punishment. Quakers were instrumental in the design and implementation of the facility, originating in Philadelphia in 1790.

Jacksonian era time period of 1829–1837, marked by the presidency of Andrew Jackson, rising industrialism, a shift from agrarian to industrial economics, and a growing division between the North and the South.

separate system prison system whereby inmates had separate cells and never interacted with other inmates or outsiders during the prison sentence (see Walnut Street Jail).

Pennsylvania System another term for the separate system since the system was created and implemented in the Walnut Street Jail and, later, in the Eastern Penitentiary, in Pennsylvania (*see* separate system).

the **Walnut Street Jail**, was erected in Philadelphia in 1790. The Walnut Street Jail "carried out incarceration as punishment, implemented a rudimentary classification system, featured individual cells, and was intended to provide a place for offenders to do penance—hence the term 'penitentiary'"(Roberts 1996, 26). The construction of penitentiaries was not undertaken on a large scale, however, until the **Jacksonian era**, between 1820 and 1830. From the outset, penitentiaries were meant to be experiments in rational, disciplined living that combined punishment and personal reform.

In the most general sense, the penitentiary was meant to be a separate and pure moral universe dedicated to the reclamation of wayward men and women. It would isolate criminals from a corrupt and corrupting world, and it would reshape their characters through the imposition of a strict routine of solitude, work, and worship. Two distinct versions of this moral universe were offered, known respectively as the separate and congregate systems.

The Separate System

The **separate system** originated in Philadelphia at the Walnut Street Jail and is sometimes called the Philadelphia or **Pennsylvania System**. The regime was one of solitary confinement and manual labor, a simple monastic existence in which the prisoners were kept separate from one another as well as from the outside world. "Locked in their cells at all times, even taking their meals alone," prisoners in the separate system "had contact only with staff members, representatives of the Philadelphia Prison Society, and chaplains." On the rare occasions prisoners did leave their cells, "They were required to wear hoods or masks." It was hoped that, "With so much solitude, prisoners . . . would spend their sentences meditating about their misdeeds, studying the Bible, and preparing to lead law-abiding lives after release" (Roberts 1996, 32–33).

Describing this system, Beaumont and de Tocqueville (1833/1964, 57) observe that its advocates

> *have thought that absolute separation of the criminals can alone protect them from mutual pollution, and they have adopted the principle of separation in all its rigor. According to this system, the convict, once thrown into his cell, remains there without interruption, until the expiration of his punishment. He is separated from the whole world; and the penitentiaries, full of malefactors like himself, but every one of them entirely isolated, do not present to him even a society in the prison.*

Prisoners served time in a manner reminiscent of the monks of antiquity or the heretics of the early Middle Ages. Sentences were formally measured in loss of freedom, but the aim of punishment was penance resulting in purity and personal reform. At issue was a fundamental change of character, a conversion. Here, the penitentiary was a place of penance in the full sense of the word. Even the prisoners' labors, essentially craft work, were intended to focus their minds on the simple things of nature, and hence to bring ever to their thoughts the image of their Maker. For the prisoners of the separate system, there was to be no escape from their cells, their thoughts, or their God. The experience of solitary confinement proved to be both oppressive and destructive, "immeasurably worse," in the words of Charles Dickens, "than any torture of the body" (1842/1996, 129).

The Congregate System

The congregate system was first introduced at Auburn Prison, and is often called simply the Auburn System. Prisoners of this system slept in solitary cells. Though they congregated in large groups for work and meals, only their bodies mingled. Silence reigned throughout the prison. "They are united," observed Beaumont and de Tocqueville (1833/1964, 58), "but no moral connection exists among them. They see without knowing each other. They are in society without [social] intercourse." There was no communication and hence no contamination. Prisoners left their cells for the greater part of each day, primarily for work and sometimes also for meals. But they carried within themselves the sharp strictures of this silent prison regime. In the congregate penitentiary,

congregate system prison system where inmates slept in single cells but were released each day to work as factory or agrarian laborers; inmates also ate and exercised together. The system originated in Auburn, New York and spread throughout the Northeast.

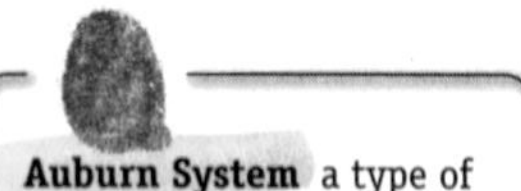

Auburn System a type of prison system that originated in Auburn, New York (*see* congregate system).

> *[Everything passes] in the most profound silence, and nothing is heard in the whole prison but the steps of those who march, or sounds proceeding from the workshops. But when the day is finished, and the prisoners have retired to their cells, the silence within these vast walls, which contain so many prisoners, is like that of death. We have often trod during night those monotonous and dumb galleries, where a lamp is always burning: we felt as if we traversed catacombs; there were a thousand living beings, and yet it was a desert solitude (Beaumont and de Tocqueville 1833/1964, 65).*

Here, too, penance and purity were sought: solitary penance by night, pure labor by day, silence broken only by the sound of machines and tools. Throughout, prisoners had time to reflect and repent. The congregate system retained the monastic features of the separate system, in its solitary cells and silent labor, but blended them with a more contemporary lifestyle. A monastery at night; by day, the congregate penitentiary was a quasi-military organization of activities (all scheduled), movement (in unison and in lockstep), eating (backs straight, at attention), and work (long hours, usually at rote factory labor). The aim of this system was to produce docile, obedient inmates. Accordingly, regimentation was the cornerstone of congregate prison life. As is made abundantly clear in Beaumont and de Tocqueville's (1833/1964, 65–66) description of the daily routine at Auburn, "The order of one day is that of the whole year. Thus one hour of the convict follows with overwhelming uniformity the other, from the moment of . . . entry into the prison to the expiration of . . . punishment."

The merits of these competing penitentiary systems were debated hotly and at great length. In the end, however, the details of the penitentiary regime and the practical definition of reform were determined as much by financial matters as by the merits of either penological perspective. Thus, the congregate system became the model for the American penitentiary at least in part because workers were in short supply in 19th-century America, and hence the deployment of prisoners at factory labor provided an affordable quarantine against the dangers and corruptions of the larger world. Elsewhere, notably in Europe, workers were in greater supply. With no appreciable demand for prison labor, the solitary system was hailed in Europe as a more pure implementation of the penitentiary ideal and became the dominant form of the penitentiary.

Women and Minorities in the Penitentiary

For the most part, women and blacks were excluded from the alleged benefits of the penitentiary. The penitentiary was considered a noble experiment in human reform; women and minorities were considered barely human—most blacks at this time were slaves, most women confined to subservient domestic roles—hence these groups were not considered fit candidates for the penitentiary's rehabilitative regime (Dodge 2002). Few women were sentenced to penitentiaries. Even fewer were exposed to the penitentiary regime.

Those who were confined to penitentiaries were warehoused, relegated to remote institutional settings such as attics, where they were often unsupervised and vulnerable to abuse (Rafter 1990, xxvi). In these barren environments, women were allowed to mingle and contaminate one another in the time-honored tradition of neglect characteristic of prisons before the advent of the penitentiary.

The early penitentiaries held few African-Americans because most were essentially incarcerated on slave plantations. Exact figures are unavailable because the early prison census figures did not even include a category for blacks (Cahalan 1979). Beaumont and de Tocqueville (1833/1964, 61) noted, that "in those states in which there exists one negro to thirty whites, the prisons contain one negro to four white persons." These prisoners were typically housed in regular, mass-confinement prisons, which made no effort at reforming prisoners and served merely to warehouse them until release. Other minorities such as immigrants were abundant in the penitentiaries, as made clear in Coffey's quote on page 23.

Paradoxically, the case can be made that women and African-Americans were inadvertently spared the considerable indignities of the penitentiary. Putting rhetoric and intention to one side, penitentiaries offered, at best, only a deceptive facade of humanity. Pain, both physical and psychological, was a central feature of the penitentiary regime. Penitentiary prisoners often went hungry; firsthand accounts report prisoners begging for food from the prison kitchen and being punished for their temerity (Johnson 2002). Diseases ran rampant among poorly nourished prisoners. Even for the healthy and well fed, life in the penitentiary was lonely and depressing and left no room whatsoever for adult autonomy. There was also the crucible of fear; from the outset, penitentiaries were maintained by the threat and practice of violence. Strict rules were routinely enforced with strict punishments, including whippings and confinement to dark cells for weeks on end. Looked at from the inside, as seen by the prisoners and not the reformers, the penitentiary was a profoundly inhumane institution.

Penitentiaries were born in a period of optimism about the prospects of reforming criminals. They reflected the **Enlightenment** faith that people entered the world as "blank slates" on which environments, including reformative prison environments, would trace individual characters. This optimism persisted for decades, even as experience proved these institutions to be unworkable. Indeed, from early on, there was evidence that penitentiaries brutalized their charges. Gradually, in the face of continuing failure, faith in the penitentiary waned.

Enlightenment time period in the 1700s when there was an explosion of art, philosophy, and science in Europe. Great thinkers such as Rousseau, Hobbes, and Locke made astute observations of the society around them, as well as the nature of human beings.

The Reformatory Era

The men's **reformatory movement**, best exemplified in the famous Elmira Reformatory, dating from 1870, kept a version of the reform-oriented prison alive after the passing of the penitentiary as a setting of reform. But this was true only for young men and only briefly, in the context of some 20 institutions developed and devoted to the discipline and rehabilitation of wayward young men (**Figure 2-1**). The reformatory movement thrived on gender stereotypes. For men, military drills formed a key feature of the reformatory regime, which sought to produce disciplined "Christian gentlemen" (Pisciotta 1983, 1994); for women, as we will see, domestic pursuits were at the heart of the reformatory regime, which in this instance sought to produce "Christian gentlewomen." The men's reformatory as a prison type proved to be a brutal, punitive penal institution, an exercise in "benevolent repression" very much like the penitentiary and no more likely to reform its inhabitants (Pisciotta 1994).

reformatory movement time period in the late 1800s when new institutions called reformatories were opened. They had a stronger emphasis on reform and targeted younger offenders.

Women's Reformatories

A notable departure from the masculine model of imprisonment for women was the reformatory. The **women's reformatory movement**, analyzed with great insight by Rafter (1990) and Freedman (1981), lasted from roughly 1860 to 1935 and produced approximately 21 institutions. Reformatories, modeled on home or domestic environments, were an explicit rejection of the male custodial model of imprisonment. These facilities were not surrounded by walls; their comparatively congenial architecture "expressed their founders' belief that women, because more tractable, required fewer constraints than men," and indeed could be housed in "cottages" featuring "motherly matrons" and a familial atmosphere rather than in traditional cell blocks run by guards. (Rafter, xxvii–xxviii).

women's reformatory movement time period in the late 1800s and early 1900s when women's reformatories were built.

The philosophy of reform that guided women's reformatories, again rejecting the male model, was premised on domestic training with an emphasis on cooking, cleaning, and waiting tables. When paroled, the women were sent to respectable families where they would work as domestic servants. Men's prisons sought to impart a tough manliness, whereas woman's reformatories preached female gentility featuring sexual restraint and domesticity. "When women were disciplined," notes Rafter (1990, xxviii), "they might be scolded and sent, like children, to their 'rooms.' Indeed, the entire regimen was designed to induce a childlike submissiveness."

Perhaps the most distinctive feature of women's reformatories was their "emphasis on propriety and decorum—on preparing women to lead the 'true good womanly life.'" Rafter draws attention to "the Thursday evening exercise and entertainment" offered at the Detroit House of Shelter in the early 1870s.

> *On this evening the whole family dress in their neatest and best attire. All assemble in our parlor . . . and enjoy themselves in conversation*

Figure 2-1 Scene from a Young Men's Reformatory. ***Source:*** *Reprinted with permission of the American Correctional Association, Lanham, Maryland.*

and needlework, awaiting the friend who week by week on Thursday evening, never failing, comes at half past seven o'clock to read aloud an hour of entertaining stories and poetry carefully selected and explained. After exchange of salutations between the "young ladies" and madam the visitor, and after the reading, tea and simple refreshments are served in form and manner the same as in refined society (quoted in Rafter 1990, 27).

Here we see what became "the hallmarks of the reformatory program: replication of the rituals of genteel society, faith in the reforming power of middle-class role models, and insistence that inmates behave like ladies" (Rafter 1990, 27). One found nothing of the sort in institutions for boys or men; as noted above, the men's reformatories were modeled on the military, not on the home.

Indeed, women from custodial institutions might well have found the domestic reformatory regime unappealing. One group of female felons, ostensibly saved from a corrupt institution for men, was reportedly angry at their new circumstances. These offenders clearly preferred the old, custodial regime, where they could trade sex for such privileges as alcohol and tobacco, to the new reformatory program, with its genteel tea parties and ladylike sociability (Rafter 1990, 32).

Women's reformatories were designed for young, minor offenders, especially those whose behavior contravened strict standards of sexual propriety (Rafter 1990). The prototypical reformatory inmate would be a young white girl of working-class background; her crime might entail little more than sexual autonomy, though this would be viewed as the earmark of prostitution. Black girls, even those convicted of minor offenses, would be routinely shunted off to custodial prisons, including the brutal custodial **plantation prisons** of the South, on the racist grounds that they were not as morally developed as white girls. As with the original penitentiaries—described by Rothman (1971/1990) as geared to reclaim "the good boy gone bad, the amateur in the trade"—reformatories were meant for novices in crime whose characters were presumed ripe for redemption. The object in both cases was to save those deemed valuable enough to warrant an investment of resources, not to reclaim hardened and essentially worthless criminals.

plantation prisons prisons in the south that were agrarian rather than industrial and utilized convict labor in the same way that earlier plantations had utilized slave labor.

- Men were leased out
- Black women worked in kitchen gardens
- white women were seldom incarcerated

Significantly, black women, who "often constituted larger proportions within female state prisoner populations than did black men within male prisoner groups" (Rafter 1990, 141), were essentially excluded from the women's reformatory movement. They were seen by reformers as too much like men to be fully adapted to the domestic model that formed the foundation of the women's reformatory. Black female offenders were sent to custodial prisons, including plantation prisons, in large numbers. In these settings, African-American women were often treated as brutally as their male contemporaries.

With the demise of the reformatory movement, reformatory institutions became filled with common felons and returned, in varying degrees, to the (male) custodial model of imprisonment. It should be noted that custodial institutions for women "were more numerous [than reformatories] even after the reformatory movement had come to fruition" (Rafter 1990, 83). These custodial institutions for women, much like those for men, "were hardly touched by the reformatory movement. They continued along lines laid down in the early 19th century, slowly growing and in some cases developing into fully separate prisons" (Rafter 1990, 83; also see Dodge 2002). Certainly it is the custodial prison, including its slave-plantation variant, which has been the main prison reserved for minorities, both men and women.

Rhetoric and the Reformatory

The rhetoric of the men's reformatory, promising differential classification and treatment but delivering heavy-handed control, had no discernible impact on the main lines of evolution followed by prisons for men (Johnson 2002). Prisons that opened at the turn of the 20th century reflected the demise of the penitentiary and reformatory. They were seen as **industrial prisons**, in which inmates labored to defray operating costs and to fill idle time; little or no attention was given to the notion of personal reform. In effect, these "fallen penitentiaries" were settings of purposeless, gratuitous pain; increasingly, they were filled with devalued minorities, mostly African-Americans. These prisons simply carried on the custodial warehousing agenda of the earliest prisons in a disciplined and regimented fashion. With the demise of prison labor in the early decades of the 20th century—due primarily to resistance from organized labor—even the industrial prison passed from the prison scene. In its wake came the "Big House," in many ways the quintessential 20th-century prison.

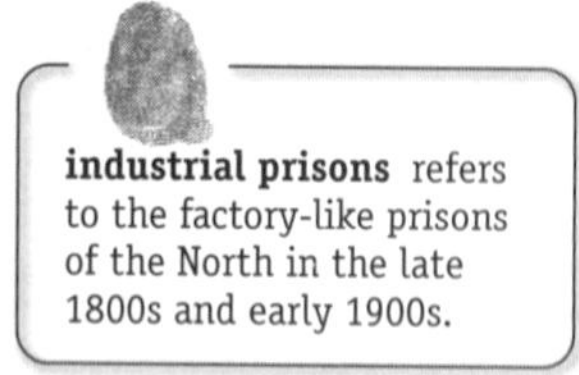

industrial prisons refers to the factory-like prisons of the North in the late 1800s and early 1900s.

The "Big House"

Maximum-security prisons throughout the first half of the 20th century were colloquially known as **Big Houses** **(Figures 2-2A, B)**. If one were to think of prisons as having lines of descent, one would say that the Big House was the primary de-

Big Houses author's term for prisons in the early 1900s.

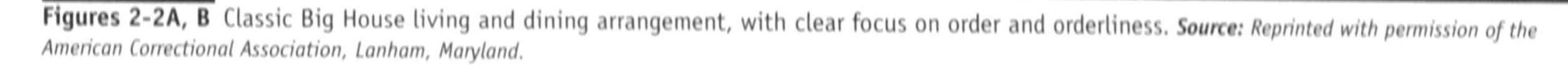

Figures 2-2A, B Classic Big House living and dining arrangement, with clear focus on order and orderliness. ***Source:*** *Reprinted with permission of the American Correctional Association, Lanham, Maryland.*

scendant and heir apparent of the penitentiary. In these prisons, a disciplined and often silent routine prevailed; prisoners worked, notably in such empty enterprises as the infamous rock pile, in which ax-wielding men broke rocks for no other reason than to show their submission to the prison authorities. The Big House prison, much like its rock piles, reflected no grand scheme or purpose; neither penance nor profits were sought. Routines were purposely empty. Activities served no purpose other than to maintain order.

The Big House's lineage was not uniform. Many southern states bypassed the penitentiary entirely. The first prisons in Texas, for example, were essentially extensions of the slave plantation (Crouch and Marquart 1989). These plantation prisons were the agrarian equivalent of the industrial prison. The object was disciplined labor of the most servile, backbreaking sort; penance was never given a second thought. From these plantation prisons, the Big Houses of the South emerged, developed primarily to provide discipline and control for inmates incapable, due to age or infirmity, of working the fields and roads of the southern states (see Rafter 1990).

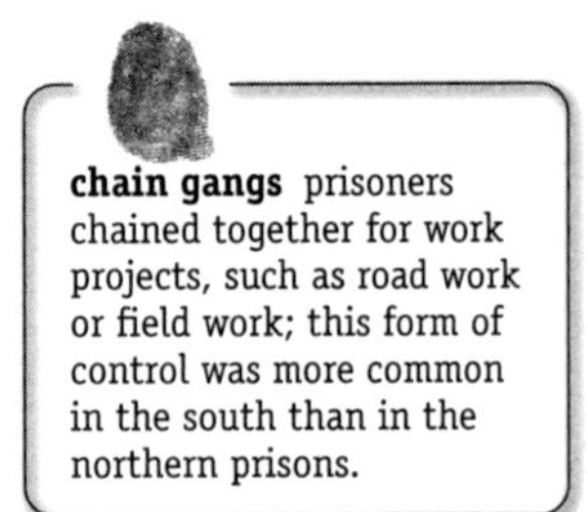

chain gangs prisoners chained together for work projects, such as road work or field work; this form of control was more common in the south than in the northern prisons.

As the name would imply, plantation prisons contained a gross overrepresentation of black prisoners, both men and women. Newly emancipated African-Americans would be incarcerated on the flimsiest pretexts and then put to hard labor in the fields of these prisons, often in **chain gangs**. Shackled groups of prisoners were also deployed to build various public works, notably roads and railroads. Other newly freed slaves would become indebted to white landowners and would be forced to work as peons to pay off debts, or to sign restrictive contracts so that they could obtain food and housing. In these various ways, vast numbers of blacks were subjected to prison or prisonlike work regimes that drew their inspiration from slavery and offered none of the hope, however illusory, associated with the penitentiary (Franklin 1989; Sellin 1976).

Significantly, southern prison chain gangs would include black female prisoners as well as black male prisoners. These black women, though few in number, were subjected to the same harsh regime as the men (Rafter 1990). Work on the chain gang and at hard field labor was generally reserved for blacks and much less often meted out to whites. Comparatively few white men, and virtually no white women, were exposed to these brutal work regimes.

Parallels between slavery and prison can be striking. Thus, southern chain gangs drew on a heritage that spanned both the original slave plantations and the lockstep march of penitentiary discipline, as revealed in first-person accounts of this brutal institution.

> *Just as day was breaking in the east we commenced our endless heart-breaking toil. We began in mechanical unison and kept at it in rhythmical cadence until sundown—fifteen and a half hours of steady toil—as regular as the ticking of a clock (Burns, a prisoner, quoted in Franklin 1989, 164–165).*

Burns, the prisoner quoted above, wrote his account of the chain gang in the 1930s. For him, the German army's goose step was the apparent inspiration for the disciplined character of the chain-gang work routine. Clearly, however, American prison officials were not borrowing from German army discipline; the lineage of this disciplined labor would be in the penitentiary lockstep, which in turn was a particular adaptation of factory discipline to the prison context (Johnson 2002). Significantly, the labor routine Burns described was unchanged from plantation-prison practices dating from the mid- to late 1800s. These practices, in turn, were rooted in slave-labor practices dating from the early 1800s, the time of the first penitentiaries. It was at this point, at the birth of the penitentiary between 1800 and 1820, that southern plantations first became formal business institutions marked by rigid discipline rather than family farming operations marked by more or less informal relations between keepers and kept (Franklin 1989). Ironically, then, the penitentiary, which originated in the North, may have found its first expression in the South in big-business plantations. Only later were facets of the penitentiary expressed in plantation prisons and custodial prisons, never reaching fruition on its own in any of the southern states.

The historical lineage of the Big House is a mixed one. Yet one can fairly conclude that the Big Houses of northern states were more than gutted penitentiaries, and the Big Houses of the South were not merely adjuncts to ersatz slave plantations. The Big House, wherever it was found, was a step forward, however modest and faltering, in the evolution of prisons. Humanitarian reforms helped to shape its inner world, though these had to do with reducing deprivations and discomforts rather than establishing a larger agenda or purpose. Thus, whereas the penitentiary offered a life essentially devoid of comfort or even distraction, the Big House routine was the culmination of a series of humanitarian milestones that made these prisons more accommodating.

The first such advance was the introduction of tobacco, which was greeted by the prisoners with great relief. Officials report, without a hint of irony, that a calm settled over the penitentiary once the "soothing syrup" of tobacco was given to the formerly irritable and rambunctious prisoners. The second reform milestone was the abolition of **corporal punishment**. In Sing Sing, a fairly typical prison of its day, corporal punishment was abolished in 1871. Prior to that time, upward of 60 percent of the prisoners would be subjected to the whip on an annual basis. Other prisons retained the practice of corporal punishment, but among prisons outside the South, whippings and other physical sanctions became an underground, unauthorized activity by the turn of the 20th century. Tragically, regimes of corporal punishment, official and unofficial, remained in place in some southern prisons for much of the 20th century (see Johnson 2002).

corporal punishment pain or punishment inflicted "to the body"; in other words, physical punishment.

The emergence of significant internal freedoms is the third and final reform milestone that paved the way for the Big House. These freedoms came in the wake of the **lockstep march**, which was abolished in Sing Sing in 1900. The daily humiliations of constrained movement implied in this shameful march soon gave way to freedom of movement in the recreation yard, first on Sundays (beginning in Sing Sing in 1912), and then, gradually over the early decades of the 20th century, each day of the week.

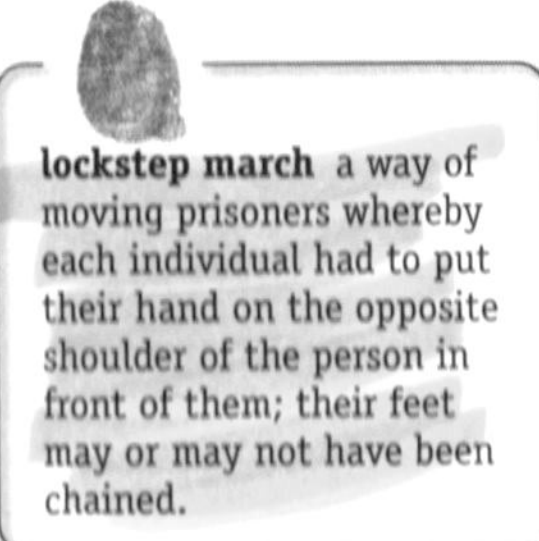

lockstep march a way of moving prisoners whereby each individual had to put their hand on the opposite shoulder of the person in front of them; their feet may or may not have been chained.

There is no doubt that the Big House was more humane than the penitentiary, but similarities between these institutions are apparent. As in the penitentiary, order in the Big House was the result of threats and force, including, in the early decades of this century, clubs and guns, which line officers carried as they went about their duties. As in the penitentiaries, rules of silence prevailed in the Big

House. Silence was both a cause and a consequence of order in the Big House and was a profound symbol of the authority of the keepers. This silence was, in the words of Lewis Lawes (1932, 34), a famous prison warden of the day, "the hush of repression."

Though the Big House was more comfortable than the penitentiary, prisoners of the Big House led spartan lives. Cells were cramped and barren; possessions were limited to bare essentials. Food was generally in good supply but was utterly uninspiring, and was, in the eyes of the prisoners, fuel for reluctant bodies and nothing more. If the dominant theme of the penitentiary was terror, the dominant theme of the Big House was boredom bred by an endlessly monotonous routine. "Every minute of the day," said Victor Nelson, a prisoner, "all the year round, the most dominant tone is one of monotony" (Nelson 1936, 15). In the extreme, the Big House could be described as a world populated by people seemingly more dead than alive, shuffling where they had once marched, heading nowhere slowly, for there was nothing of any consequence for them to do. In Nelson's words, "All about me was living death: anemic bodies, starved souls, hatred and misery: a world of wants and wishes, hungers and lusts" (Nelson, 4). In the Big House, as in the penitentiary, the prison was a world circumscribed by human suffering.

Big House prisons contained an overrepresentation of minorities, though no accounts seem to exist describing the distinctive experience and adjustment of minority prisoners in these highly structured milieus. Certainly Big House prisons, like the larger society, were racially segregated—by policy in the early years, and later by custom. Ethnic segregation of a voluntary sort was no doubt quite extreme, just as it was—and to some degree still is—in and out of prison (Carroll 1988). Early sociological discussions of northern prisons proceeded as if African-American prisoners did not exist at all within Big House walls, though, of course, that is entirely untrue. Minority prisoners, invisible to white social scientists and even to white convicts, must have formed a world of their own, apart from that of white prisoners and white officials. Fictional accounts, written by black convicts, suggest that black inmates of Big House prisons led a more materially impoverished life than their white contemporaries.

In one story, a white prisoner stumbles on an enclave of black prisoners far from the main prison living area in an area labeled "Black Bottom." In the story, it is as if the black prisoners were buried within the prison, residing at its bottom, left to suffer greatly in isolation from the larger white prison society (Himes 1934). In the typical Big House, it is white prisoners who rise to positions of considerable influence and even comfort due to their connections with the white power structure; few, if any, blacks have such an opportunity. Accounts from southern prisons, which during this era were of the plantation type, suggests that the harshest and most restrictive conditions within these prisons, particularly relating to labor, were reserved for blacks.

Big House prisons existed for women as well as for men. The origins of Big House prisons for women, like those of their male counterparts, can be traced to the penitentiary. As the numbers of female penitentiary prisoners grew, separate units within men's penitentiaries were developed for women. Eventually, these units were moved off the men's prison grounds to become completely separate and autonomous institutions. Most of these new separate institutions for women were run on a custodial model, which Rafter (1990) convincingly argues is an inherently masculine model of imprisonment. Confinement in custodial regimes was hard on the women, who were uniquely vulnerable in such settings. "Probably lonelier and certainly more vulnerable to sexual exploitation, easier to ignore because so few in number, and viewed with distaste by prison officials, women in custodial units were treated as the dregs of the state prisoner population" (Rafter, 21).

Accounts by inmates of women's custodial prisons highlight the diversity of populations within these institutions (similar to the diversity within men's prisons). This 1930 description of entry into a women's prison offers a glimpse of a motley and diverse crew of women—petty thieves, addicts, prostitutes of varying ages and nationalities, young and old, diseased and healthy. When they meet the warden, he promptly proceeds to fondle the women as a part of their orientation to the regime, warning them that those who do not submit to their superiors will be punished:

> *"The first law of this prison," he continued, putting his hand on my shoulder, and gradually running it down my side, a smirk of sensual pleasure playing upon his leather-like countenance ". . . is to obey at all times . . . to obey your superiors . . . to fit in to your surroundings without fault-finding or complaint. . . ." His hand had now progressed below my skirt, and he was pressing and patting my naked thigh. . . . "because unruly prisoners are not wanted here and they are apt to get into trouble. . . ." (quoted in Franklin 1989, 171–172).*

Sexual abuse was disturbingly common in custodial prisons run by men. Women might well be molested at intake, as in the preceding excerpt, then later raped in their cells by their male keepers. On other occasions, guards would make sport of their sexual encounters with their female captives (see, for example, Anonymous 1871). The impression one gains from this literature is that the women relegated to custodial institutions, from the penitentiary onward, had little or no choice but to submit to the predations of their keepers. In the harsh assessment of a 19th-century observer, criminal women were "if possible, more depraved than the men; they have less reason, more passion and no shame. Collected generally from the vitiated sewer of venality, they are schooled in its depravity, and practiced in its impudence" (Coffey 1823, 61). Many early 20th-century observers shared Coffey's views, at least through the period when men ran custodial prisons for women, and female offenders were routinely called "whores and thieves of the worst kind" (see Dodge 2002). As outright moral pariahs, female offenders in these prisons were presumed to be spoiled goods, there for the taking by their male keepers.

The Correctional Institution

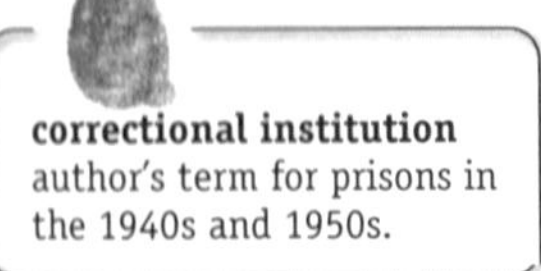

correctional institution author's term for prisons in the 1940s and 1950s.

The correctional institution emerged gradually from the Big House, with the first stirrings of this new prison type manifesting themselves in the 1940s and 1950s.

In correctional institutions, harsh discipline and repression by officials became less-salient features of prison life. The differences between Big Houses and correctional institutions were real: daily regimes at correctional institutions were typically more relaxed and accommodating. But the benefits of correctional institutions are easily exaggerated. The main differences between Big Houses and correctional institutions are of degree rather than kind. Correctional institutions did not correct. Nor did they abolish the pains of imprisonment. They were fundamentally more-tolerable human warehouses than the Big Houses they supplanted, less a departure than a toned-down imitation. Often, correctional institutions occupied the same physical plants as the Big Houses. Indeed, one might classify most of these prisons as Big Houses "gone soft."

Correctional institutions were marked by a less intrusive discipline than that found at the Big Houses. They offered more yard and recreational privileges; more-liberal mail and visitation policies; more amenities, including an occasional movie or concert; and more educational, vocational, and therapeutic programs, though these various remedial efforts seemed to be thrown in as window dressing. These changes made life in prison less oppressive. Even so, prisoners spent most of their time in their cells or engaged in some type of menial work. They soon discovered that free time could be "dead" time; like prisoners of the Big Houses before them, prisoners in correctional institutions often milled about the yard with nothing constructive to do. Boredom prevailed, though it was not the crushing boredom born of regimentation as in the Big House. Gradually, considerable resentment developed: officials had promised programs but had not delivered them. The difficulty was that officials, however well intended they might have been, simply did not know how to conduct a correctional enterprise. Nor did they have the resources or staff to make a serious attempt at that task. The correctional institution promised to transform people—a claim reminiscent of the penitentiaries—but mostly these institutions simply left prisoners more or less unchanged.

In the 1950s, Trenton State Prison in New Jersey was a fairly typical correctional institution for men, merging the disciplined and oppressive climate of the Big House with a smattering of educational, vocational, and treatment programs. Gresham Sykes's classic study *The Society of Captives* (1958/1966) describes Trenton State Prison. Significantly, Sykes describes the dominant reality at Trenton as one of pain. "The inmates are agreed," he emphasized, "that life in the maximum security prison is depriving or frustrating in the extreme" (Sykes 1958/1966, 63). To survive, the prisoner turned not to programs or officials but to peers. In essence, Sykes concluded that the prisoners must reject the larger society and embrace the society of captives if they were to survive psychologically. The prison society, however, promoted an exploitative view of the world. Weaker inmates were fair game for the strong. At best, prisoners "do their own time," to use an old prison phrase, and leave others to their predations, turning a deaf ear to the cries of victims.

Trenton State contained a substantial overrepresentation of minority offenders, no doubt a source of some conflict in the prison community. This salient fact is mentioned only in passing by Sykes (1958/1966, 81), who observes, "The inmate population is shot through with a variety of ethnic and social cleavages" that kept prisoners from acting in concert or maintaining a high degree of solidarity. Similarly, Irwin (1970) makes clear that during the 1950s, Soledad Prison, also a correctional institution, was populated largely by groups called **tips**, or cliques, that were defined largely in racial and ethnic terms. Conflict simmered below the surface of daily life, erupting only occasionally, suppressed in large measure out of a vain hope that all inmates might benefit from correctional programs. In fact, however, treatment and the prospect of mature interpersonal relations were, at best, a footnote to the Darwinian ebb and flow of daily life in the prison yard of the correctional institution. The violence would come later, after the demise of the correctional institution.

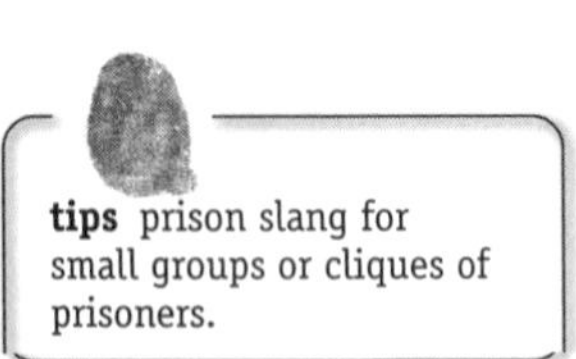

tips prison slang for small groups or cliques of prisoners.

Life in Trenton State Prison was grim. The plain fact is that prisons—whether they are meant to house men, women, or adolescents—are built for punishment, and hence are meant to be painful. The theme of punishment is nowhere more evident than in the massive walls that keep prisoners both out of sight and out of circulation. Many of our contemporary prisons are built without those imposing gray walls, though these institutions usually feature barbed wire that, ironically, is often a shade of gray. Almost all prisons feature a dull gray or other drab-colored interior environment. Colorful prisons—so-called **pastel prisons**, some built to resemble college dormitories—are few in number and are reserved for prisoners judged to pose little threat to one another, to staff, or to the public.

pastel prisons refers to prison architecture that softens the custodial aspects of the institution.

To the extent that such pastel prisons exist, they are likely to be reserved for women (Rafter 1990). Women's penal institutions more often resemble college campuses than prison compounds. Dorm rooms often replace cells; it is not uncommon to find vases of flowers in the rooms of confined female felons. Yet the ostensible comforts of women's prisons are belied by the custodial realities of daily life in these institutions, which are experienced by their inhabitants as prisons that, at best, offer too many rules and too few program opportunities. Those programs that exist, moreover, still follow stereotypical gender lines, focusing on domestic skills rather than job skills. The continuing theme is one of sexism and neglect.

Contemporary Prisons

Most of today's prisons are still formally known as correctional institutions, but this label can be misleading. One problem has been that, with the passing of the disciplined and repressive routines of penitentiaries and Big Houses, today's prisons are marked by more inmate violence than at any other time since the advent of the penitentiary. This is most apparent in men's prisons. Prison uprisings, including such debacles as the infamous Attica and Santa Fe prison riots, occur with disturbing regularity. So, too, is inmate-on-inmate and inmate-on-staff violence more common today than was the case in earlier prisons (see generally Johnson 2002). While some staff members still abuse inmates, this grossly unprofessional behavior is considerably less in evidence in today's prisons than in earlier prisons, where staff-on-inmate violence was a routine feature of daily life.

Racial and ethnic imbalances, little-noted facts of prison history, are today more pronounced than ever, and often give rise to inmate-on-inmate violence, again particularly in men's prisons. Beginning in the correctional institution, when discontent with failed programs often followed racial lines, race forms perhaps the key fault line in today's prison community (Johnson 1976; Jacobs 1977; Carroll 1988; and McCall 1995). Prisons are balkanized along racial and ethnic lines; groups and gangs defined in terms of race and ethnicity are increasingly central sources of violence in today's prisons. It is only in the contemporary prison, dating from roughly 1965, in which a black prisoner might say with confidence, "I was in jail, the one place in America that black men rule" (McCall, 149). It is, to be sure, an exaggeration to say that African-American men rule today's prisons, but minority groups—African-American, Latin American, and Native American—wield disproportionate power behind bars. Too often, that power is used to dominate and abuse whites, the despised minority group in many men's prisons. Racial and ethnic relations have been and remain more pacific in women's prisons, though anecdotal evidence from practitioners suggests that racial tensions may be rising in some women's prisons.

Over the past two or three decades, a fair number of American prisons for men have seemed out of control, with inmate violence reaching frightening proportions. Some evidence suggests that the worst of today's prison violence may be a thing of the past. The statistical trend in prisoner assaults and killings is down, at least over the past five years or so. This suggests that nonviolent accommodations are finally being worked out in our prisons, within and perhaps between inmate groups and the officials who run the prison (Johnson 2002).

The evolution of women's prisons has shown a more marked change in recent decades. Women's correctional institutions often had a relaxed climate, sometimes set in a campuslike environment. Niantic Prison in Connecticut is a case in point. During the correctional era, one correctional officer observed, "It was nice and cozy then, every inmate had her own room. She knew everybody's name and the names of her kids, boyfriends, husband, mother, father, brothers, sisters, and friends" (Rierden 1997, 2). In the correctional era, this officer "spent most of her time with a handful of inmates trying to decide whether to take them fishing or to sit with one of them and have a nice long chat" (Rierden, 2–3).

Those days are gone in Niantic—and indeed, in most, if not all, women's prisons. With the War on Drugs and the explosion of prison populations—an explosion that hit women's prisons especially hard—Niantic gradually became a "repository more than a reformatory," holding inmates with a daunting array of medical and social problems (Rierden 1997, 3). In the wake of the transition from a correctional setting to a setting of containment, the culture of the institution changed. Discipline problems became rampant. "Everybody's getting on everybody's nerves. There are more drugs, more assaults, and now AIDS and gangs" (Rierden, 13). Niantic was described as "a once snug little town steeped in tradition and culture that had been forced to undergo urbanization" (Rierden).

Rierden (1997) documents serious emerging problems at Niantic, but the worst was yet to come. In a few short years, a huge concrete-and-steel "confinement model" prison was built in the middle of the prison grounds. This new institution, called the York Correctional Institution, named, with some irony, after a warden known for her commitment to rehabilitation, is modeled on high-security men's prisons.

Each living unit was a replica of the next and each cell conformed to a standard intolerant of deviation. Unlike "old" Niantic, where inmates could add their own small touches—a crocheted pillow, an embroidered picture frame—expressions of individuality were now taboo. All personal clothing was surrendered, replaced by inmate uniforms. Inmates were now addressed by their last names only (Griffith, in Lamb 2003, 342)

Entry into this institution is accompanied by a ritual sexual humiliation reminiscent of women's prisons of the Big house era, as seen in this description provided by a York prisoner:

"Take everything off," she ordered.

"Even my bra and panties?"

"What did I just say, you stupid bitch?"

I undressed and stood naked before her.

"Now, turn around, bend over, spread your butt cheeks, and cough."

There I stood, a woman who had been too inhibited to appear naked before her husband unless it was in the dark, now facing this hostile stranger under the glare of fluorescent lights. Ashamed, I obeyed her because I had no choice.

"Okay," she said. "Now hold out your hands, palms up." She poured a thick yellow liquid into my cupped hands. "Rub this stuff in your pubic hair and the hair on your head," she commanded (Adams, in Lamb 2003, 71).

The repressive regime at York, moreover, came to influence the social climate at the other units within Niantic, units called cottages and long run as low-security dorms. "Now *all* inmates lived under tightly enforced maximum-security regulations. Many of the small, incentive-building privileges and humanizing gestures extended to low-risk inmates were surrendered during this transition" (Griffith, in Lamb 2003, 343). As so often has happened in institutions for men, the regime developed to contain and constrain the worst offenders has been extended to affect the daily lives of all prisoners, creating tensions and resentments that often have a pernicious influence on daily life. We are told that programs still abound at Niantic, and even at the York facility, but it is not clear how well these programs have survived the substantial regime changes that have taken place at these institutions.

Conclusions

George Bernard Shaw once described the prison as "a horrible accidental growth" that was made worse, rather than better, by reform efforts. (Shaw 1946, 104) There is some truth in this observation, but it is nevertheless the case that, in some important respects, conditions in today's prisons are notably better than was the case in earlier prisons. Prisons can never be returned to the days when officials ruled with an iron hand and prisoners marched, silently docile, at the command of their keepers. These regimes were themselves acts of violence, and no doubt inflicted harms in excess of today's penal institutions. It is widely recognized that prisoners today are no longer slaves of the state to be worked at will, often to the point of injury or death. Prisoners, no matter how serious their crimes, retain basic civil and human rights that were unheard of in earlier prisons. Accordingly, arbitrary or violent disciplinary practices are, with regrettable exceptions, relics of a long-dead correctional past. Similarly, involuntary treatment programs, chosen for inmates by experts, are also a thing of the past.

We know more about prison life and prison reform than ever before, and we can point to successes on a number of discrete fronts (Lin 2000; Johnson 2002). And even where the prison fails inmates, sometimes inmates find a way to change for the better. As one African-American female inmate told Paula Johnson, "Prison—now, you might think this is crazy—but prison has brought out the best in me. It has brought out the best in me, because it makes me resourceful" (Johnson 2003, 63). In the words of another woman studied by Johnson, "Prison has made me a better woman." Explaining, she observed:

> *I could have been this bigoted person. I was bigoted when I first came. I didn't want them telling me nothing. If they said something to me, I had something to say back. I wanted to have the last word. What I learned is that this is not what this is about here. They didn't put me here. They are only here for care, custody, and control. . . . The place doesn't make you or break you. You make or break yourself, depending on how you live with yourself (Johnson 2003, 105–106).*

Growth through adversity is a central component of change in prison, we believe. Yet too often our prisons squander human potential for growth and change because they are overcrowded and underfunded. Our penal institutions house, more than at any other time in prison history, excessive numbers of minorities, mostly African-Americans and Latin Americans. Incarceration of women is growing at an alarming rate, considerably above that of men; again, this is particularly true for women of color (Johnson 2003). We are, moreover, experiencing a minor rebirth of the worst excesses of Big House discipline in the form of supermax prisons. Though few in number, these brutal institutions—maximizing control, minimizing autonomy—are on the rise and may inadvertently serve as models for prisons meant to be run at lower levels of security.

Two reform strategies suggest themselves. On the one hand, we can work to implement management strategies that accommodate the legitimate human needs of our captive criminals at reasonable levels of security. On the other, we can work to reduce policies that promote the overuse of prisons, particularly among minorities. Social justice—at a minimum, color-blind use of imprisonment—is a necessary if not sufficient condition of prison reform. History teaches us that, in the absence of conscious, explicit, and continuing efforts at reform, our prisons all too readily degenerate into warehouses for our least valued and most vulnerable fellow citizens.

KEY TERMS

Auburn System—a type of prison system that originated in Auburn, New York (*see* congregate system).

Big Houses—author's term for prisons in the early 1900s.

chain gangs—prisoners chained together for work projects, such as road work or field work; this form of control was more common in the South than in the northern prisons.

congregate system—prison system where inmates slept in single cells but were released each day to work as factory or agrarian laborers; inmates also ate and exercised together. This system originated in Auburn, New York and spread throughout the northeast.

corporal punishment—pain or punishment inflicted "to the body"; in other words, physical punishment.

correctional institution—author's term for prisons in the 1940s and 1950s.

Enlightenment—time period in the 1700s when there was an explosion of art, philosophy, and science in Europe. Great thinkers such as Rousseau, Hobbes, and Locke made astute observations of the society around them, as well as the nature of human beings.

industrial prisons—refers to the factory-like prisons of the North in the late 1800s and early 1900s.

Jacksonian era—time period of 1829–1837, marked by the presidency of Andrew Jackson, rising industrialism, a shift from agrarian to industrial economics, and a growing division between the North and the South.

lockstep march—a way of moving prisoners whereby each individual had to put their hand on the opposite shoulder of the person in front of them; their feet may or may not have been chained.

pastel prisons—refers to prison architecture that softens the custodial aspects of the institution.

penitentiary—institution designed for offenders to meditate upon their crimes and, through penitence, achieve absolution and redemption.

Pennsylvania System—another term for the separate system since the system was created and implemented in the Walnut Street Jail and, later, in the Eastern Penitentiary, in Pennsylvania (*see* separate system).

plantation prisons—prisons in the South that were agrarian rather than industrial and utilized convict labor in the same way that earlier plantations had utilized slave labor.

reformatory movement—time period in the late 1800s when new institutions called reformatories were opened. They had a stronger emphasis on reform and targeted younger offenders.

separate system—prison system whereby inmates had separate cells and never interacted with other inmates or outsiders during the prison sentence (*see* Walnut Street Jail).

tips—prison slang for small groups or cliques of prisoners.

Walnut Street Jail—the first institution that followed penitentiary ideals; i.e., single cells, individual handcrafts, isolation from temptation, classification of prisoners, a mission of reform rather than simple punishment. Quakers were instrumental in the design and implementation of the facility, originating in Philadelphia in 1790.

women's reformatory movement—time period in the late 1800s and early 1900s when women's reformatories were built.

REVIEW QUESTIONS

1. What was the purpose of the penitentiary? Distinguish the *separate system* from the *congregate system*.
2. How did the *congregate system* come to dominate American corrections?
3. How was reform conceptualized during the reformatory era, and how did it differ for men and women?
4. What is the *custodial model*?
5. Describe the original prototype of the reformatory inmate, and discuss how it differs from today's prototype.
6. List three reform milestones that paved the way for the Big House.
7. Describe some of the similarities between the penitentiary and the Big House.
8. When did the correctional institution emerge, and how did it differ from the Big House?
9. Are contemporary corrections more humane than early penitentiaries?
10. List three parallels between slavery and today's prisons.
11. How did race relations among inmates evolve through time?

FURTHER READING

Dodge, M. 2002. *Whores and Thieves of the Worst Kind: A Study of Women, Crime, and Prisons, 1835–2000*. De Kalb, IL: Northern Illinois University Press.

Foucault, M. 1977. *Discipline and Punish: The Birth of the Prison*. New York: Pantheon.

Freedman, E. 1981. *Their Sister's Keepers: Women's Prison Reform in America, 1830–1930*. Ann Arbor: University of Michigan Press.

Johnson, R. 2002. *Hard Time: Understanding and Reforming the Prison*. Belmont, CA: Wadsworth.

Rafter, N. H. 1990. *Partial Justice: Women, Prisons, and Social Control*, 2d ed. Boston: Northeastern University Press.

REFERENCES

Adams, C. A. 2003. "Thefts." In W. Lamb (Ed.), *Couldn't Keep It to Myself: Testimonies from Our Imprisoned Sisters*, pp. 65–93. New York: HarperCollins.

Anonymous. 1871. *An Illustrated History and Description of State Prison Life, by One Who Has Been There. Written by a Convict in a Convict's Cell* [Prison Life, 1865–1869]. New York: Globe.

Beaumont, G. D., and A. de Tocqueville. 1833/1964. *On the Penitentiary System in the United States and Its Application to France*. Carbondale: Southern Illinois University.

Cahalan, M. 1979. "Trends in Incarceration in the United States Since 1880." *Crime and Delinquency* 25, 1: 9–41.

Carroll, L. 1988. "Race, Ethnicity, and the Social Order of the Prison." In R. Johnson and H. Toch (Eds.), *The Pains of Imprisonment*, pp. 181–203. Prospect Heights, IL: Waveland.

Coffey, W. A. 1823. *Inside Out: Or, An Interior View of the New York State Prison; Together with Bibliographic Sketches of the Lives of Several of the Convicts*. New York: printed for the author.

Crouch, B. M., and J. W. Marquart. 1989. *An Appeal to Justice: Litigated Reform of Texas Prisons*. Austin: University of Texas Press.

Dickens, C. 1842/1996. American Notes: Philadelphia and its Solitary Prison. New York: Modern Library.

Dodge, M. 2002. *Whores and Thieves of the Worst Kind: A Study of Women, Crime, and Prisons, 1835–2000*. DeKalb: Northern Illinois University Press.

Franklin, H. B. 1989. *Prison Literature in America*. New York: Oxford University Press.

Freedman, E. 1981. *Their Sister's Keepers: Women's Prison Reform in America, 1830–1930*. Ann Arbor: University of Michigan Press.

Garland, D. 1990. *Punishment and Modern Society: A Study in Social Theory*. Chicago: University of Chicago Press.

Griffith, D. 2003. "Bad Girls." In W. Lamb (Ed.), *Couldn't Keep It to Myself: Testimonies from Our Imprisoned Sisters*, pp. 335–350. New York: HarperCollins.

Himes, C. 1934. "To What Red Hell?" *Esquire* 2 (October): 100–101, 122, 127.

Irwin, J. 1970. *Prisons in Turmoil*. Boston: Little, Brown.

Jacobs, J. B. 1977. *Stateville: The Penitentiary in Mass Society*. Chicago: University of Chicago Press.

Johnson, P., 2003. *Inner Lives: Voices of African American Women in Prison*. New York: New York University Press.

Johnson, R. 1976. *Culture in Crisis in Confinement*. Lexington, MA: D. C. Heath.

Johnson, R. 2002. *Hard Time: Understanding and Reforming the Prison*. Belmont, CA: Wadsworth.

Lawes, L. E. 1932. *Twenty Thousand Years in Sing Sing*. New York: Ray Long and Richard R. Smith, Inc.

Lin, A. C. 2000. *Reform in the Making: The Implementation of Social Policy in Prison*. Princeton, NJ: Princeton University Press.

McCall, N. 1995. *Makes Me Wanna Holler: A Young Black Man in America*. New York: Vintage.

Merlo, A. V., and J. M. Pollock. 1995. *Women, Law, and Social Control*. Boston: Allyn and Bacon.

Nelson, V. E. 1936. *Prison Days and Nights*. New York: Garden City Publishing Company.

Pisciotta, A. W. 1983. "Scientific Reform: The 'New Penology' at Elmira." *Crime and Delinquency* 29, 4: 613–630.

Pisciotta, A. W. 1994. *Benevolent Repression: Social Control and the American Reformatory-Prison Movement*. New York: New York University Press.

Pollock, J. M. 2002. *Women, Prison, and Crime*, 2d ed. Belmont, CA: Wadsworth.

Rafter, N. H. 1990. *Partial Justice: Women in State Prisons, 1800–1935*. Boston: Northeastern University Press.

Rierden, A. 1997. *The Farm: Life Inside a Women's Prison*. Amherst: University of Massachusetts Press.

Roberts, J. W. 1996. *Reform and Retribution: An Illustrated History of American Prisons*. New York: United Book Press, Inc.

Rothman, D. 1971/1990. *The Discovery of the Asylum: Social Order and Disorder in the New Republic*. Boston: Little, Brown.

Sellin, T. 1976. *Slavery and the Penal System*. New York: Elsevier.

Shaw, G. B. 1946. *The Crime of Imprisonment*. New York: Greenwood Press.

Sykes, G. 1958/1966. *The Society of Captives*. Princeton, NJ: Princeton University Press.

3

Sentencing Trends and Incarceration

Alida V. Merlo
Indiana University of Pennsylvania

Chapter Objectives

- Know the incarceration rate of this country as it compares to other countries.
- Be aware of the incarceration rate differences between states and regions within the country.
- Be able to articulate some of the reasons for the high incarceration rate.
- Know the "front-end strategies" and the "back-end strategies" for dealing with overcrowding.
- Understand the effect that sentencing practices and the incarceration rate have on other social services.

From all indications, Americans love prisons. We willingly support legislation to increase their use and to extend the lengths of incarceration, with little regard for the resulting construction and maintenance costs, opportunity costs, or effects on other components of the system. We reported in the first edition of this book that there were 1.1 million offenders incarcerated in federal and state prisons (*Corrections Digest* 1994; *Corrections Today* 1996, 20). By 2003, there were more than 1.4 million offenders in state and federal prisons and approximately 760,000 offenders incarcerated in jails (Harrison and Beck 2004; Harrison and Karberg 2004, 1; Sentencing Project 2003). In short, more than 2 million men and women were incarcerated in the United States in 2003. We are in the midst of a cycle that shows little sign of abatement, although the prison population grew by only 2 percent in 2003, which is somewhat less than the average yearly growth since 1985 of 3.4 percent (Harrison and Beck, 1). This chapter examines some of the reasons for prison overcrowding, the various strategies that have been attempted to address it, and the costs in both economic and social terms of a correctional policy that is heavily influenced by incapacitation and deterrence.

capacity prison capacity is the number of beds minus a small percentage set aside for necessary transfers and emergency uses.

incarceration rate a display of the number of people incarcerated by the population, so that different size population groups can be compared.

Prisons in the United States are often overcrowded. Both the state and the federal prison systems are operating at or above their **capacity**. In 2002, "state prisons were operating between 1 percent and 17 percent above capacity, and the federal prisons were 33 percent above their rated capacity" (Harrison and Karberg 2004, 1). The **incarceration rate** displays how many adults are in prison for every 100,000 in the population. Rates are used so that we can compare the pattern of imprisonment across time or between different countries or states with different populations. The data indicate that the national prison incarceration rate in the United States in 2003 was 482 per 100,000 residents. As **Table 3-1** shows, 11 states had incarceration rates that were higher than the national average. Most notably, Louisiana was the leader with 801, followed by Mississippi with 768, Texas with 702, and Oklahoma with 636. Conversely, there were 9 states that ". . . had rates that were less than half the national rate" (Harrison and Beck 2004, 3).

Inmates in prison in 2003 were likely to be serving longer sentences than their predecessors. In 1995, the average length of incarceration was 23 months. In 2001, the average length was 30 months (Butterfield 2004, A14). These longer sentences are the result primarily of changes in sentencing laws that include mandatory minimum sentences, truth-in-sentencing laws, and various three-strikes laws that increase the prison term for repeat offenders (Butterfield, A14). As Vaughn noted in 1993, overcrowding continues to pose a serious challenge to correctional administrators (Vaughn 1993, 12). When Vaughn surveyed correctional administrators from each of the 50 states in 1990, he found that overcrowding was the number-one problem in 48 states (Vaughn, 15). At least 44 administrators linked overcrowding to four factors: increased sentence length, the drug problem, the public's desire to get tough on crime, and the legislative response to that demand (Vaughn, 15–16). These factors continue to affect the prison population and the attendant overcrowding today. Other factors such as **mandatory sentences**, increases in minimum sentence length, and more-effective law-enforcement procedures and techniques were also noted by the respondents, but not with the same level of consensus.

mandatory sentences type of sentencing whereby for a particular crime, there is no judicial discretion; the decision whether to imprison or not and/or the amount of prison time is set by statute.

The overcrowding problem occurs in the federal system as well. On June 30, 2003, there were more than 170,000 inmates in the federal prison system; this represented a 5.4 percent increase from the previous year, and more than a fifth of the growth in the U.S. prison population for 2003 (Harrison and Karberg 2004, 3; Sentencing Project 2004). The exponential increase in the federal system is evidenced by the fact that the population of federal inmates has almost doubled in the past decade with a 93 percent increase (Sentencing Project, 1). The majority of current federal inmates are incarcerated for drug rather than violent offenses. Only 13 percent of federal inmates were convicted of a violent crime, and 55 percent of the federal system population have drug-offense convictions (Sentencing Project, 1).

Prison Inmates

Unfortunately, there are two distinct groups that have been particularly affected by the increased use of imprisonment: minorities and women. Their incarcera-

Table 3-1 Prisoners Under the Jurisdiction of State or Federal Correctional Authorities, by Region and Jurisdiction, Year-end 2002–2003

Region/jurisdiction	Total			Percentage change from	
Region and jurisdiction	**12/31/03**	**6/30/03**	**12/31/02**	**12/31/02 to 12/31/03**	**6/30/03 to 12/31/03**
U.S. Total	**1,470,045**	**1,457,884**	**1,440,144**	**2.1%**	**0.8**
Federal	**173,059**	**170,461**	**163,528**	**5.8**	**1.5**
State	**1,296,986**	**1,287,423**	**1,276,616**	**1.6**	**0.7**
Northeast	**173,330**	**175,753**	**175,907**	**21.5%**	**21.4**
Connecticut[a]	19,846	20,525	20,720	−4.2	−3.3
Maine	2,013	2,009	1,900	5.9	0.2
Massachusetts	10,232	10,511	10,329	−0.9	−2.7
New Hampshire	2,434	2,483	2,451	−0.7	−2
New Jersey	27,246	28,213	27,891	−2.3	−3.4
New York	65,198	65,914	67,065	−2.8	−1.1
Pennsylvania	40,890	40,545	40,168	1.8	0.9
Rhode Island[a]	3,527	3,569	3,520	0.2	−1.2
Vermont[a]	1,944	1,984	1,863	4.3	−2
Midwest	**247,388**	**247,478**	**245,303**	**0.8%**	**0**
Illinois	43,418	43,186	42,693	1.7	0.5
Indiana	23,069	22,576	21,611	6.7	2.2
Iowa[c]	8,546	8,395	8,398	1.8	1.8
Kansas	9,132	9,009	8,935	2.2	1.4
Michigan	49,358	49,524	50,591	−2.4	−0.3
Minnesota	7,865	7,612	7,129	10.3	3.3
Missouri	30,303	30,649	30,099	0.7	−1.1
Nebraska	4,040	4,103	4,058	−0.4	−1.5
North Dakota	1,239	1,168	1,112	11.4	6.1
Ohio	44,778	45,831	45,646	−1.9	−2.3
South Dakota	3,026	3,059	2,918	3.7	−1.1
Wisconsin	22,614	22,366	22,113	2.3	1.1
South	**587,814**	**578,865**	**575,048**	**2.2%**	**1.5**
Alabama	29,253	28,440	27,947	4.7	2.9
Arkansas	13,084	12,378	13,091	−0.1	5.7
Delaware[a]	6,794	6,879	6,778	0.2	−1.2
Florida[b]	79,594	77,316	75,210	5.8	2.9

continued

Table 3-1 Prisoners Under the Jurisdiction of State or Federal Correctional Authorities, by Region and Jurisdiction, Year-end 2002–2003, continued

Region/jurisdiction	Total			Percentage change from	
Georgia[c]	47,208	47,004	47,445	−0.5	0.4
Kentucky	16,622	16,377	15,820	5.1	1.5
Louisiana	36,047	36,091	36,032	0	−0.1
Maryland	23,791	24,186	24,162	−1.5	−1.6
Mississippi	23,182	20,542	22,705	2.1	12.9
North Carolina	33,560	33,334	32,832	2.2	0.7
Oklahoma	22,821	23,004	22,802	0.1	−0.8
South Carolina	23,719	24,247	23,715	0	−2.2
Tennessee	25,403	25,409	24,989	1.7	0
Texas	166,911	164,222	162,003	3	1.6
Virginia	35,067	34,733	34,973	0.3	1
West Virginia	4,758	4,703	4,544	4.7	1.2
West	**288,454**	**285,327**	**280,358**	**2.9%**	**1.1**
Alaska[a]	4,527	4,431	4,398	2.9	2.2
Arizona[c]	31,170	30,741	29,359	6.2	1.4
California	164,487	163,361	161,361	1.9	0.7
Colorado	19,671	19,085	18,833	4.4	3.1
Hawaii[a]	5,828	5,635	5,423	7.5	3.4
Idaho	5,887	5,825	5,746	2.5	1.1
Montana	3,620	3,440	3,323	8.9	5.2
Nevada	10,543	10,527	10,478	0.6	0.2
New Mexico	6,223	6,173	5,991	3.9	0.8
Oregon	12,715	12,422	12,085	5.2	2.4
Utah	5,763	5,594	5,562	3.6	3
Washington	16,148	16,284	16,062	0.5	−0.8
Wyoming	1,872	1,809	1,737	7.8	3.5

Source: Harrison, P., and A. J. Beck. 2004. *Prisoners in 2003: Bureau of Justice Statistics Bulletin*, Table 3, p. 3. Washington, DC: U.S. Department of Justice.

Note: As of year-end 2001, the transfer of responsibility for sentenced felons from the District of Columbia to the Federal Bureau of Prison was completed. The District of Columbia is no longer eligible to participate in NPS.

[a]Prisons and jail from one integrated system.

[b]Population figures are based on custody counts (see jurisdiction notes).

[c]Jurisdiction counts reported by Florida totaled 82,012 on 12/31/03 and 80,352 on 6/30/03.

tion rates, in both the state and federal systems, have been especially dramatic. About 44 percent of all inmates sentenced to one year or more of incarceration in 2003 were black men, and 19 percent were Hispanic. By midyear 2003, there were more than 100,000 women in federal and state prisons, and this represents a 5 percent increase from 2002 (Harrison and Karberg 2004, 5). Although white female inmates outnumbered black and Hispanic female inmates in 2003, the incarceration rate for black women was substantially higher. For black women, the incarceration rate was 185 per 100,000 residents, while for whites it was 38 per 100,000 (Harrison and Beck 2004, 10). When jails are included in the incarceration rate, the differences are even more apparent: almost 6 in 10 inmates in local jails in 2003 were racial or ethnic minorities (Harrison and Karberg, 8).

There were approximately 900,000 African-Americans—or one of every eight African-American men between the ages of 25 and 29—in prison or jail in the United States in 2003 (Sentencing Project 2004, 3). When that same age group is examined for all groups in the United States, we see that 12.8 percent of African-Americans, 3.7 percent of Hispanics, and 1.6 percent of whites were in jail or prison in 2003 (Harrison and Karberg 2004, 11). Clearly, young African-American men are adversely affected by their disproportionate incarceration rates.

Of all the minority groups incarcerated in the United States, Hispanics are the fastest-growing group that is being sentenced to prison. They made up more than 15 percent of all state and federal inmates in 2001 (Sentencing Project 2004). Four percent of Hispanic men in their 20s and early 30s were in jail or prison. By comparison, 1.8 percent of white men in that age range were in jail or prison in the United States (Sentencing Project 2003). In Connecticut and Pennsylvania, the rate of incarceration for Hispanics was seven times higher than the rate for whites (Sentencing Project 2003, citing Beck, Karberg, and Harrison 2002).

Table 3-2 illustrates that women are continuing to experience the rapid increase in incarceration that was documented in the first edition of this book. In 1994, Gilliard and Beck (1994) noted that women made up 5.8 percent of all inmates in the United States at the end of 1993, but that their numbers increased annually at a faster rate (9.6 percent) than men (7.2 percent). For 2003, the number of female inmates increased 3.6 percent compared to a 2.0 percent increase for male inmates (Harrison and Beck 2004, 4). By 2003 female inmates made up almost 7 percent of all inmates (Harrison and Beck, 4). From 1995 to 2003, the number of male inmates grew by 29 percent. By contrast, the number of female inmates during that same time period increased by 48 percent (Harrison and Beck). There is no empirical evidence to suggest that an increase in women's criminal behavior can completely explain the surge in their imprisonment (Pollock 2002).

On the contrary, the national data on incarcerated women found that the proportion of women in prison for violent offenses actually dropped in the 1980s and the 1990s with the exception of assault (Merlo and Pollock 2005; Chesney-Lind and Pollock 1995, 159). When comparing 1986 arrest data with 2000 arrest data, the Justice Policy Institute staff found that the number of arrests of women decreased by 3 percent, "... but that their admission to state prisons had increased by 170 percent," and "... two out of three of these women were convicted of nonviolent drug and property offenses" (Justice Policy Institute 2004, para. 6). State statistics that have been compiled indicate that the War on Drugs has adversely

War on Drugs the term coined to describe the governmental response to drug use, starting during President Reagan's presidency. This "war" is characterized by more resources devoted to interdiction and control rather than treatment and prevention.

Table 3-2 Women Under the Jurisdiction of State or Federal Correctional Authorities, Year-end 1995, 2002, and 2003

	Number of female inmates			Percentage change		
Region and jurisdiction	2003	2002	1995	2002–2003	Average 1995 to 2003[a]	Incarceration rate, 2003[b]
U.S. Total	101,179	97,631	68,468	3.6%	5 %	62
Federal	11,635	11,234	7,398	3.6%	5.8%	7%
State	89,544	86,397	61,070	3.6	4.9	56
Northeast	9,108	9,381	8,401	−2.9%	1 %	28
Connecticut	1,548	1,694	975	−8.6	5.9	46
Maine	124	90	36	37.8	16.7	18
Massachusetts[c]	708	704	656	0.6	1	12
New Hampshire	117	144	109	−18.8	0.9	18
New Jersey	1,517	1,586	1,307	−4.4	1.9	34
New York	2,914	2,996	3,615	−2.7	−2.7	29
Pennsylvania	1,823	1,821	1,502	0.1	2.5	29
Rhode Island	222	214	157	3.7	4.4	10
Vermont	135	132	44	2.3	15	27
Midwest	15,682	15,306	10,864	2.5%	4.7%	47
Illinois	2,700	2,520	2,196	7.1	2.6	42
Indiana[c]	1,758	1,583	892	11.1	8.9	56
Iowa	716	703	425	1.8	6.7	48
Kansas	629	537	449	17.1	4.3	46
Michigan[c]	2,198	2,267	1,842	−3	2.2	43
Minnesota	435	455	217	−4.4	9.1	17
Missouri	2,239	2,274	1,174	−1.5	8.4	76
Nebraska	323	352	211	−8.2	5.5	35
North Dakota	113	103	29	9.7	18.5	34
Ohio	2,897	2,929	2,793	−1.1	0.5	49
South Dakota	269	227	134	18.5	9.1	69
Wisconsin	1,405	1,356	502	3.6	13.7	47
South	43,389	41,801	27,366	3.8%	5.9%	74
Alabama	2,003	1,697	1,295	18	5.6	82
Arkansas	887	854	523	3.9	6.8	63

Delaware	508	542	358	-6.3	4.5	53
Florida[c]	5,068	4,595	3,660	10.3	4.2	58
Georgia	3,145	3,129	2,036	0.5	5.6	71
Kentucky	1,411	1,269	734	11.2	8.5	63
Louisiana	2,405	2,398	1,424	0.3	6.8	104
Maryland	1,248	1,264	1,079	-1.3	1.8	42
Mississippi	2,163	2,082	791	3.9	13.4	134
North Carolina[c]	2,256	2,173	1,752	3.8	3.2	37
Oklahoma	2,320	2,338	1,815	-0.8	3.1	127
South Carolina	1,576	1,671	1,045	-5.7	5.3	68
Tennessee[c]	1,826	1,735	637	5.2	14.1	61
Texas	13,487	13,051	7,935	3.3	6.9	98
Virginia	2,681	2,641	1,659	1.5	6.2	71
West Virginia	405	362	129	11.9	15.4	42
West	21,365	19,909	14,439	7.3%	5%	61
Alaska	392	349	243	12.3	6.2	55
Arizona	2,656	2,428	1,432	9.4	8	85
California[c]	10,656	9,987	9,082	6.7	2	57
Colorado	1,736	1,566	713	10.9	11.8	77
Hawaii	685	669	312	2.4	10.3	68
Idaho	592	592	212	0	13.7	86
Montana	419	345	112	21.4	17.9	91
Nevada	880	851	530	3.4	6.5	79
New Mexico	576	518	278	11.2	9.5	56
Oregon	883	812	465	8.7	8.3	49
Utah	427	371	161	15.1	13	35
Washington	1,288	1,254	793	2.7	6.3	41
Wyoming[c]	175	167	106	4.8	6.5	70

Source: Harrison, P., and A. J. Beck. 2004. *Prisoners in 2003: Bureau of Justice Statistics Bulletin*, p. 5. Washington, DC: U.S. Department of Justice.

[a]The average annual percentage increase from 1995 to 2003.

[b]The number of female prisoners with sentences of more than 1year per 100,000 female U.S residents.

[c]Growth from 1995 to 2003 may be slightly overestimated due to a change in reporting from custody to jurisdiction counts.

affected women (Chesney-Lind and Pollock 1995, 158; McShane and Williams 2005; Owen 2005). National data indicate that women who are currently in prison in the United States are more likely than men to be incarcerated for a drug charge (30 percent versus 20 percent) (Sentencing Project 2004, 2).

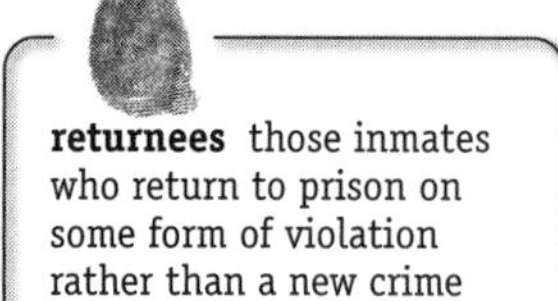

returnees those inmates who return to prison on some form of violation rather than a new crime conviction.

Prison admission statistics consist primarily of two categories of offenders: new court commitments and **returnees** (parole violators, mandatory-release offenders, or others released under some level of supervision). Both categories of admissions have increased dramatically since the early 1990s. However, admissions have continued to outpace releases, and this has contributed to the overcrowding problem. According to Harrison and Karberg (2004, 6), new court commitments to prison have continued to rise since 1990 from 323,069 to 392,717 in 2002. There were 133,870 offenders returned to prison because of parole violations in 1990, and 207,251 offenders returned in 2002 (Harrison and Karberg, 6). These offender readmission statistics help explain the burgeoning prison population in the United States. However, new court commitments have increased 13 percent since 1998 while parole violators' commitments have increased by only 1 percent (Harrison and Karberg).

State courts continue to rely heavily on incarceration in felony cases. In 2002, 69 percent of all the felons who were convicted in state courts were sentenced to a term of incarceration. For these offenders, 41 percent were sentenced to state prisons, and 28 percent were sentenced to local jails (Durose and Langan 2004, 2). With respect to offenders convicted of drug offenses, 66 percent of those convicted were sentenced to a term of incarceration in jail or prison in the state system (Durose and Langan, 3).

When comparing 1990 and 2002 sentencing data, it is apparent that judges utilized probation slightly more frequently in 2002. In 1990, straight probation was the sentence imposed on 29 percent of convicted felons. In fact, it was used most often for fraud and embezzlement; more than half of these offenders were sentenced to probation (Langan, Perkins, and Chaiken 1994, 5). In 2002, judges imposed a sentence of probation on 31 percent of felons who were convicted in state courts (Durose and Langan 2004, 1). These data suggest that part of the prison-overcrowding problem may be due to the incarceration of property offenders (especially burglars and thieves).

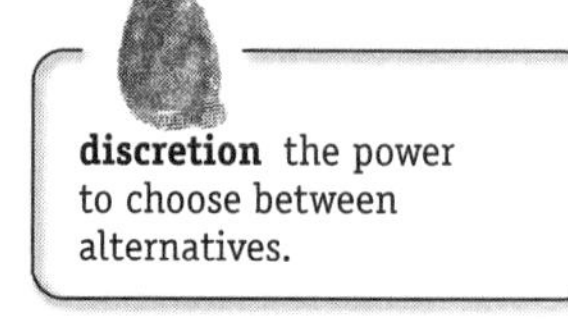

discretion the power to choose between alternatives.

indeterminate sentencing a form of sentencing where the amount of time served is not set at sentencing, but rather is determined by some other body such as a parole board.

Sentencing Reforms

In trying to understand the rationale for our increased reliance on prison as the primary method of crime control, it is useful to examine the sentencing changes that have occurred in the past 50 years. The federal government and the states employ various sentencing approaches. According to the U.S. Department of Justice (1988, 91), "the basic difference in sentencing systems is the apportioning of **discretion** between the judge and parole authorities." Discretion is the power to decide; in this case, the power to decide how much time the offender will serve in prison.

Shortly after World War II, the predominant approach was some type of **indeterminate sentencing**. These sentences meshed with the notion of prisons as correctional institutions (Irwin and Austin 1994, 9). In this kind of sentencing scheme, the judge imposed a minimum and a maximum sentence length. However, some authority other than the judiciary—most typically, the parole authority in the various states—determined the actual time served.

In some jurisdictions, **partially indeterminate sentencing** statutes existed. These statutes empowered the judge to set the maximum sentence that the offender could serve, but there was no minimum stipulated. Once again, the ultimate decision to release was in the hands of the parole authority (Bureau of Justice Statistics). These kinds of sentencing approaches reflected the rehabilitative ideal. Decisions to release an inmate were to be individualized and tailored to the specific offender's progress within the institution. They were very popular in the United States for most of the 1950s and 1960s.

partially indeterminate sentencing a form of sentencing whereby judges may set either the minimum or the maximum or some percentage of total time that must be served and the paroling authority would have the remaining power to set the actual time served.

judicial discretion in reference to sentencing, this term means that judges make the determination of how long the sentence will be.

sentencing disparity the phenomenon of offenders who commit similar crimes but receive different sentences.

determinate sentencing a form of sentencing that eliminates judicial discretion; sentences are set by statute for each crime.

Beginning in the 1970s, the indeterminate sentence came under attack from both liberals and conservatives. As discussed in Chapter 1, liberals, advocating for the rights of offenders (including war resisters and civil-rights activists), contended that indeterminate sentencing adversely affected inmates and treated then unjustly and coercively (Goodstein and Hepburn 1985, 14). They also criticized excessive **judicial discretion** and **sentencing disparity** (Goodstein and Hepburn, 18). Simultaneously, conservatives criticized the treatment ideal and found some empirical support for their position that rehabilitation was a failure (Goodstein and Hepburn, 15). Additionally, conservatives were anxious to make sentences tougher by restricting the judges' ability to impose lenient sentences (Tonry and Hamilton 1995, 3). These two diverse groups agreed on at least one thing: the indeterminate sentence was unfair, albeit for very different reasons.

The **determinate sentencing** approach began to be viewed as more impartial and a greater deterrent to crime. In those states that adopted such statutes, the central feature was a fixed period of incarceration from which there could be no deviation. An offender served his full term, less whatever good-time credits he accrued while in prison. These kinds of sentences effectively ended any parole discretion (Bureau of Justice Statistics). Proponents of determinate-sentencing statutes contend that they physically restrain offenders from engaging in crime while deterring them from future criminal tendencies.

Determinate-sentencing statutes affect overcrowding in three ways. First, offenders who previously might have been placed on probation or in other community-based settings are now sent to prison. Second, the average sentence length is increased so that offenders stay in prison longer. Third, some states abolished parole when they established determinate-sentencing models, thereby eliminating a potential release mechanism to cope with overcrowding (Holten and Handberg 1994, 228; Singer 1993, 174).

Federal and state governments rely on a number of sentencing options in addition to the three basic approaches just discussed (indeterminate, partially indeterminate, and determinate) (Bureau of Justice Statistics). For example, most states now have mandatory-sentencing laws (Bureau of Justice Statistics). The central feature of these laws is that judges are required to impose fixed periods of incarceration for offenders convicted of certain specific crimes. These very popular laws severely restrict the judges' discretion in sentencing.

Some states have adopted a determinate-sentencing structure called **presumptive sentencing**. This system requires judges to sentence offenders convicted

presumptive sentencing a form of determinate sentencing that allows for a small increase or decrease of a set sentence depending on stated mitigating or aggravating circumstances that must be proved in court.

of certain offenses to a specified period of incarceration. Judges can deviate from these set terms if there are extenuating or mitigating circumstances, but judges must still conform to the legislative boundaries established. Any deviation usually requires the judge to stipulate in writing the rationale for the change in the sentence authorized (Holten and Handberg 1994, 222; U.S. Department of Justice 1988, 91).

The Supreme Court and the Sentencing Process

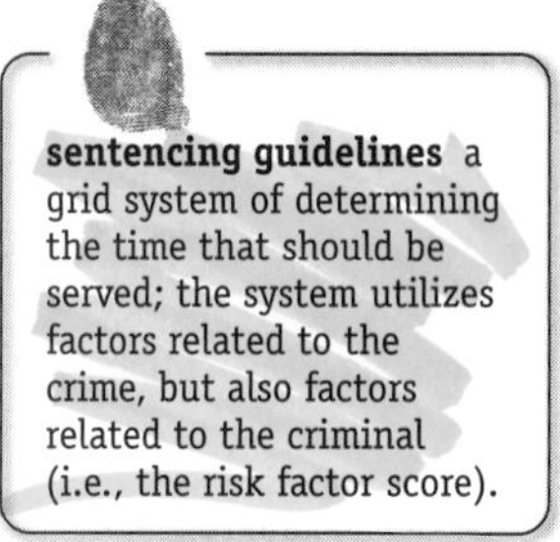

sentencing guidelines a grid system of determining the time that should be served; the system utilizes factors related to the crime, but also factors related to the criminal (i.e., the risk factor score).

good-time credits time taken off of a prison sentence, usually earned for good behavior and/or for participating in prison programs.

Two other sentencing changes have occurred in the past 20 years. The first involves the use of <u>sentencing guidelines</u>, which usually are drafted by a separate sentencing commission. They can be prescriptive, stipulating whether the judge is required to impose a prison sentence, or they can be advisory, providing information to the judge but not requiring a specific sentence (U.S. Department of Justice 1988, 92). These guidelines can be either incorporated in the statute or used selectively by the various states and the federal government (U.S. Department of Justice, 91). For example, when Congress enacted the Comprehensive Crime Control Act of 1984, it included the Sentencing Reform Act of 1984 (Kennedy 1985, 113). This legislation established the independent U.S. Sentencing Commission, which developed guidelines for all federal judges to use in the sentencing of convicted federal offenders (Kennedy, 119). It abolished parole and mandated that offenders could only accrue <u>good-time credits</u> not to exceed 15 percent of their total sentence or 54 days per year (Kennedy, 119; Langan, Perkins, and Chaiken 1994, 8). This "truth-in-sentencing" law applies to offenders who committed crimes after November 1, 1987 (Langan, Perkins, and Chaiken).

There is considerable controversy about federal sentencing guidelines that are used to sentence approximately 60,000 offenders each year (Cohen 2004, A1). Two recent Supreme Court decisions, *U.S. v. Booker* (No. 04-104) and *U.S. v. Fanfan* (No. 04-105), challenged the federal sentencing guidelines (McGough, 2004, A-6). In January of 2005, the justices determined that the federal sentencing guidelines were in violation of a defendant's right to a trial by jury when the guidelines authorized judges to increase an offender's sentence without having to prove the elements (charges) relied upon to increase the sentence beyond a reasonable doubt. In addition, the justices ruled that the guidelines should be used as advisory rather than mandatory when federal judges are imposing sentences on convicted offenders. The justices left open the possibility of appeals for defendants who questioned the "reasonableness" of the disposition (Greenhouse 2005, A1; *U.S. v. Booker* [No. 04-104]). However, the justices determined that only those cases in the appellate process could be affected by their decision. Cases that have already been reviewed cannot be revisited in light of this ruling (Cohen and Fields 2005, A4). The discretion that federal judges have regained in the sentencing process may not last too long. Congress is already discussing ways to revise the sentencing guidelines to continue the mandatory sentences designed to get tough with offenders (Cohen and Fields, A1).

The federal sentencing guidelines were considered by the Court after its ruling in *Blakely v. Washington* (124 S. Ct. 2531, 159 L. Ed. 2d 403) in June of 2004. By a 5-4 decision, the justices struck down the sentencing requirements under Washington State law that permitted a judge to increase an offender's prison term if there were aggravating circumstances associated with the crime. For example, if an offender used "deliberate cruelty" or played a leadership role when he or she committed the crime of kidnapping, then the judge could impose a longer sentence (McGough 2004, A6; Cohen 2004, A1).

In the *Blakely* decision, the justices determined that the state's statute was invalid because it permitted judges rather than juries to increase an offender's sentence on the basis of these aggravating factors (McGough 2004, A6). In order for a judge to continue the practice of increasing a sentence, the justices determined that the offender would have to have admitted to the aggravating factor, or the jury would have to have found the exacerbating factor true beyond a reasonable doubt (Cohen 2004, A1). Since the federal sentencing guidelines also permit judges rather than juries to enhance the offender's sentence if there are aggravating circumstances, it was widely speculated that the U.S. Supreme Court would apply its decision in *Blakely* to the federal system (McGough, A6). In the interim between *Blakely* and the Court's decision in January 2005, federal district-court judges were sentencing offenders with an eye toward a possible decision from the U.S. Supreme Court. These strategies ranged from requiring a defendant to sign a waiver that he/she would not contest the sentence imposed if the U.S. Supreme Court overturned the federal sentencing guidelines enhancement provisions, to deferring the imposition of the sentence, although some judges sentenced the defendant as if nothing had changed or was likely to change (Cohen, 1).

sentence-enhancement statutes statutes designed to increase the total sentence if certain stated factors are proven to be present.

Sentence-enhancement statutes authorize judges to impose longer sentences or to remove the parole possibility for chronic offenders (U.S. Department of Justice 1988, 91). Since the sentencing guidelines took effect in 1987, more than 44 percent of sentenced offenders have had their sentences enhanced (Cohen and Fields 2004, A1). In the federal system, offenders can have their sentences increased by a variety of strategies. It is important to note that these enhancements can occur without ever having proved the allegations that are made. In one instance, Carla Lyn Clifton was studying criminology when she agreed to allow her cousin's boyfriend to use her name to get a cell phone. He allegedly had bad credit. The boyfriend used the phone to sell crack cocaine. Although Ms. Clifton initially told the police that he was the only one using the phone, she changed her story when she testified before the federal grand jury. At that time, she said that she was the only one using the phone. Ms. Clifton was convicted of perjury, and under the sentencing guidelines, she probably would have been sentenced to a year in prison. However, the probation officer pointed out that if someone commits perjury associated with a crime, then that person should be sentenced according to that crime. In short, Ms. Clifton should be sentenced as a cocaine trafficker. Under that formula, Ms. Clifton would have to be sentenced to 10 years in prison. Although the judge was shocked, he eventually sentenced her under the cocaine-trafficking statute, but as a minor player. That netted her a sentence of three years and five months (Cohen and Fields 2004, A5).

three-strikes laws statutes that allow for an increased sentence if the offender has been previously convicted of two crimes.

Chronic-Offender Legislation

The second change that has occurred involves the proliferation of statutes known as chronic-offender, or **three-strikes law** in the federal and state systems. These

Found constitutional by Supreme Court $ voters rejected changes to it statutes have become increasingly popular in the United States. For example, the Violent Crime Control and Law Enforcement Act of 1994 authorized mandatory life imprisonment for offenders convicted of a third violent or drug felony (Benekos and Merlo 1995, 4). Half of the states have joined the federal system in passing these laws (King and Mauer, 2001, 3). The irony of these laws is how many times they are applied to nonviolent offenders. For example, in California as of March of 2001, approximately 58 percent of the offenders convicted of a "third strike" were for nonviolent offenses, and 69 percent of those convicted of a "second strike" were for nonviolent offenses (King and Mauer, 8–9).

These laws are popular for a variety of reasons. First, they are a manifestation of the public's penchant for the "get tough" rhetoric that has characterized criminal-justice policy in recent years. Second, the baseball slogan is appealing because of its sound-bite appeal. Politicians are able to convey their message simply and quickly. Third, the laws are perceived as a deterrent to potential criminals. Last, these laws purport to take dangerous offenders off the streets for good, providing the public with a sense of safety and security.

In November 2004, California voters had the opportunity to approve Proposition 66, a referendum to revise the state's three-strikes law and require that the first and/or second strike be a violent or serious crime as opposed to any felony offense. The referendum was defeated when 52.7 percent of the voters indicated their disapproval of Proposition 66 and their interest in maintaining the status quo with the three-strikes law (Institute of Governmental Studies 2004). The referendum results suggest that voters desire to keep the law as it was originally enacted. It also reflects a continuation of the "get tough" sentiment that helped fuel the enactment of this law in 1994.

Although the shibboleth is novel, the legislative action is not. Habitual-offender laws were first created in New York in the late 1700s (Turner, Sundt, Applegate, and Cullen 1995, 17). Other states followed New York's lead by enacting similar popular laws. Even as recently as 1968, 23 states had statutes that authorized life imprisonment for habitual offenders who had previously been convicted of certain specified offenses (Turner et al.).

Despite their appeal, these sentencing laws exacerbate already overcrowded prison conditions while simultaneously creating more problems for state and federal governments, criminal-justice agencies, and prison administrators. The hidden costs of such policies will be borne ultimately by the public, whose penchant for the "quick fix" does not generally include a long-term commitment of public funding. Most important is the high cost that is associated with housing older inmates sentenced under such statutes and who are required to remain in prison long after their criminality has receded and when their health-care demands are accelerating.

Federal prisons are becoming overcrowded with drug and other offenders who increasingly are being prosecuted and sentenced in the federal courts. Using a variety of legislative strategies such as the 1946 Hobbs Act and the Violent Crime Control and Law Enforcement Act of 1994, prosecutors are able to bring felons into the federal court system. In addition, Congress has enacted more criminal laws during the past 30 years, and it is estimated that there are more than 3,500 such criminal statutes now. Federal judges are sentencing these offenders to longer terms of incarceration than they would have received under the state statutes (Fields 2004c, 1). For example, James McFarland was sentenced to 97 years in a federal prison after he was convicted under the Hobbs Act of four robberies of convenience stores. The conviction could have resulted in a sentence of 13 years in prison if his case had been retained in the state system. The Hobbs Act was ". . . originally designed to keep racketeers from obstructing interstate commerce" (Fields, 1). The federal prosecutor reasoned that the money McFarland netted in the robberies might have been used to buy out-of-state goods by the Fort Worth convenience stores (Fields, A10). The most dramatic escalation of federal prosecution and incarceration, however, has been related to the War on Drugs.

Drug Offenders and Prison Overcrowding

In his survey of prison administrators, Vaughn (1993) found that, in addition to the increase in the length of sentences, the practice of incarcerating drug offenders was perceived as one of the explanations for prison overcrowding. This trend continues, particularly in the federal system. In 2002, 91 percent of offenders convicted of drug offenses in the federal system were sentenced to a term of incarceration in a prison or jail, compared to 66 percent of felons convicted of drug offenses in state courts (Durose and Langan 2004, 3). To more clearly understand how the incarceration of drug offenders affects prison populations, it is useful to examine a few recent trends regarding criminal-justice policy and drugs.

According to Zimring and Hawkins (1995, 162), two categories of offenders are responsible for the majority of all sentences to prison in the United States: (1) drug users who have been convicted of drug or property offenses, and (2) recidivist property offenders. In examining new court commitments, Durose and Langan (2004, 3) found that the number of offenders admitted to prison or jail totaled 224,617 for drug offenses. The second-largest category was the 214,632 property offenders who were sentenced to a term of incarceration in prison or jail in state courts in 2002 (Durose and Langan). Clearly, there are more drug and property offenders being convicted and sentenced to prison than violent offenders.

In their analysis of 2002 federal and state court data, Durose and Langan (2004, 3) found that of the 15 offenses they studied, drug-trafficking offenses were the most numerous. They accounted for 21 percent of the total number of convictions in the United States. Not surprisingly, drug crimes accounted for 31 percent of the state-prison sentences that were imposed, and 45 percent of the federal sentences imposed in 2002 (Durose and Langan). Clearly, American drug policy and law-enforcement efforts have played a part in augmenting prison populations.

Zimring and Hawkins (1995) contend that, unlike other categories of offenders, the number of offenders arrested and convicted of drug offenses varies dramatically over time as policies change. These kinds of variations are not found for other serious crimes (Zimring and Hawkins, 134–136). Not only has the number of drug arrests continued to increase, but "the proportion of prisoners being punished for drug crimes has increased more rapidly since the mid-1980s than ever before in United States history" (Zimring and Hawkins, 162).

There is ample evidence that the War on Drugs initiated during the Reagan administration targeted cocaine rather than heroin. Cocaine first appeared to be the drug of wealthy Americans in the 1970s, but during 1985–1986 a new form of cocaine known as crack appeared (Kappeler, Blumberg, and Potter 1996, 153). It was this new "smokable" cocaine that was relatively inexpensive and widely available that prompted the criminal-justice system to respond. With strong public support, the government aggressively investigated, apprehended, and incarcerated drug offenders.

For the most part, these policies have continued. Irwin and Austin (1994) contend that the war the government has waged against drugs has not only prompted a movement toward more-punitive sentencing for drug offenders, but has focused on crack cocaine, which is sold and available primarily in inner-city neighborhoods. The residents of these neighborhoods, African-Americans and Hispanics who use the drug, have been disproportionately incarcerated. In fact, two groups have been significantly impacted by the War on Drugs: minorities and women.

With the Sentencing Reform Act of 1984 and the implementation of the U.S. Sentencing Commission's guidelines, the disproportionate representation of African-Americans in prisons has increased significantly. Fields (2004a) elaborates on the current racial disparity versus the preguideline days. Today, minorities receive longer sentences than white offenders for offenses such as drug trafficking. Furthermore, there are significant differences in the type of trafficking for African-American and white offenders. This accounts for most of the variation in sentencing: African-Americans are more likely to be convicted of trafficking crack cocaine, and whites are more likely to be convicted of trafficking powdered cocaine (McDonald and Carlson 1994, 9).

When Congress enacted the Anti-Drug Abuse Act of 1986, it established minimum sentences for trafficking crack and powdered cocaine. However, the punishments for selling crack are greater than those for selling powdered cocaine. In short, "the punishments for selling or possessing crack (50 grams) are the same as those for selling 100 times these amounts of powdered cocaine" (McDonald and Carlson 1994, 9). For example, if an offender is convicted of possessing or selling as little as 5 grams of crack (approximately a teaspoon), the sentence is a minimum of five years in prison (Fields, 2004a, A4; Isikoff 1995, 77). By contrast, possession of 500 grams of powdered cocaine results in a five-year sentence. According to Fields, "The average crack defendant is sentenced to 115 months, compared to 77 months for defendants involved in powder-cocaine offenses."

The sentencing differences are especially problematic in view of prospective users. Crack is more likely to be used by African-Americans than by whites. As McDonald and Carlson (1994) point out, it may not be the case that African-Americans prefer crack to powdered cocaine, but crack is more heavily promoted in minority neighborhoods, and it is cheaper. Research shows that about 13 percent of African-Americans use drugs, but more than 60 percent of narcotics convictions are African-Americans (Pollock 2004, 41–42).

For women, the effects of the War on Drugs are even more pronounced. According to Pollock, "Drug offenses account for the largest percentage of women in prison" (Pollock 2002, 53). Bloom, Chesney-Lind, and Owen (1994) found that one out of three women in prison in the United States was serving a sentence for drug offenses in 1993. Even though the drug-enforcement strategies target major drug dealers, Bloom et al. (1994, 1) found that more than 35 percent of the women who were incarcerated in prison for drug convictions were there for the crime of "possession." In the 2002 Arrestee Drug Abuse Monitoring Program (ADAM) study, researchers found that even though women made up 20 percent of those arrested in 2000, 63 percent of the women arrested tested positive for using illegal drugs compared to 64 percent of men. In the sites that were tested, women used cocaine, marijuana, opiates, and methamphetamines (Taylor, Newton, and Brownstein [N.I.J.], 2003, 47; Davis, Merlo, and Pollock 2005). The available data lend support to the research suggesting that female drug use is increasing and that it is a factor involved in women's criminal behavior (Davis, Merlo, and Pollock).

These trends are particularly disturbing when the paucity, duration, and intensity of drug-treatment programs in prison are well documented (Wellisch, Prendergast, and Anglin 1994). More women are being convicted and imprisoned for drug offenses than previously (one manifestation of mandatory-sentencing laws), but there is little emphasis directed toward treating these women in prison. For example, Covington found that 60 percent of the women in prison in California were there for crimes related to drugs, but only about 3 percent had the option of participating in drug or alcohol programs (Covington, 1998, cited in Pollock 2002, 103). Rather than treatment, some drug companies have actually provided grants to local law-enforcement agencies to facilitate their investigative and arrest strategies. For example, Purdue Pharma. L.P., a drug company that produces OxyContin, has provided more than $1.7 million to police and sheriff departments throughout the United States since 2002 (Tesoriero 2004, B3). Unfortunately, mandatory-sentencing laws also foreclose the possibility of community treatment.

Responding to Overcrowding

The research on prison overcrowding and its relationship to inmate violence, health, stress, prison disciplinary problems, and recidivism is not conclusive, probably for several reasons. First, there is the difficulty in finding comparable control groups. Although some jurisdictions are reportedly operating below 100 percent of their highest capacity, they are still experiencing some aspects of overcrowding. Therefore, this current correctional climate restricts opportunities for finding institutions to evaluate. Second is the difficulty in comparing one inmate population with another. One institution might house more-recalcitrant or more-violent offenders, thus affecting the impact of overcrowding (Durham 1994). Last, living in overcrowded conditions may have long-term effects that are not detectable until several years later (Durham).

Despite these limitations, some research suggests "that prisons housing significantly more inmates than a design capacity based on sixty square feet per inmate are likely to have high assault rates" (Gaes 1985, 95). Lawrence and Andrews (2004) confirmed earlier research when they found that prison crowding is related to an increase in stress and arousal among inmates, and that it reduces their psychological well-being. Overcrowding affects both the mentally ill as well as inmates who are in good mental condition. As Pollock (2004, 125) notes, "Overcrowding stresses even the most even-tempered and mentally healthy inmate." For inmates who have already been diagnosed with mental illness, overcrowding can lead to more-serious psychological problems (Kupers 1999). In their research, Clayton and Carr (1984, 1987) also found a strong relationship between crowding and inmate rule infractions, but not to recidivism.

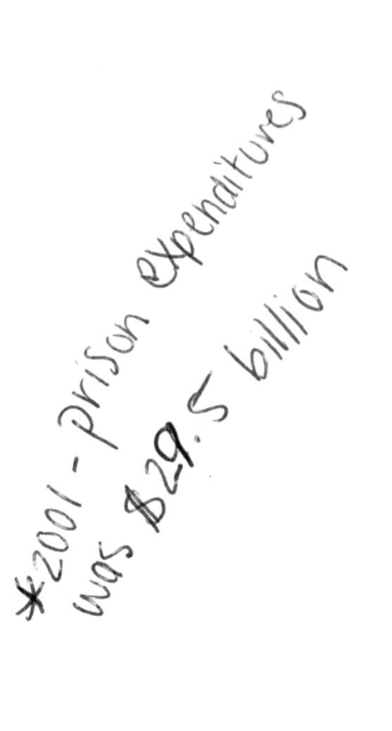

Although overcrowding may not be a direct cause of stress, Gaes (1985, 141) contends that it may exacerbate the many other sources of stress in prison, such as separation from family, loss of freedom, and fear of assault. Normally, crowding may intensify the effects of stress; however, when it reaches some as-yet-undetermined level, crowding may act as a direct stress, rather than merely as a stress elevator. There is little dispute among correctional staff. They contend that densely populated prisons are more difficult to control and to supervise.

At the end of 2003, ". . . 22 states and the Federal system reported operating at 100 percent or more of their highest capacity" (Harrison and Beck 2004, 7). Although not all the states reported that they were at such peak levels in terms of overcrowding, these data suggest that the states and the federal government are experiencing difficulties in finding enough prison beds, staff, programs, and services to accommodate the demand.

Previous research has indicated that inmates who live in open areas (such as dormitories) tend to use medical services more often and to have higher blood-pressure levels than offenders in single cells (Crouch, Alpert, Marquart, and Haas 1995, 66). In addition, there has been increased publicity about the incidence of HIV and tuberculosis infections among inmate populations. Although the risk of contracting HIV while in prison is well known, the risk of spreading tuberculosis is another serious concern (Durham 1994, 54). Tuberculosis is highly contagious, and it, too, can be fatal. Institutions that require inmates to be bunked in densely populated cell blocks or dormitories may increase the likelihood of spreading the disease (Pollock 2004, 121).

Strategies to Fight Overcrowding

States employ a variety of strategies to deal with overcrowding. They range from constructing new prisons or increasing the size of existing prisons, through expansion projects or double bunking, to the increased use of local jails and community-based correctional services. Historically, the most prevalent response has been to construct new facilities (Vaughn 1993, 16).

Overcrowding typifies the current correctional dilemma. Correctional administrators are in the unenviable position of trying to locate or construct additional space to cope with sentencing changes after they have been enacted by the legislature. Typically, these kinds of "get tough" sentencing laws are enacted without the accompanying funding for the additional prison space that will be required (Durham 1994, 45).

Court intervention compounds the dilemma. While the courts are concerned about crowding issues as they relate to violations of Eighth Amendment provisions regarding protection against cruel and unusual punishment, elected officials who fear being portrayed as "soft on crime" do not share their zeal. Historically, the result has been a somewhat acrimonious relationship in which the courts decree population limits, the remedies to be utilized, and the penalties (usually fines) to be invoked if the state refuses to comply (Cole and Call 1992).

The quickest method of dealing with overcrowding is the acquisition of space from within the institution. For example, the state may opt to convert a gymnasium into a dormitory. Simultaneously, additional facilities (either temporary or permanent) can be constructed. None of these space-expansion projects comes cheaply in economic or social costs. During the 1980s, prison budgets in many states increased more rapidly than any other portion of the state budget (McGarrell 1993, 16). That trend continues. In state budgets in 2001, there was an average increase of 6.4 percent for prisons, and the increased costs associated with adult incarceration surpassed those for health care, education, and natural resources (Stephan 2004, 2).

In fiscal year 2001, the total for prison expenditures by all states was $29.5 billion. Of all the states, California spent the most for its prisons, $4.2 billion (Stephan 2004, 2). Overall, prison-construction costs declined from 1996, but still resulted in $860,954,000 in expenditures. When land, equipment, and other devices were included, expenditures surpassed $1.1 billion in 2001 (Stephan, 5).

Contracting with a private provider for the incarceration of inmates is another option that states have utilized. In 30 states and the federal system, private providers hold more than 95,000 inmates in their facilities. Overall, 5.7 percent of all state inmates and 12.6 percent of all federal inmates (or 6.5 percent of all inmates) were in privately operated facilities in 2003. In terms of reliance on private facilities, Texas was the leader, with more than 16,000 inmates in institutions operated by private corporations in 2003 (Harrison and Beck 2004, 5–6).

For those states that opt to contract with private providers, there are substantial costs associated with increases in the prison population. In Tennessee, more than $150 million was utilized to hold offenders in private facilities and in local jails in fiscal year 2001. Contracting with private vendors to house inmates is one alternative to a state's reluctance to build additional prisons. In some instances, inmates can be shuttled between states when there are private facilities that have vacant cells in one state and overcrowding in the inmate's home state (see Pollock 2004, 63–64).

States have continued to rely on local jails to help with the overcrowding problems. For example, at the end of 2003, 32 states and the federal system utilized local jails to house more than 73,000 inmates. These inmates made up 5 percent of all incarcerated offenders in 2003 (Harrison and Beck 2004, 6). Some states utilize local jails more than others. In Louisiana, Tennessee, and Kentucky, more than a fifth of their state inmate population was held in local jails (Harrison and Beck).

Texas and California: Leaders in Incarceration

To attempt to assess the costs of overcrowding, it is useful to examine the experiences of Texas and California. On December 31, 2003, the Texas prison system was the largest in the country, with a population of 166,911. California was a close second, with a prison population of 164,487 (Harrison and Karberg 2004, 3).

Together, these two states account for 25 percent of all incarcerated inmates, or more than one in five inmates in the United States. Currently, California spends more money for prison expenditures than any other state. In 2001, the state of California spent $4.2 billion for its prisons (Stephan 2004, 3).

The enormous expenses incurred due to prison construction and to longer incarceration periods might be worthwhile if they resulted in reduced crime rates. Unfortunately, California has not demonstrated such a savings. Ehlers, Schiraldi, and Ziedenberg (2004, 23) contend that the three-strikes legislation that voters so enthusiastically supported in 1994 and endorsed again in 2004 has cost or will cost the residents of California an extra $10.5 billion in prison and jail expenditures. These California expenditures include $6.2 billion for offenders who were convicted of nonviolent offenses, but who are serving or will serve longer prison sentences under the terms of the law. These same nonviolent offenders will spend an extra 164,314 years in prison than they would have before the three-strikes legislation was enacted (Ehlers et al. 2004, 23). If deterring crime were the goal, the legislation has not had the effect it intended. States without three-strikes legislation, such as New York, have experienced a more dramatic drop in index offenses including violent offenses than has California (Schiraldi, Colburn, and Lotke 2004, 10).

The Texas experience is not much different. Although that state reports one of the lowest average annual operating costs for each inmate ($13,808), Texas spent more than $2.3 billion on prisons in fiscal year 2001 (Stephan 2004, 3). The situation is remarkably similar, only on a somewhat smaller scale, in most other states. In terms of national growth, between 1986 and 2001, state-prison expenses for each resident of the United States went from $49 per resident to $104 per resident (Stephan, 2). Clearly, the costs of incarceration are staggering.

States routinely increase their correctional budgets, and these increases allow prison systems to surpass other state budgets. As previously noted, the average annual increase for state prison was 6.4 percent—while health-care spending went up 5.8 percent, spending on education increased by 4.2 percent, and spending on natural resources increased by 3.3 percent in 2001 (Stephan 2004, 2). Even though state spending on corrections continues to escalate, if the current incarceration trends continue, there is little optimism for the likelihood of constructing enough prison space to satisfy most of the states' demands. The search for alternative strategies has intensified.

Front-End Strategies to Fight Overcrowding

In addition to new construction, the majority of states have utilized front-end strategies to deal with overcrowding (Vaughn 1993). Typically, these include increased use of probation and other kinds of intermediate sanctions such as intensive probation, house arrest, electronic monitoring, and shock probation or split sentences (Clark 1994, 105). Recently, an increased reliance on drug courts and treatment in lieu of incarceration for drug offenders has gained popularity. There is little doubt that the primary reason for the search for alternatives to incarceration is based primarily on economic rather than humanitarian motives.

Minnesota has been a leader in community corrections beginning with its Community Corrections Act, in 1973. That legislation authorizes annual state subsidies totaling approximately $31 million to local counties to help them create corrections programs including community service, electronic monitoring, and intensive supervision. These programs have flourished, seemingly without increased risk to the community (Wood 1996, 54).

Intensive supervision-probation programs in Georgia and intensive supervision-parole programs in New Jersey first became publicized in the early 1980s. They soon developed in every state. Similarly, house arrest and electronic monitoring were also products of the 1980s. In the intervening years, these programs and shock-incarceration programs have become increasingly utilized (Tonry and Hamilton 1995, 4–5).

Alternatives to incarceration are perceived as especially useful for nonviolent offenders. In 2003, 49 percent of offenders sentenced to probation were convicted of a felony. When a judge imposes a sentence of probation, he or she can do so by directly sentencing an offender to probation, by suspending a prison sentence in lieu of probation, or by employing a **split sentence**, in which an offender serves part of the sentence incarcerated and part of the sentence on probation. In recent years, judges have utilized probation by itself less often, and probation with some kind of incarceration more than in the past. For example, in 1995, judges used incarceration with probation in 13 percent of the probation cases, but in 2003, judges sentenced 22 percent of probation offenders to a split sentence of incarceration and then probation (Glaze and Palla 2004, 4). Of all the offenders sentenced to probation in 2003, 25 percent had been convicted of a drug-law violation, 17 percent had been convicted of driving while they were intoxicated or under the influence of alcohol, 12 percent had been convicted of larceny/theft, and 9 percent had been convicted of assault, which excludes domestic violence and sexual assault (Glaze and Palla). Probationers were less likely to have been convicted of domestic violence (7 percent), minor traffic offenses (6 percent), burglary (5 percent), fraud (4 percent), or sexual assault (3 percent). These data indicate that almost 70 percent of the men and women sentenced to probation in 2003 were nonviolent offenders (Glaze and Palla).

split sentence a form of sentence that combines probation and some form of incarceration, either in a work release facility, jail, or prison.

Using probation or some other type of community-based program for property offenders and drug offenders, particularly if it were coupled with some type of drug treatment and community service, might satisfy the public's demand to "get tough" while simultaneously saving much-needed capital for other public-service projects. As in any community-based program, screening appropriate offenders is imperative.

There is some apprehension, however, that these kinds of alternatives will produce a **net-widening effect**. Having these alternatives available may prompt judges to take offenders who previously would have been diverted from the system or subject to a fine, and instead sentence them to intensive probation or shock probation. When this occurs, these alternatives simply create more correctional clients rather than serve as alternatives to incarceration.

net-widening effect the idea that some sentence alternatives are used, not for those who would have received a harsher sentence, but for those who would have received a lesser sentence, thereby increasing the total number of correctional clients.

A second concern is related to funding for probation and parole-officer positions. The federal government has been proposing a freeze on spending for the judicial branch, accompanied by additional cuts to the current budget. Federal probation departments took a 7 percent salary cut in 2004, and in some jurisdictions, there have not been any replacements for probation officers who have left. For example, in New Mexico, there are 27 vacancies, and the caseload for those officers has to be absorbed by other probation staff (Fields 2004b, B1). Without adequate probation staff to supervise and assist clients, the imposition of a sentence to probation may lose its effectiveness.

Back-End Strategies to Fight Overcrowding

Using back-end strategies to cope with prison overcrowding generally necessitates the acceleration of early release, either through good-time credits or parole (Clark 1994, 105). These release procedures are extremely controversial. Consider the 1993 case of Polly Klaas, the 12-year-old girl who was abducted and murdered by Richard Allen Davis, a parolee who had served 8 years of a 16-year sentence for kidnapping and who had previous kidnapping, assault, and robbery convictions (*New York Times* 1993; Benekos and Merlo 1995). Media sensationalism and the fear of victimization by criminals who are prematurely released from prison discourage consideration of these kinds of approaches. Unfortunately, these kinds of cases prompt the public to demand tougher sentences, including life without possibility of parole.

The get-tough era has had an effect on parole decisions in the United States. Parole boards are less inclined to parole inmates today than in previous years. For example, of those offenders eligible for parole in 2003, 39 percent of these offenders were granted parole releases compared to 51 percent in 1995 (Glaze and Palla 2004, 6). The more likely scenario is that offenders are being released under supervision because of their mandatory-release dates. More than 50 percent of the parolees in 2003 were paroled due to a mandatory release compared to 45 percent in 1995 (Glaze and Palla).

Nonetheless, the parole population is growing in the United States; in 2003, it grew by more than 23,000 offenders (Glaze and Palla 2004, 1). Texas and California have the largest number of offenders on parole and probation, and, together, they account for more than 1 million offenders under one of these types of community supervision (Glaze and Palla, 2). The increased use of parole in 2003 represented the largest increase in the population of parolees since 1995 (Glaze and Palla, 2). Notably, the largest category included those convicted of drug offenses; 36 percent of inmates paroled in 2003 were incarcerated for drug offenses (Glaze and Palla, 6).

Typically, states will authorize emergency-release mechanisms (for example, adding on extra good-time credits) when they are under some type of court decree not to exceed a certain population ceiling. However, emergency-release mechanisms have become less necessary as states have expanded their incarceration capabilities. States can and do utilize county jails, private facilities, and neighboring states when the population increases occur. These strategies suggest that there is less likelihood that emergency-release procedures are being mandated today than in the past.

Part of the reluctance to utilize early-release programs may be due to the limited research on their effectiveness. Lane (1986) reported that the National Council on Crime and Delinquency's (NCCD's) evaluation of an early-release program used by Illinois proved that the program did not jeopardize public safety or increase crime.

Future Directions

One of the more troubling aspects of the overcrowding phenomenon is whether most states will ever be able to overcome the problem. The public and elected officials seem to have an insatiable appetite for the increased use of incarceration. The three-strikes statute that Californians embraced in 1994 and reaffirmed in 2004 costs billions of dollars.

California is not the only state with a three-strikes law (Benekos and Merlo 1995). However, the state's application of the law is unique. The federal government and more than half of the states have enacted similar statutes, but they do not utilize them to the same degree as California. In that state, more than 50,000 offenders had been sentenced under the three-strikes legislation by 2001. In June of 2003, more than one of every four California inmates (over 27 percent) was either a second-strike or a third-strike offender (Ehlers, Schiraldi, and Ziedenberg 2004, 6). According to Schiraldi, Colburn, and Lotke (2004, 4–5), California incarcerated almost four times as many offenders under its three-strikes law than all the other states that have three-strikes legislation combined. By contrast, in Georgia, there were 942 offenders, and in Florida 116 offenders were similarly sentenced (King and Mauer, 2001; Pollock 2004). In some instances, they are called habitual-offender laws rather than three-strikes legislation. California has certainly felt the effects of the legislation—with the second-largest prison system in the United States, and a budget that is more than $5.3 billion a year (Institute of Governmental Studies, 2004).

One of the more alarming aspects of these kinds of laws is their inability to deter crime. By contrast, these laws may serve to exacerbate the commission of certain kinds of crime, most notably homicide. Marvell and Moody (2001) found that states with the laws experienced a 10 to 12 percent increase in homicides when compared to states that did not enact three-strikes legislation. When Kovandzic, Sloan, and Vieraitis (2002, 213) used panel data from 188 cities from 1980 to 1999, they found that ". . . cities in states with the three strikes laws had a 13 to14 percent increase in homicide rates in the short term and a 16 percent to 24 percent increase in the long term." In sum, these studies suggest that the three-strikes legislation exemplifies the "unintended consequences" that Kovandzic et al. (2002) describe. The three-strikes legislation not only failed to reduce the incidence of serious crime, but, in fact, increased it (Kovandzic, Sloan, and Vieraitis 2004, 213).

In a later analysis, Kovandzic et al. (2004) examined the effects of three-strikes legislation on other crimes by analyzing data from 1980 through 2000 in 188 cities with a population of more than 100,000 in states that had enacted such laws. Their findings indicate that "consistent with other studies, ours finds no credible statistical evidence that passage of three-strikes laws reduced crime by deterring potential criminals or incapacitating repeat offenders" (2004, 234). If the intent of these laws is to deter crime, they have not succeeded. What remains is expensive incarceration without any of the benefits that the law intended.

Skolnick's research (1995, 9) suggested that the federal-prison population would increase by 50 percent by the year 2000. In fact, the federal-prison population has grown from 83,663 in 1995 to 151,919 in 2003, an 82 percent increase (Harrison and Beck 2004, 4). Furthermore, Skolnick predicted the majority of incarcerated offenders would be there for drug offenses (Skolnick). Today, drug offenders are grossly overrepresented in the federal-prison population. This kind of selective incapacitation might be useful if it deterred drug-related crimes. However, there is no evidence that the public demand for cocaine and heroin has abated as a result of these criminal-justice policies. In fact, there are an estimated 2 million hard-core abusers of these two drugs in the United States (Clark 1994, 103). Would incarcerating all of these abusers eliminate the demand for drugs in America?

Another distressing aspect of the laws concerns the likely targets. African-Americans and Hispanics are overrepresented in the California prison system. With respect to the three-strikes legislation, African-Americans, who are 6.5 percent of the state population, constituted 36 percent of the offenders who were sentenced under the second-strike law, and 45 percent of those convicted and sentenced under the third-strike law (Ehlers, Schiraldi, and Lotke 2004, 2). There are also variations by county. Of the 2,897 nonviolent criminals sentenced in Los Angeles County to life under three strikes, 1,613, or 56 percent, were African-Americans (Stewart Wheeler 2004, para. 14). An offender who lives in Los Angeles County is more likely to be incarcerated under three-strikes legislation than an offender in San Francisco County. These data suggest that any perceived uniformity in the disposition of sentences that the legislators had intended is nonexistent (Ehlers et al. 2004, 19). Although California has a record of utilizing the three-strikes legislation more than other states, the pattern is remarkably similar in other states. For example, in Washington state, African-Americans make up less than 4 percent of the state's population, but 37 percent of offenders sentenced under Washington's three-strikes law are African-Americans (King and Mauer 2001, 13). Just as the War on Drugs has increased the incarceration of minorities disproportionately, Mauer's (1994, 23) early concern that these three-strikes laws would have a similar effect appears to have been realized.

The pessimism concerning minority overrepresentation and three-strikes laws is augmented by the overwhelming expenditures that could be put to better use. Finally, there are the subtle effects of the three-strikes legislation that King and Mauer (2001) have noted. Although California's three-strikes legislation is just slightly over 10 years old, it has the potential to be even more costly than originally realized due to the fact that it will require the incarceration of an aging prison population with their attendant health-care and special-needs requirements. King and Mauer (2001, 4) estimate that by 2026, there will be approximately 30,000 inmates who were sentenced under the three-strikes provisions serving a sentence of 25 years to life. The onerous consequences of housing a population that is elderly include exorbitant costs and the concomitant realization that these inmates are probably the safest to parole and least likely to return to a life of crime. Just how much will it cost to house 30,000 inmates under the prescribed legislation? Based on existing costs of incarceration for inmates (not the costs associated with elderly-inmate incarceration), research suggests it will cost $750 million each year (King and Mauer, 5). Nonetheless, they will remain incarcerated.

An aging inmate population will affect both the state and federal systems. Currently, 1 of every 23 incarcerated offenders in prison is 55 years of age or older, and this represents an 8.5 percent increase from 1995. There were approximately 127,000 inmates serving a life sentence in the United States in 2003, and one quarter of them were serving life without the possibility of parole. The number of offenders sentenced to life imprisonment has increased by 83 percent since 1992, and it likely to continue for the next decade (Sentencing Project 2004, 3). These older inmates will potentially strain the existing overcrowded institutions with their special needs and health issues.

Social Costs of Prison Expansion

There is little evidence that the massive expansion of federal and state prison beds has reduced crime. Given that fact and the enormous expenditures that these policies necessitate, there is little to justify such a position, especially when the opportunity costs that these policies represent are considered. Zimbardo (1994) contended that, factoring in the interest paid for a construction-bond debt, it costs $333 million to build one new prison in California. To build and maintain more prisons, states and the federal government have to forgo other programs. Funding has to come from other sources of public revenue, and these consequences affect everyone long after the prison has been constructed, staffed, and filled with inmates.

Prisons are costly. Total expenditures for state prisons skyrocketed between 1985 and 1995, jumping to $27 billion from $13 billion (Cose 2000, 42). Consider the growth of California's corrections budget. In order to incarcerate one felon in California, it costs taxpayers $26,606 each year. It is estimated that it will cost $1.5 million for an elderly person to be incarcerated for the minimum 25 years that is stipulated in the statute (King and Mauer 2001, 12).

The trade-off is even more troubling when one considers the less obvious losses that occur in the way of human connections. In the United States, 2 percent of our children in 2000 had to go to a prison to see a parent (Cose 2000, 43). When the state institutions are overcrowded and the parent has to be transferred to another state, parenting becomes a virtual nightmare. This is especially traumatic for women, who tend to be the children's caregivers and who are trying to continue their role as mothers (Pollock 2002, 191). In the African-American community, more than 12 percent of men between the ages of 25 and 29 are incarcerated, and there is a significant effect on the quality of life for the children who are not only deprived of their father, but also of any economic and emotional stability that he may have provided the family. Although Californians' share of these kinds of costs have been well documented, there are other costs that California and other states will have to address as a result of these policies. It makes more sense to adopt a proactive strategy and invest in early intervention in poor children's lives than to persist with a reactive strategy, incarcerating large numbers of people and expending vast amounts of money after they have been involved in crime.

Some research that looks at African-American men between the ages of 18 to 67 has found that there are more African-American men in prison than in college in the United States (Justice Policy Institute 2002). Although there are more African-American men attending college than in prison in the 18- to 24-year-old age group, it is a dramatically smaller ratio than that of white men. For African-American men, the ratio is 2.6:1; for white men, it is 28:1 (Hocker 2002, 1, cited by Pollock 2004, 83). Even more troubling is the reality of states such as Ohio. In 2000, there were more African-American men in prison than enrolled in colleges and universities in Ohio (Collins 2002). These differences are significant, and merit closer evaluation along with alternatives to prison sentences.

Conclusions

It is unlikely that the excessive reliance on incarceration will completely dissipate in the next 10 years. Building prisons helps to enhance the illusion of their "incapacitative effect" (Sieh 1989, 49). As Zimring and Hawkins (1995, 156) contend, incapacitating offenders is perceived as desirable because "restraint directly controls the behavior of the offender rather than leaving him or her any choice in the matter." It is the ability to control an offender rather than simply to influence him or her that makes incarceration of criminals so appealing to the public (Zimring and Hawkins, 157).

There are, however, some indications emerging that portend the possibility for change. The incidence of violent crime, particularly murder, has decreased dramatically. The FBI data from 1994 to 2003 indicate a 16 percent drop in violent-crime arrests—including a 36 percent drop in arrests for murder and a 25 percent decrease in robbery arrests (Butterfield 2004, A14). In New York, Chicago, and a number of other major cities, the number of homicides again decreased in 2004 (Dewan 2004, A17). Although some public officials might quickly conclude that incarceration is the primary reason for the decline in the crime rate, there is no empirical evidence to support such a claim.

No single reason can be given for the precipitous drop in the murder rates in large cities or for the decrease in crime in general in the United States. Researchers include increased incarceration as one of a number of reasons that include demographic changes, economic changes, improved job market, drug-market changes, and improved policing (King and Mauer 2001, 7; Schiraldi, Colburn, and Lotke 2004, 10). In New York City, Commissioner Raymond F. Kelly attributes the drop to Operation Impact, which uses computer technology to identify troubled areas, and then deploys large numbers of uniformed officers to patrol those areas (Dewan 2004, A17). Although incarceration policies are a component of the change, they do not solely explain the drop in crime. In fact, King and Mauer (2001, 7) cite a report conducted by researchers at the California Department of Justice Statistics Center who were unable to find any objective evaluations of ". . . 'get tough' laws that were able to empirically verify that they exercise a direct negative effect on crime rates."

Interestingly, other states also experienced dramatic decreases in crime during this time period, most notably New York, but without any three-strikes legislation. Without three-strikes legislation, New York experienced a greater decline in index crimes and in violent index crimes between 1993 and 2002, and it had a 5.7 percent decline in its incarceration rate when contrasted with California.

During the same time, California increased its incarceration rate by 17.7 percent (Schiraldi, Colburn, and Lotke 2004, 10). In short, there is no empirical evidence that can show that the drop in crime has been *substantially* caused by the use of incarceration.

Although the decline in violent crime did not previously prompt legislators to demand that prison populations be reduced or sentencing laws revised, there is some recent evidence of a softening in their approach to certain crimes. In 2004, there are indications that the punitive stances for drug crimes may be slightly eroding on the state level. For example, legislators in New York recently reviewed and revised the 30-year-old drug laws in their state. Although the new legislation is far from lenient, it will reduce the sentences for those who are first-time offenders and are convicted of drug offenses. Previously, these first-time offenders had to receive a mandatory sentence of 15 years to life; under the new law, they will usually receive terms of less than 8 years. For inmates who are serving very long sentences, the legislation will enable them to petition the courts to review their terms of incarceration and have the terms reduced to the new levels (Eaton and Baker 2004). In Pennsylvania, legislators recently enacted a statute that makes it possible for the Department of Corrections to determine if offenders are eligible for alternative sentencing for drug offenses, sidestepping the mandatory-sentencing laws for drug offenders now in place (Collis 2004, 1, 13). Legislators in Arizona and Kansas also have revised their statutes and utilize community-based alternatives for drug offenders (Collis, 13).

In California, voters enacted Proposition 36, the Substance Abuse and Crime Prevention Act, in 2000. As a result, offenders who are convicted of drug possession are eligible for drug treatment as opposed to incarceration. The law is also applicable to three-strikes-eligible offenders who have been out of prison for five years (Ehlers, Schiraldi, and Ziedenberg 2004, 4). This legislation has the potential to reduce the growth in the prison population that California has been experiencing, especially considering the fact that there are more third-strike offenders serving a sentence of 25 years to life for drug possession (672) than there are third-strike offenders serving a sentence for second-degree murder (62) (Ehlers et al. 2004, 8).

There is also some indication that public opinion may be softening with respect to offenders who receive prison terms. According to Schiraldi et al. (2004), public-opinion polls demonstrate a shift in public opinion toward a more balanced approach to crime. In 2001, 32 percent of those polled by Hart and Associates favored a tougher approach to crime with an emphasis on more-punitive sentences, increased use of capital punishment, and a decrease in the use of parole, compared to 48 percent who had expressed those sentiments in 1994 (Schiraldi et al., 11). These data suggest that the public may be weary of the approaches and expenses that characterized criminal-justice policy in the 1990s. These data also may influence Congress as it prepares to draft new legislation as a result of the recent Supreme Court decision in *U.S. v. Booker* (No. 04-104).

Unfortunately, our desire to control offenders will cost all of us. We will lose a variety of services—from early-education programs such as Head Start, to medical services for the poor, to college educations. According to Malik Russell of the Justice Policy Institute, "They [Californians] are spending more to imprison non-violent people with three-strikes than they are spending on people in state colleges" (Stewart Wheeler 2004, 2). Fiscal restraint will necessitate curtailing road construction and repair, social services, health care, and immunization programs. The costs associated with prison construction will continue to escalate, and if the demand for stricter punishment does not waver, these costs will necessitate deep budget cuts in all states and the federal government. It is against this backdrop that we look at what is occurring inside the nation's prisons.

KEY TERMS

capacity—prison capacity is the number of beds minus a small percentage set aside for necessary transfers and emergency uses.

determinate sentencing—a form of sentencing that eliminates judicial discretion; sentences are set by statute for each crime.

discretion—the power to choose between alternatives.

good-time credits—time taken off of a prison sentence, usually earned for good behavior and/or for participating in prison programs.

incarceration rate—a display of the number of people incarcerated by the population, so that different size population groups can be compared.

indeterminate sentencing—a form of sentencing where the amount of time served is not set at sentencing, but rather is determined by some other body such as a parole board.

judicial discretion—in reference to sentencing, this term means that judges make the determination of how long the sentence will be.

mandatory sentences—type of sentencing whereby for a particular crime, there is no judicial discretion; the decision whether to imprison or not and/or the amount of prison time is set by statute.

net-widening effect—the idea that some sentence alternatives are used, not for those who would have received a harsher sentence, but for those who would have received a lesser sentence, thereby increasing the total number of correctional clients.

partially indeterminate sentencing—a form of sentencing whereby judges may set either the minimum or the maximum or some percentage of total time that must be served and the paroling authority would have the remaining power to set the actual time served.

presumptive sentencing—a form of determinate sentencing that allows for a small increase or decrease of a set sentence depending on stated mitigating or aggravating circumstances that must be proved in court.

returnees—those inmates who return to prison on some form of violation rather than a new crime conviction.

sentencing disparity—the phenomenon of offenders who commit similar crimes but receive different sentences.

sentence-enhancement statutes—statutes designed to increase the total sentence if certain stated factors are proven to be present.

sentencing guidelines—a grid system of determining the time that should be served; the system utilizes factors related to the crime, but also factors related to the criminal (i.e., the risk factor score).

split sentence—a form of sentence that combines probation and some form of incarceration, either in a work release facility, jail, or prison.

three-strikes laws—statutes that allow for an increased sentence if the offender has been previously convicted of two crimes.

War on Drugs—the term coined to describe the governmental response to drug use, starting during President Reagan's presidency. This "war" is characterized by more resources devoted to interdiction and control rather than treatment and prevention.

REVIEW QUESTIONS

1. What is an incarceration rate? How is it computed? What is the incarceration rate for your state? How does your state's rate compare to the country's average rate?
2. How many people are in prison today in the United States? What are the projections for the future?
3. What are the increases in the incarceration rate and prison populations attributed to?
4. Define indeterminate, partially indeterminate, and determinate sentencing.
5. What has been the effect of the various forms of determinate sentencing on prison populations?
6. What seems to be the relationship between crime rates and incarceration rates?
7. How have drug laws and commitments of drug offenders affected prison populations?
8. What does the research show regarding the effects of overcrowding on inmates?
9. What are some front-end strategies to reduce overcrowding? What are some back-end strategies?
10. What are the costs associated with the increase of prison commitments?

FURTHER READING

Ehlers, S., V. Schiraldi, and E. Lotke. 2004. "Racial Divide: An Examination of the Impact of California's Three Strikes Law on African-Americans and Latinos." Justice Policy Institute. Retrieved September 23, 2005, from www.justicepolicy.org/.

Ehlers, S., V. Schiraldi, and J. Ziedenberg. 2004. "Still Striking Out: Ten Years of California's Three Strikes Law." Justice Policy Institute. Retrieved September 23, 2005, from, www.justicepolicy.org/.

Kovandzic, T. V., J. Sloan, III, and L. M. Vieraitis. 2002. "Unintended Consequences of Politically Popular Sentencing Policy: The Homicide Promoting Effects of 'Three Strikes' Laws in U.S. Cities (1980–1999)." *Criminology and Public Policy* 1: 399–424.

Kupers, T. 1999. *Prison Madness: The Mental Health Crisis Behind Bars and What We Must Do About It*. San Francisco: Jossey-Bass.

Sentencing Project. 2004. "New Incarceration Figures: Rising Population Despite Falling Crime Rates." Washington, DC: Sentencing Project.

REFERENCES

Beck, A., J. Karberg, and P. Harrison. 2002. *Prison and Jail Inmates at Mid-year, 2001*. (April 2002). Washington, DC: U.S. Dept. of Justice.

Benekos, P. J., and A. V. Merlo. 1995. "Three Strikes and You're Out!: The Political Sentencing Game." *Federal Probation* 59 (March): 3–9.

Bloom, B., M. Chesney-Lind, and B. Owen. 1994. "Women in California Prisons: Hidden Victims of the War on Drugs." *In Brief* (May): 1–11. San Francisco: Center on Juvenile and Criminal Justice.

Bureau of Justice Statistics. 1988. *Report to the Nation on Crime and Justice—Sentencing and Corrections*, 2d ed. Washington, DC: U.S. Department of Justice.

Butterfield, F. 2004. "Despite Drop in Crime, An Increase in Inmates." *New York Times* (November 8): A14.

Chesney-Lind, M., and J. M. Pollock. 1995. "Women's Prisons: Equality with a Vengeance." In A. V. Merlo and J. M. Pollock (Eds.), *Women, Law, & Social Control*, pp. 155–175. Needham Heights, MA: Allyn and Bacon.

Clark, C. S. 1994. "Prison Overcrowding." *Congressional Quarterly Researcher* 4, 5 (February): 97–120.

Clayton, O., Jr., and T. Carr. 1984. "The Effects of Prison Crowding upon Infraction Rates." *Criminal Justice Review* 9: 69–77.

Clayton, O., Jr., and T. Carr. 1987. "An Empirical Assessment of the Effects of Prison Crowding upon Recidivism Utilizing Aggregate Level Data." *Journal of Criminal Justice* 15: 201–210.

Cohen, L. P. 2004. "Double Standard: In Wake of Ruling, Disarray Plagues Federal Sentencing." *Wall Street Journal* (December 28): A1, A4.

Cohen, L. P., and G. Fields. 2004. "Reasonable Doubts: How Unproven Allegations Can Lengthen Time in Prison." *Wall Street Journal* (September 20): A1, A5.

Cohen, L. P., and G. Fields. 2005. "New Sentencing Battle Looms After Court Decision." *Wall Street Journal* (January14): A1, A4.

Cole, R. B., and J. E. Call. 1992. "When Courts Find Jail and Prison Overcrowding Unconstitutional." *Federal Probation* 56 (March): 29–39.

Collins, M. 2002. "Prison Spending Outpaces Higher Ed." *Cincinnati Post* (August 28). Retrieved January 3, 2005, from www.policymattersohio.org/media/CP_cell.htm.

Collis, A. 2004. "Sentencing Alternatives to Incarceration: Senate Bill 217 Passes." *Correctional Forum* (December): 1,13.

Corrections Digest. 1994. "Senate Crime Bill Will More Than Double American Prison Population by Year 2005." *Corrections Digest* (March 9): 1–4.

Corrections Today. 1996. "U.S. Prison Population Grows at Record Rate." *Corrections Today* 58, 1 (February): 20.

Cose, E. 2000. "The Prison Paradox." *Newsweek* (November 20): 43–49.

Covington, S. 1998 "Women in Prison: Approaches in the Treatment of Our Most Invisible Population." In J. Harden and M. Hill, *Breaking the Rules: Prison and Feminist Therapy*, pp. 141–155. New York: Harrington Park Press Books.

Crouch, B., G. P. Alpert, J. W. Marquart, and K. C. Haas. 1995. "The American Prison Crisis: Clashing Philosophies of Punishment and Crowded Cellblocks." In K. C. Haas and G. P. Alpert, *The Dilemmas of Corrections*, 3d ed., pp. 64–80. Prospect Heights, IL: Waveland Press.

Davis, S., A.V. Merlo, and J. M. Pollock. 2005. "Female Criminality: Ten Years Later." In A. V. Merlo and J. M. Pollock (Eds.), *Women, Law, & Social Control*, 2d ed. In press. Needham Heights, MA: Allyn and Bacon.

Dewan, S. K. 2004. "New York Murder Rate Falls Again, But Has It Hit Bottom?" *New York Times* (December 24): A17.

Durham, A. M. 1994. *Crisis and Reform: Current Issues in American Punishment*. Boston: Little, Brown.

Durose, M. R., and P. A. Langan. 2004. *Felony Sentences in State Courts, 2002. Bureau of Justice Statistics Report*. Washington, DC: U.S. Department of Justice.

Eaton, L., and A. Baker. 2004. "Changes Made to Drug Laws Don't Satisfy Advocates." *New York Times* (December 9). Retrieved January 4, 2005, from www.justicepolicy.org/article.php?id=465.

Ehlers, S., V. Schiraldi, and E. Lotke. 2004. *Racial Divide: An Examination of the Impact of California's Three Strikes Law on African-Americans and Latinos*. Washington, DC: Justice Policy Institute. Retrieved September 23, 2005, from www.justicepolicy.org/.

Ehlers, S., V. Schiraldi, and J. Ziedenberg. 2004. *Still Striking Out: Ten Years of California's Three Strikes Law.* Washington, DC: Justice Policy Institute. Retrieved September 23, 2005, from www.justicepolicy.org/.

Fields, G. 2004a. "Commission Finds Racial Disparity in Jail Sentences." *Wall Street Journal* (November 24): A4.

Fields, G. 2004b. "Federal Sentencing Changes Could Strain Probation System." *Wall Street Journal* (September 20): B1, B4.

Fields, G. 2004c. "Sentencing Shift: In Criminal Trials, Venue Is Crucial but Often Arbitrary." *Wall Street Journal* (December 30): A1, A10.

Gaes, G. G. 1985. "The Effects of Overcrowding in Prison." In M. Tonry and N. Morris (Eds.), *Crime and Justice*, Vol. 6, pp. 95–146. Chicago: University of Chicago Press.

Gilliard, D. K., and A. Beck. 1994. *Prisoners in 1993: Bureau of Justice Statistics Bulletin* Washington, DC: U.S. Department of Justice.

Glaze, L. F., and S. Palla. 2004. *Probation and Parole in the United States, 2003: Bureau of Justice Statistics Bulletin*. Washington, DC: U.S. Department of Justice.

Goodstein, L., and J. Hepburn. 1985. *Determinate Sentencing and Imprisonment*. Cincinnati, OH: Anderson.

Greenhouse, L. 2005. "Supreme Court Transfers Use of Sentence Guidelines: Discretion in Federal Cases Is Given Back to Judges." *New York Times* (January 13): A1, A27.

Harrison, P., and A. J. Beck. 2004. *Prisoners in 2003: Bureau of Justice Statistics Bulletin*. Washington, DC: U.S. Department of Justice.

Harrison, P. M., and J. C. Karberg. 2004. *Prison and Jail Inmates at Midyear 2003: Bureau of Justice Statistics Bulletin*. Washington, DC: U.S. Department of Justice.

Hocker, C. 2002. "More Brothers in Prison Than College?" Blackenterprise.com. Retrieved October 10, 2002, from www.blackenterprise.com/ExclusivesOpen.asp?source=Articles/10082002ch.html.

Holten, N. G., and R. Handberg. 1994. "Determinant Sentencing." In A. Roberts (Ed.), *Critical Issues in Crime and Justice*, pp. 217–231. Thousand Oaks, CA: Sage.

Institute of Governmental Studies at the University of California. 2004. Report retrieved December 27, 2004, from igs.berkeley.edu/library/htThreeStrikes Prop66.htm.

Irwin, J., and J. Austin. 1994. *It's About Time: America's Imprisonment Binge*. Belmont, CA: Wadsworth.

Isikoff, M. 1995. "Crack, Coke, and Race." *Newsweek* 126, 19 (November 6): 77.

Justice Policy Institute. 2002. *Cellblocks or Classrooms in Ohio*. Washington, DC: Justice Policy Institute. Retrieved September 22, 2004, from www.justice policy.org/.

Justice Policy Institute. 2004. *Prison Numbers Increase to Nearly 1.5 million Despite Decade-long Drop in Crime*. Washington, DC: Justice Policy Institute. Retrieved September 22, 2004, from www.justicepolicy.org/.

Kappeler, V. E., M. Blumberg, and G. W. Potter. 1996. *The Mythology of Crime and Criminal Justice*. Prospect Heights, IL: Waveland.

Kennedy, E. M. 1985. "Prison Overcrowding: The Law's Dilemma." *Annals of the American Academy* 478 (March): 113–122.

King, R. S., and M. Mauer. 2001. *Aging Behind Bars: 'Three Strikes' Seven Years Later.* Washington, DC: Sentencing Project. Retrieved September 22, 2004, from www.sentencing project.org.

Kovandzic, T. V., J. Sloan, III, and L. M. Vieraitis. 2002. "Unintended Consequences of Politically Popular Sentencing Policy: The Homicide Promoting Effects of 'Three Strikes' Laws in U.S. Cities (1980–1999). *Criminology and Public Policy* 1: 399–424.

Kovandzic, T. V., J. Sloan, III, and L. M. Vieraitis. 2004. "'Striking Out' as Crime Reduction Policy: The Impact of 'Three Strikes' Laws on Crime Rates in U.S. Cities." *Justice Quarterly* 21, 2: 207–239.

Kupers, T. 1999. *Prison Madness: The Mental Health Crisis Behind Bars and What We Must Do About It*. San Francisco: Jossey-Bass.

Lane, M. 1986. "A Case for Early Release." *Crime and Delinquency* 32, 4: 399–403.

Langan, P. A., C. A. Perkins, and J. M. Chaiken. 1994. *Felony Sentences in the United States, 1990: Bureau of Justice Statistics Bulletin*. Washington DC: U.S. Department of Justice.

Lawrence, C., and K. Andrews. 2004. "The Influence of Perceived Prison Crowding on Male Inmates' Perception of Aggressive Events." *Aggressive Behavior* 30, 4: 273–284.

Marvell, T. B., and C. E. Moody. 2001. "The Lethal Effects of Three Strikes Laws." *Journal of Legal Studies* 30: 89–106.

Mauer, M. 1994. *Americans Behind Bars: The International Use of Incarceration, 1992–1993*. Washington, DC: Sentencing Project. Retrieved from www.sentencingproject.org.

McDonald, D. E., and K. E. Carlson. 1994. "Drug Policies Causing Racial and Ethnic Differences in Federal Sentencing." *Overcrowded Times* 5, 6: 1, 8–10.

McGarrell, E. F. 1993. "Institutional Theory and the Stability of a Conflict Model of the Incarceration Rate." *Justice Quarterly* 10, 1: 7–28.

McGough, M. 2004. "Sentencing Guideline Ruling Delayed." *Pittsburgh Post-Gazette* (December 20): A6.

McShane, M., and F. Williams. 2005. "Women Drug Offenders." In A. V. Merlo and J. M. Pollock (Eds.), *Women, Law, & Social Control*, 2d ed., pp.211–276. Needham Heights, MA: Allyn and Bacon.

Merlo, A. V., and J. M. Pollock. 2005. *Women, Law, & Social Control*, 2d ed. Needham Heights, MA: Allyn and Bacon.

New York Times. 1993. "Hunt for Kidnapped Girl, 12, Is Narrowed to Small Woods." *New York Times* (December 3): A22.

Owen, B. 2005. "The Context of Women's Imprisonment." In A. V. Merlo and J. M. Pollock (Eds.), *Women, Law, & Social Control*, 2d ed., pp. 251–270. Needham Heights, MA: Allyn and Bacon.

Pollock, J. M. 2002. *Women, Prison, & Crime*. Belmont, CA: Wadsworth.

Pollock, J. M. 2004. *Prisons and Prison Life*. Los Angeles: Roxbury Publishing Company.

Schiraldi, V., J. Colburn, and E. Lotke. 2004. *Three Strikes and You're Out: An Examination of the Impact of 3-Strike Laws 10 Years After Their Enactment*. Washington, DC: Justice Policy Institute. Retrieved September 22, 2004, from www.justicepolicy.org/.

Sentencing Project. 2003. *Hispanic Prisoners in the United States*. Washington, DC: Sentencing Project. Retrieved September 22, 2004, from www.sentencing project.org.

Sentencing Project. 2004. *New Incarceration Figures: Rising Population Despite Falling Crime Rates*. Washington, DC: Sentencing Project. Retrieved September 22, 2004, from www.sentencingproject.org.

Sieh, E. W. 1989. "Prison Overcrowding: The Case of New Jersey." *Federal Probation* 53, 3: 41–51.

Singer, R. 1993. "Sentencing." In C. W. Eskridge (Ed.), *Criminal Justice: Concepts and Issues*, pp. 172–174. Los Angeles: Roxbury.

Skolnick, J. H. 1995. "What Not to Do About Crime—The American Society of Criminology 1994 Presidential Address." *Criminology* 33, 1: 1–13.

Stephan, J. J. 2004. *State Prison Expenditures, 2001. Bureau of Justice Statistics, Special Report*. Washington DC: U.S. Department of Justice.

Stewart Wheeler, S. 2004. "Jim Crow Constitution Upheld in Alabama." BlackAmericaWeb.com. Retrieved January 4, 2005, from www.justicepolicy.org/article.php?id=460.

Taylor, B. G., P. J. Newton, and H. H. Brownstein. 2003. *2000 Arrestee Drug Abuse Monitoring: Annual Report, Part 1, Chapter V—Drug Use Among Female Arrestees*. Washington DC: U.S. Department of Justice. Retrieved August 2, 2004, from www.ncjrs.org/txtfilesl/nij/193013.txt.

Tesoriero, H. 2004. "Drug Maker Helps the Police Fight Abuse." *Wall Street Journal* (September 20): B3, B4.

Tonry, M., and K. Hamilton. 1995. *Intermediate Sanctions in Overcrowded Times*. Boston: Northeastern University Press.

Turner, M. G., J. L. Sundt, B. K. Applegate, and F. T. Cullen. 1995. "Three-Strikes and You're Out Legislation: A National Assessment." *Federal Probation* 59, 3: 16–35.
Vaughn, M. S. 1993. "Listening to the Experts: A National Study of Correctional Administrators' Responses to Prison Overcrowding." *Criminal Justice Review* 18 (Spring): 12–25.
Wellisch, J., M. L. Prendergast, and M. D. Anglin. 1994. *Drug-Abusing Women Offenders: Results of a National Survey: Research in Brief*. Washington, DC: U.S. Department of Justice.
Wood, F. W. 1996. "Cost-Effective Ideas in Penology—The Minnesota Approach." *Corrections Today* 58, 1: 52, 54, 56.
Zimbardo, P. G. 1994. *Transforming California's Prisons into Expensive Old Age Homes for Felons: Enormous Hidden Costs and Consequences for California's Taxpayers*. San Francisco: Center on Juvenile and Criminal Justice.
Zimring, F. E., and G. Hawkins. 1995. *Incapacitation: Penal Confinement and the Restraint of Crime*. New York: Oxford University Press.

CASES CITED

Blakely v. Washington, 124 S. Ct. 2531, 159 L. Ed. 2d 403 (2004)
U.S v. Booker (No. 04-104, 2005)
U.S. v. Fanfan (No. 04-105, 2005)

II

The Social World of the Prison

On the Yard

Muggers, rapists, robbers, and thieves mill about aimlessly, or so it seems, among small-time junkies and dealers with big-time dreams, in occasional conferences, bargaining, overheard but not understood by the nut cases, men in their own orbit, on their own highs, in turn beset by the retarded, who think the crazies are conduits to the gods, think their hallucinations are harbingers of things to come, signs that hold answers to their own muttered, stuttered pleas for guidance, direction, relief from the chaos that envelops their days and nights in the prison world, a PlayStation® fantasy place where folks play for keeps and scripts get written and rewritten all the time.

Tattoos form a crazy quilt of sick art, rendered on human flesh, pointing the way to the various and sundry constellations on planet prison, one sadder or madder than the next.

This way to gangs and girls (or a reasonable facsimile); that way to muscles with mothers perched on biceps; watch out for guns half hidden by boxer shorts, peeking out at the waist, as if in a holster; beware devils and goblins and serpents, medieval creatures loose on the sagging skin of bearded, ponderous, dangerous men, folk you watch out for, can't befriend.

Follow the yellow brick road but don't show yellow, fellow, or it's a long, long way from Kansas to where you'll be heading, a piece of meat, somebody's bedtime treat. . . . Primitive, Primeval. Just plain Evil.

But there it is, there you are, far from home, trying to find a home on the prison range, where life is downright strange, and ain't nobody free.

Robert Johnson

4

The Social World of the Prisoner

Chad Trulson
North Texas State University

Chapter Objectives

- Be able to describe the prison population.
- Discuss the ways to study prisoner subcultures.
- Understand the changes that have taken place in the subculture over the years.
- Describe the differences between men's- and women's-prison subcultures.
- Discuss the deprivation theory versus the importation theory of subculture development.

Some propose that prisons are a microcosm, or simply a smaller version of the larger free society. If this is true, then society is a strange place, because the world that prisoners inhabit turns upside down the values that most of us live by. It is probably more accurate to say that the prisoner world is one with similarities to the larger free society, but that free-world actions have a very different meaning in the context of a prison. This chapter focuses on the prison subculture in this universe—parallel to, but very different from, the one in which most of us live.

Who Is in Prison?

Prisons have a unique organization that, when compared to the free society, seems to be backward. In most prisons, racial and ethnic minorities represent the majority. The youngest and least educated of this society emerge as top leaders. Our nation's most violent individuals, who are vilified in the free world, gain status in the prison setting, because being a predator is held in higher esteem than being the prey. This section examines select characteristics of those who inhabit this world.

Overall Population

As was discussed in Chapter 3, the incarcerated population in America exceeds 2 million people (Harrison and Karberg 2004). The prison population grew 84 percent from 1990 to year-end 2002—to 1,367,856 prisoners from 743,382 in just 12 years (Harrison and Beck 2003). Although there are signs that the massive prison-population binge of the 1980s and 1990s may be slowing as we move further into the 21st century, the growth in the incarcerated population over the past 10 to 20 years has had a significant impact on the social arrangements in prisons today.

Race and Gender

Prison populations contain disproportionate numbers of racial minorities relative to their proportion in the general population. This is not a new or recent development, however. Not only are racial minorities disproportionately represented in prison—they now constitute the majority of prison inmates. As **Figure 4-1** shows, African-Americans represent a much higher percentage of the prison population than their percentage of the general population would predict. Such disproportionate minority-group representation in the prison population is reflected in the rate at which minority members of American society are sent to prison. As discussed in Chapter 3, at the end of 2002 the incarceration rate for African-American men was almost eight times higher than the rate for white men. There were 3,437 African-American male inmates per 100,000 African-American male

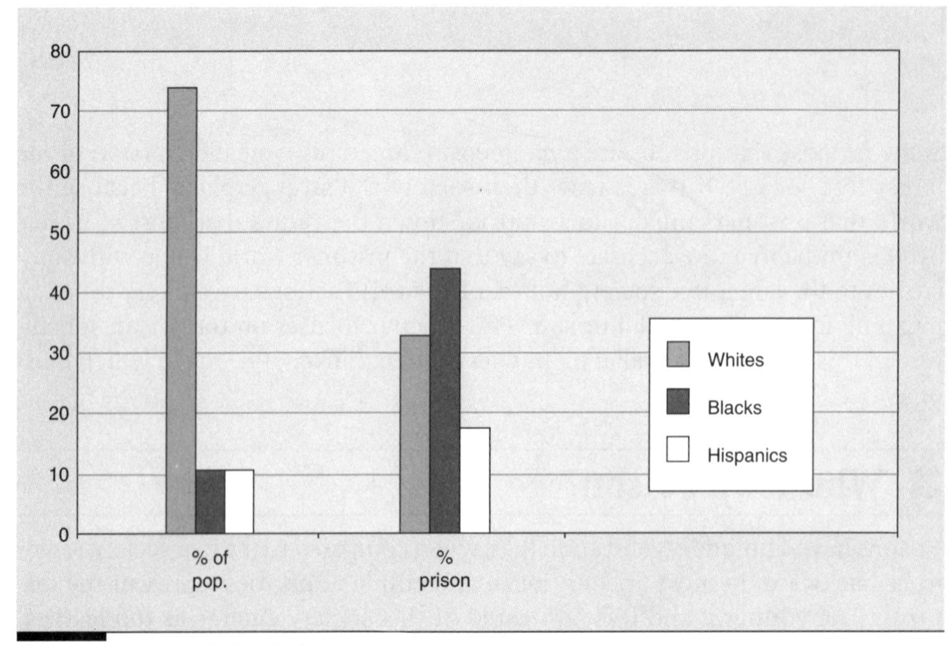

Figure 4-1 Prison Populations by Race. ***Source:*** *Created by author from Harrison and Beck 2003; U.S. Census Bureau 2004.*

residents compared to 450 white male inmates per 100,000 white male residents. Hispanic men were incarcerated at a rate of 1,176 per 100,000 Hispanic male residents, or almost three times the incarceration rate of white males (Harrison and Beck 2003).

At year-end 2002, 39 percent of all female prisoners were white, 40 percent were African-American, and 17 percent were Hispanic. The incarceration rate per 100,000 residents for African-American women was five times higher than the incarceration rate of white female residents. There were 191 African-American women incarcerated per 100,000 African-American female residents compared to 35 white women incarcerated per 100,000 white female residents. Hispanic women were incarcerated at a rate in between African-American women and white women (80 per 100,000 Hispanic female residents) (Harrison and Beck 2003). Although racial proportions are not as lopsided among female prisoners as they are for men, disproportionate incarceration rates for minorities and women are the most obvious factor in any description of this nation's prisons.

Men make up just over 93 percent of the prison population, and the overall rate of incarceration for men is 912 per 100,000 male residents, almost 15 times higher than that of women. However, the number of female inmates has increased at a faster rate than for men in the past two decades. From 1995 to 2002, for example, the total number of male prisoners grew 27 percent, while the total number of female prisoners in this same time period grew 42 percent (Harrison and Beck 2003). Caution should be noted, however, as percentage changes can be misleading when describing growth rates among female prisoners. This is because percentage changes of women are calculated from a much smaller base rate than for males, and the absolute number of female prisoners still remains much lower than the number of male prisoners. For example, from 1995 to 2002, the number of male prisoners in the United States grew by roughly 286,000 inmates (or 27 percent), while the number of women prisoners grew by only 29,000 inmates (or 42 percent). Despite this caution, the incarceration of women is no longer being viewed as an anomaly not worthy of attention, and is now being viewed as a significant issue in corrections.

People in Prison by Type of Crime Committed

What about the types of crimes for which inmates are in prison? Of all state-prison inmates at year-end 2002, 49 percent were in prison for a violent offense—including, but not limited to, murder, rape/sexual assault, robbery, and assault. Property offenders accounted for 19 percent of all state-prison inmates, drug offenders accounted for just over 20 percent of inmates, and the remaining inmates were in prison for public-order offenses such as drunken driving or weapons-possession charges (Harrison and Beck 2003).

In the six-year period from 1995 to 2001, the percentage of the total number of property and drug offenders in prisons declined (to 19.3 percent from 22.9 percent for property offenders, and to 20.4 percent from 21.5 percent for drug offenders), while the percentage of the total number of violent and public-order offenders in prison increased (to 49.3 percent from 46.5 percent for violent offenders, and to 10.8 percent from 8.7 percent for public-order offenders) (Harrison and Beck 2003). From 1995 to 2001, the absolute number of violent offenders grew by roughly 131,000 inmates, or 63 percent of the total growth of the prisoner population (total growth was 207,300 inmates); the number of property offenders in prison grew by 3,600 inmates, or 1.7 percent of total growth; the number of drug offenders grew by roughly 31,000 inmates, or 14.8 percent of the total growth; and the number of public-order offenders in prison grew by 42,400 inmates, or roughly 20 percent of the total growth in the prisoner population during these years. The bottom line today is that almost half of the prisoner population consists of violent offenders, and most of the total growth in the prison population in the past several years is also attributed to these offenders.

Caution should be noted because the six-year trend from 1995 to 2001 does not show the entire picture of change in the correctional population. For a more complete picture, a longer-term look at correctional populations is needed. As **Figure 4-2** shows, the percentage of the prison population consisting of those offenders incarcerated for violent offenses has fluctuated through the years, and has yet to reach its former high level.

The fact that violent offenders have been and continue to be the most represented group in prison, and that their proportion of growth in the total prison population from 1995 to 2001 is larger relative to other offender groups examined, is not surprising. Perhaps the main reason for this finding is that violent offenders receive longer sentences, serve larger portions of their sentences in prison, and are denied parole at a greater rate than most property, drug, and public-order offenders. As a result, violent offenders tend to "stack up" in prison, whereas other offenders are revolving in and out of prison (Pollock 2004). It is entirely possible that some violent offenders sent to prison in 1980 are still there, whereas it would be rare to find a public-order or a property offender still incarcerated for a crime they had been sent to prison for in 1980. Thus, violent offenders continue to accumulate, while other groups of offenders are rotated out of prison.

"Truth-in-sentencing" initiatives and sentencing practices in the 1990s that called for even longer sentences for violent offenders ensured that these criminals served a greater proportion of their sentence in prison as opposed to parole in the community. Some evidence for this is found in the average time served in prison

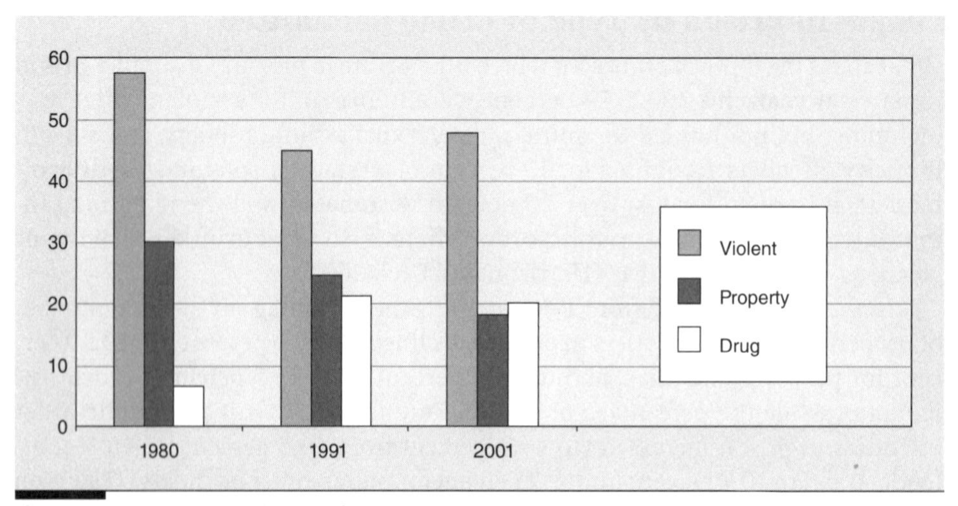

Figure 4-2 Prison Population by Type of Crime. ***Source:*** *Author's adaptation of data from Bureau of Justice Statistics 2004.*

for a violent offense in the 1990s. From 1990 to 1999, for example, the average time served in prison for violent offenders increased to 55 percent of their sentence from 44 percent. For a 20-year sentence, this would equal an additional 2 years in prison. Although the average time served in prison increased for most types of offenders in the 1990s, increases were most pronounced for violent offenders (Bureau of Justice Statistics 2001). **Figure 4-3** displays the percentage of violent offenders in prison in a slightly different manner from the previous figure.

The foregoing discussion is a national look at prison populations and trends, and there is sure to be variation in individual states. Prison-population demographics and the types of offenders in prison differ among a number of factors. For example, one factor is gender. Although violent offenders have contributed the most to the growth in male- and female-prisoner populations in the past several years (64 percent of the growth in prisons for men versus 49 percent in prisons for women from 1995 to 2001), property and public-order offenders have accounted for 38 percent of the growth in the women's-prison population since 1995 compared to only 21 percent in men's prisons (Harrison and Beck 2003).

Sentence Length and Time Served

How long are sentences, and how long do offenders actually serve in prison? Information on sentence length in the past several years shows that while the

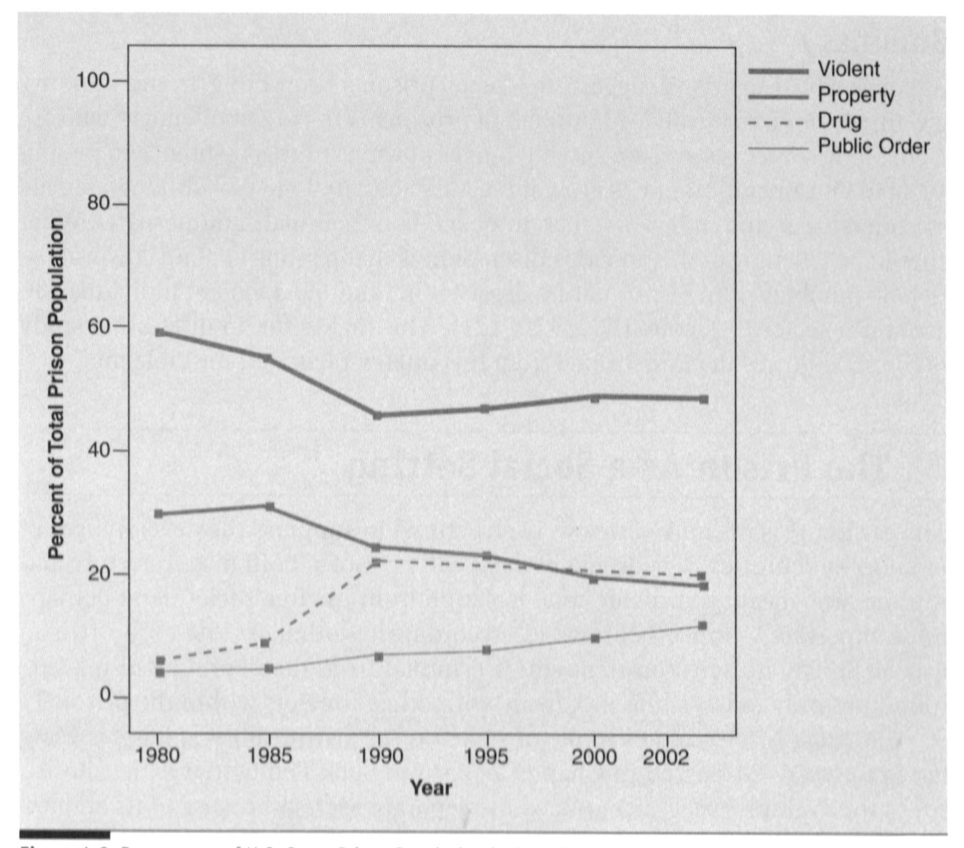

Figure 4-3 Percentages of U.S. State-Prison Population in Custody by Most Serious Offense, 1980–2002.
Source: *Author's adaptation of data from Bureau of Justice Statistics 1997 and Bureau of Justice Statistics 2003.*

average length of sentence for prisoners has *decreased*, the proportion of the time that inmates will serve in prison has *increased*. From 1990 to 1999, the average sentence length decreased to 65 months from 69 months, while the percentage of sentence served increased to 49 percent from 38 percent among all offenses. Although these figures vary considerably by specific crime, the indication is that prisoners will be serving a longer portion of their sentence in prison. This translates to 34 months, or about three years for the "average" prisoner (Bureau of Justice Statistics 2001).

Age and Education

Prisoners are also likely to be young and undereducated. Fifty-eight percent of prisoners at year-end 2002 were 34 years old and under, while 40 percent were under age 30 (Harrison and Beck 2003). Approximately 41 percent of prisoners have less than a high-school education (this rises to 65 percent if considering those only with a GED) (Harlow 2003). The percentage of prison inmates without a high-school education has remained stable from the early 1990s. There is some variation when comparing education levels among male and female prisoners and among minority and nonminority prisoners. In general, female prisoners are more likely to have a high-school diploma than are male prisoners. Minority prisoners overall are generally less educated than white prisoners, as determined by completion of high school or GED program (Harlow).

Summary

A look at who is in prison suggests that many prisoners are young, from a minority group, and undereducated. About half of prisoners are violent offenders who are facing long sentences and are serving longer portions of their sentence in prison. Women's incarceration rate is growing relative to that of men's—although female prisoners are found in fewer actual numbers than their male counterparts. What should be taken from the previous discussion is that prisons contain "disproportionate numbers of the least mature, least stable, and most violent individuals in American society" (Jacobs 1982, 120–121). This makes for an interesting study of the subcultures that are formed from this often violent and unstable mix.

The Prison As a Social Setting

The earliest prisons in America were structured to suppress the development of prisoner subcultures. The design of these early prisons, both in architecture and routine, was meant to provide total isolation from the free society, and perhaps more importantly, from other inmates. According to Rothman (1990, 85), "[t]here was obviously no sense to removing a criminal from the depravity of his surroundings only to have him mix freely with other convicts within the prison."

The most recognized example of such a **total institution** was Eastern State Penitentiary. As discussed in Chapter 2, Eastern State Penitentiary (the successor to the Walnut Street Jail) utilized the **separate system** because of its emphasis on the total physical and social isolation of prisoners. The rationale for the separate system of confinement was that total isolation would guarantee that all

total institution Erving Goffman's term for an institution where the boundaries of work, play, and sleep are eliminated.

separate system type of prison system, like Eastern Penitentiary in Pennsylvania, where inmates were totally isolated and spent their entire sentence in their cell and private exercise yard. Also called Pennsylvania System or Philadelphia System.

inmates would avoid being contaminated by other prisoners—such isolation would give ample time and an environment for a prisoner to "reflect on the error of his ways" (Rothman 1990, 85).

By the mid-1800s, Eastern State was one of only a few prisons in America that attempted the complete physical and social isolation of prisoners. Other prisons dabbled with efforts to socially isolate prisoners by enforcing strict rules of silence, although in these prisons, inmates were allowed to congregate together for work and at mealtimes (**Figure 4-4**). Called the congregate system, the most recognized examples of this form of social isolation by silence were adopted in New York at the Auburn Prison and at Ossining Prison (or Sing Sing), which operated at the same time as Eastern State. Instead of physical isolation from other prisoners, to enforce the strict rules of silence, administrators at Auburn and Sing Sing relied on harsh regimes of corporal punishment for the smallest of infractions (Rothman 1990, 87).

congregate system type of prison like Auburn Penitentiary where inmates slept in separate cells but were released each day to work and eat together; also called New York System or Auburn System.

Attempts to isolate prisoners through complete physical separation, as in Eastern State, or to socially isolate them through strict rules of silence, as in Auburn or Sing Sing, eventually eroded. Penitentiary administrators found that inmates were masters at subverting prison control mechanisms, and they soon realized that no setting, no matter how coercive, could prevent the development of a semblance of an inmate subculture. Inmates at Eastern State, for example, developed elaborate tapping methods on sewage pipes—a sort of prison Morse code called the rapping alphabet (Johnston 1994, 50). Prisoners also perfected what

"rapping alphabet" the type of communication inmates at Eastern Penitentiary developed by tapping on their sewer pipes as a type of early Morse code.

Figure 4-4 Mess Hall, typical correctional institution. ***Source:*** *Reprinted with permission of the American Correctional Association, Lanham, Maryland.*

"kite" prison notes that are thrown to their recipient; today a "kite" is known as any form of written communication.

today is sometimes referred to as a kite—by throwing notes weighted with pebbles over the walls of separate recreation yards (which did not have covered roofs). Still other inmates at Eastern State found they could communicate with cells directly above them through ventilation shafts or through skylights in the roof of the cells (which were eventually boarded up in 1852) (Johnston). Much was the same in the congregate systems of Auburn and Sing Sing. According to Rothman (1990, 87), the routine in the congregate institutions ". . . diabolically tempted the convicts. They were to sit together at mess tables and workbenches, and yet abstain from talking—an unnecessarily painful situation." As a consequence, guards at Auburn and Sing Sing were not able to totally suppress conversations among inmates, and prisoners also developed elaborate hand gestures as an early form of prison sign language (Rothman).

Possibly the greatest barrier to the physical and social isolation of prisoners in early American prisons was burgeoning prison populations. Later, as discussed in Chapter 2, the reforms of the Big House and the correctional institution allowed prisons to develop as quasi-social settings, leading to prisoner solidarity and subcultures replete with a unique code, slang, roles, and values. It was during this time, roughly the 1930s to 1950s, that American sociologists and others became interested in the workings of the prison as a "social community" (Clemmer 1958). From these early American sociologists we received the first systematic look at prison life in their description of a prisoner subculture—a *sub rosa* system of power and exchange that includes the special rules, norms, values, and behavior patterns of prisoners.

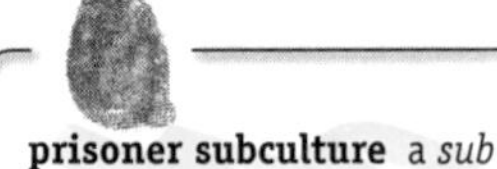

prisoner subculture a *sub rosa* system of power and exchange that includes the special rules, norms, values, and behavior patterns of prisoners.

Researching the Prisoner Subculture

closed system organization where there is little or no permeability in communication with the outside world.

While sociologists and others have been interested in prisons since their inception, a new era where prison researchers and others attempted to study the norms, values, and behavior patterns of prisoners began in the 1940s. Prior to this time, prisons were generally closed systems, and invisible to the public. Once removed from society, the prisoner and his or her social world were forgotten. As researchers and others became interested in the unique world of the confined, prisons became pristine research environments for the interested few that chose to venture into the prison walls.

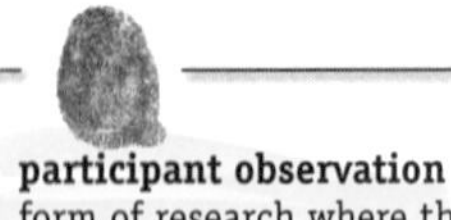

participant observation form of research where the researcher lives or works with, or simply observes, the target population.

The earliest research on prison subcultures used participant observation. This method involves the researcher's living or working with a population of prisoners for a length of time, participating to some degree in the daily activities of prisoners while they record and observe what they see and experience (Hagan 2003). The advantage of participant observation, as opposed to surveys and interviews, is that the researcher does not have to rely on secondhand accounts from prisoners or others—methods that require trusting people who are sometimes not so trustworthy. The general disadvantage of participant observation is that prisoner behavior may be influenced by the researcher's presence, if his or her identity is known—or, in the most extreme cases, the researcher's observations may be biased if he or she "goes native" and begins to sympathize or align with the inmate subjects under study (Hagan).

Donald Clemmer, for example, was a sociologist on the mental-health staff at Menard Prison in Illinois. His employment resulted in one of the seminal books on the prisoner subculture, *The Prison Community*, first published in 1940. Gresham Sykes was allowed access to the New Jersey State Prison in Trenton because the warden, Dr. Lloyd W. McCorkle, was familiar with one of Sykes's mentors. The resulting work, Sykes's *The Society of Captives* (1958), came from nearly three years of participant observation at Trenton. Similarly, Erving Goffman's book *Asylums* (1961)—which was not about a prison, but was important nonetheless—was in part facilitated by his role as the assistant to the athletic director at St. Elizabeth's Hospital in Washington, D.C. (Goffman).

A more contemporary example of participant observation in prison came from Leo Carroll, who worked as a prison guard. This employment helped him gain entrance to a Rhode Island maximum-security prison, and eventually resulted in his important contribution to understanding race relations in prisons, titled *Hacks, Blacks, and Cons*, in 1974. James Jacobs, although not employed by the prison, was a participant observer at Stateville Penitentiary in Joliet, Illinois. This experience culminated in the publication of *Stateville* in 1977. James W. Marquart became employed as a prison guard for almost three years in Texas, which made possible his access to the inner sanctum of prisoner control meted out by a group of co-opted prisoners (called Building Tenders) in the Texas prison system (Marquart 1986; Crouch and Marquart 1989). Mark Fleisher was certified as a full-time federal correctional staffer and studied the United States Penitentiary at Lompoc, California, in the mid-1980s to form his work on prison violence called *Warehousing Violence* (Fleisher 1989). Rose Giallombardo studied Alderson, a federal women's prison, and published one of the first accounts of the prison subculture of women in her book *Society of Women* in 1966. A more contemporary example of research in a women's prison is by Barbara Owen, who studied the Central California Women's Facility and wrote *In the Mix* (1998). Still other accounts are less sociological and more journalistic, but give an indication of prison life and subcultures nonetheless. For example, Ted Conover's book *Newjack* (2000) details the experiences he had while employed as a correctional officer at Sing Sing prison in New York, and Jennifer Wynn examined the incarceration experience at New York's Riker's Island as a journalist and participant teacher for five years, which culminated in her book *Inside Riker's* (2000). This is just a small sampling of researchers, who either through employment or other contact, were able to study the prison firsthand through participant observation.

Some excellent works about prison life and prisoner subcultures come from personal accounts or autobiographies of inmates who wrote while in prison or once they were released. These authors might be said to be the ultimate participant–observer as incarcerated felon. Some early examples include Eldridge Cleaver's *Soul on Ice* (1968) and George Jackson's *Soledad Brother* (1970). Although these two works were generally political essays and quasi-autobiographies, they do hint at prison life and subcultures. John Irwin, a former California inmate who later earned a doctorate degree and became a renowned prison sociologist, wrote *The Felon* (1970) and *Prisons in Turmoil* (1980). Jack Henry Abbott, a convicted bank robber and murderer (for killing another inmate), wrote about prison in his book, *In the Belly of the Beast*, published in 1981. Edward Bunker (a.k.a. "Mr. Blue" in the movie *Reservoir Dogs*) was incarcerated in California for various stints of time and wrote his first nonfiction book about prison, *Education of a Felon*, in 2000. He also wrote several fictional accounts based on his prison experiences, including *The Animal Factory* (1977), recently made into a movie of the same name.

Major publishers have released several other accounts of prison life in the past few years, including *Behind a Convict's Eyes* (2004) by "K. C. Carceral" (a pseudonym of a prisoner) and *About Prisons* (2004) by Michael G. Santos, both of whom remain incarcerated. *You Got Nothing Coming: Notes from a Prison Fish* (2002) by Jimmy Lerner is an especially powerful observation of prison life by a middle-class author who was incarcerated for manslaughter in a Nevada prison. Other good examples would include Victor Hassine's *Life Without Parole* (1999) and *Committing Journalism: The Prison Writings of Red Hog* (1993) by Dannie Martin and Peter Sussman. Moreover, a separate subfield (or subculture) of criminology called **convict criminology**, of the prison has emerged in recent years. Convict criminologists are generally composed of current or former prisoners, many of whom are professors now, who utilize their firsthand experiences in prison along with academic analysis to describe prison life (Ian Ross and Richards 2003).

convict criminology a developing sub-field comprised of researchers and professors who served time and offer their unique perspective to the study of prisons and crime.

The aforementioned works are only a sample of many forms of participant observation in prison, and engender a long history that has examined prison life and the prisoner subculture. Through various participant-observation methods, these individuals were able to penetrate the prison walls to provide a description of prison life at a level that might not be obtained by other methods. This is how we "know" about prison life and subcultures. The remainder of this chapter will draw on the important contributions of these and other authors as the social world of the prisoner is examined.

The Process of Imprisonment

Some of the earliest descriptions of the social world of the prison dealt with the experiences of inmates as they first entered the prison walls. These examinations focused on the special ceremonies and deprivations that new prisoners endured as they were transformed from free citizens into prisoners. These processes were thought to contribute to the development of the prison subculture as inmates responded and adapted to the ceremonies and deprivations of being imprisoned. Later works offered alternative explanations for the development of prisoner subculture.

Degradation Ceremonies

Prisons have been described as "forcing houses for changing persons," and entrance into the prison setting might best be characterized as the social process by which once-free citizens are "made" into inmates (Goffman 1961, 12). As Goffman (1961, 16) noted: ". . . being squared away the new arrival allows himself to be shaped and coded into an object that can be fed into the administrative machinery of the establishment, to be worked on smoothly by routine operations."

The terms used to describe the inmate-making process portray the inmate as a lifeless automaton to be "shaped" and "coded" into a compliant prison object. To shape them into inmates, new commitments to prison must pass through elaborate procedures of the "administrative machinery" designed to "strip," "trim," and "program" them into compliance with institutional regimes. These **degradation ceremonies** underscore the concept of a "civil death"—a leaving of one life as a free citizen and the beginning of another as a convict (Goffman 1961, 16). Strip searches, photographing, weighing, disinfecting, and fingerprinting are standard degradation ceremonies as inmates enter the prison world. Another ceremony includes the creation of dossiers (or social- and criminal-history files) detailing the most intimate aspects of the inmate's life history, in which staff members have full access and where information becomes power. The giving of a number instead of a name symbolizes the loss of an old identity and the adoption of an official prison identity for inmates. Heads are shaved in many prison systems (at least among men) to remove any individual personality based on outward appearance, and prison garb is issued to formally solidify an inmate's position as different from those in the outside world.

degradation ceremonies procedures in a total institution toward new entrants that depersonalize and strip them of identity and self esteem.

Two of the most debasing rituals include forced nakedness and the landslide of intricate and paternal rules for inmates to follow. Nakedness is both a symbolic and actual degradation in that the practice relays to inmates that they are completely vulnerable and powerless in the prison environment. It is akin to Nazi Germany's concentration camps whereby prisoners would be marched naked in single file to shower areas to be deloused. This is coupled with the massive number of paternalistic rules that govern the simplest of activities for inmates. Through these and other rituals, inmates are transformed from adult men and women into little boys and girls who are assumed to need maximum supervision and guidance for even the simplest of activities (Abbott 1981; Goffman 1961). For all intents and purposes, then, the new arrival to prison has begun life over as a child—friendless, fundless (a term used in a television interview with Caryl Chessman pending his execution in California in 1960), and with no identity.

Such degradation ceremonies serve to make a strong statement from the prison administration that everyone in prison is like everyone else—and that no one with a number instead of a name is running anything within the prison walls. Whether every inmate is like any other inmate is debatable, but what is not debatable is that these ceremonies are meant to reprogram and orient the inmate to his or her new life as a state convict—something different from the life that he or she had known previously.

Getting "Cliqued On" and "Ho Checked"

The initial moments of prison socialization may be characterized by occasions whereby staff members administer "tests" or "checks" to an inmate to gauge their level of obedience. An inmate who becomes defiant in response to staff orders may receive immediate and sometimes harsh retribution, "which increases until he openly 'cries uncle' and humbles himself" (Goffman 1961, 17). An interesting example of the "checking" process is described by Marquart (1986, 351–352), based on his study of physical coercion and prisoner control in a southern penitentiary:

> *I [hall officer] had a hard time in the North Dining Hall with an inmate who budged in line to eat with his friend. Man, we had a huge*

> *argument right there in the food line after I told him to "Get to the back of the line." I finally got him out [of the dining hall] and put him on the wall. I told my supervisor about the guy right away. Then the inmate yelled "Yea, you can go ahead and lock me up [solitary] or beat me if that's how you get your kicks." Me and the supervisor brought the guy into the Major's office. Once in the office, this idiot [inmate] threw his chewing gum in a garbage can and tried to look tough. One officer jumped up and slapped him across the face and I tackled him. A third officer joined us and we punched and kicked the sh—— out of him. I picked him up and pulled his head back by the hair while one officer pulled out his knife and said "You know, I ought to just go ahead and cut your lousy head off."*

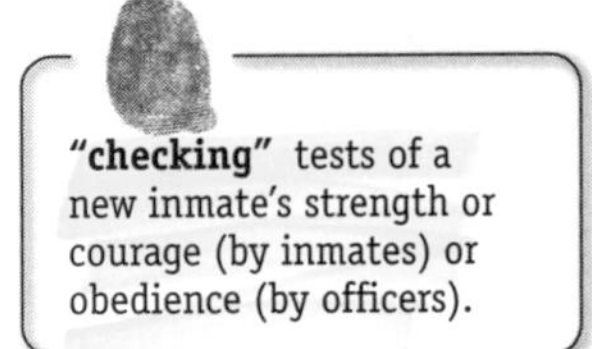

"checking" tests of a new inmate's strength or courage (by inmates) or obedience (by officers).

"Checking" is not limited to staff. Inmates participate in this process as well to get a gauge on an inmate's "pedigree," sometimes with the blessing of staff members (Lerner 2002; Santos 2004, 100). One inmate maintained that getting "ho checked" or "cliqued on" is a daily feature of prison life, designed to test the new arrival.

> *. . . A new Anglo will report to work in a hoe squad, and the blacks and Hispanics will perform what is referred to as a "ho check," and will, one at a time, drop whatever implements they are using, step out of the work line, and ask the field boss if they can "take a look" at the new guy. This means they are calling the man out to fight. The boss will approve, and the Anglo must step out of line and fight. This may happen to that new Anglo a dozen times that day, and not one of those incidents will be reported. The field boss will not allow the Anglo to be beaten too bloody, as that would necessitate a trip to the infirmary and accompanying paperwork. What often follows is that the new man, beaten and fearful, has then received a taste of what he must face on the blocks, and when two or three blacks or Hispanics then fall into his cell, will be much more likely to capitulate and "catch a ride," or agree to be their bank or prostitute or whatever. (Inmate correspondence to Trulson and Marquart, 2004)*

The "checking" process suggests that new arrivals to prison will be tested by both staff and inmates. As they try to navigate the prison maze, the neophytes' response may indicate how they will survive in the prisoner subculture—attack, submit, align, divert, bargain, exploit, or avoid.

Deprivation, Importation, and the Development of Prisoner Subcultures

All inmates are subject to what Sykes (1958) called the deprivations of prison life. The earliest descriptions of the prisoner subculture focused on these deprivations and explained that the prison subculture developed as a response to them (Sykes; Sykes and Messinger 1960). According to Sykes (1958, 63–83), there are five deprivations, or "pains," of imprisonment. These are described in **Box 4-1**.

Box 4-1
Deprivations of Prison

Liberty—Includes restrictions on the freedom of movement, as well as deprivation of contact with family members and friends and others from the free world.

Goods and services—Inmates' inability to obtain these reduce them to an infantile state whereby they must depend on staff to provide everything from socks and underwear to shampoo and toothpaste.

Heterosexual relationships—Refers to the deprivation of "normal" sexual contact in prison settings.

Autonomy—Refers to the fact that inmates are not free to make decisions about their own lives; they are not free to do what they want, when they want, and how they want. They are told when to eat, sleep, and work by being treated like a child.

Security—Inmates are in a constant state of anxiety over their safety, and there is little that they can do to avoid danger (Sykes 1958, 78).

Sykes (1958) concluded that prisoner subcultures are developed and maintained as a response to the deprivations of imprisonment. For example, the deprivation of heterosexual relationships may result in a prisoner subculture that places little or no stigma on homosexual rape, at least for the aggressor. In the free society, the homosexual rapist might be stigmatized, but in the context of prison deprivations, being a predatory rapist garners status as a "man" willing to take what he wants. Negative stigma attaches to the victims, as they are referred to as "punks." The deprivations mentioned above are thought to lead to the formation of a prisoner subculture—which includes, but is not limited to, an inmate code (a set of behavioral rules), a special language (prison slang), defined prisoner roles (the wolf and the punk, among others), and a set of prisoner values (trust, loyalty, violence, and strength) (Welch 2004).

inmate code an informal but powerful set of rules for behavior for inmates.

A competing perspective on the development of prisoner subcultures explains that they do not develop from the deprivations of imprisonment, but instead are brought into or imported to prison from the outside (Irwin and Cressey 1962). This perspective is called the "importation hypothesis." Researchers subscribing to this view believe that a prisoner subculture's special codes, language, roles, and values are derived from the preprison characteristics of inmates rather than from the deprivations of prison. In simpler terminology, prisons and their routines do not make the prisoner subculture; instead, the people entering prisons come with a ready-made culture of their own (Irwin and Cressy).

There has been research support for both the importation and the deprivation perspectives. To test these competing hypotheses, researchers have examined whether preprison characteristics such as gang membership, age at admission, sentencing offense, and marital status are more helpful in explaining prison behavior than are deprivation factors such as sentence length, living arrangements, or custody level in prison. Generally, if preprison characteristics are found to be more influential on prison behavior, this indicates more support for importation as opposed to deprivation explanations. Jiang and Fisher-Giorlando (2002) found support for both the importation and the deprivation models in an explanation of forms of inmate violence. They found that the deprivation model was more important in explaining inmate violence against correctional staff, while the importation model was more helpful for explaining inmate-on-inmate assaults. Paterline and Peterson (1999) found more support for deprivation-model factors than for importation-model factors in explaining adaptation to the prison regime. Sorenson, Wrinkle, and Gutierrez (1998) found that age and race were significant predictors of prison violence, suggesting support for the importation model. In explaining prison disciplinary actions, Cao, Zhao, and Van Dine (1997) found more support for preprison characteristics than for prison influences (deprivation model), again suggesting that prison behavior may be more influenced by what is brought in than by a response to prison deprivations.

One of the strongest pieces of evidence that prison itself changes people was an experiment conducted with college students. The Stanford Prison Experiment, conducted in 1973, shed light on the powerful influence of a prisonlike environment on individual behavior and adaptations. Twenty-four average male college students were paid $15 a day to participate in a two-week study. In this study, half of the subjects were randomly assigned as prisoners to be confined in a mock prison built in the basement of Stanford University's psychology building. The others were randomly assigned as guards. These were normal, mentally and physically healthy college students. They did not import into the prison a background of violence and intimidation, gang membership, or generally irresponsible behavior. The results of this study, which had to be aborted after only six days, were stunning and were never duplicated again in an academic environment. As Haney and Zimbardo (1998, 709) recalled 25 years after the experiment:

> *The outcome of our study was shocking and unexpected to us, our professional colleagues, and the general public. Otherwise emotionally strong college students who were randomly assigned to be mock-prisoners suffered acute psychological trauma and breakdowns. Some of the students begged to be released from the intense pains of less than a week of merely simulated imprisonment, whereas others adapted by becoming blindly obedient to the unjust authority of the guards.*

This seminal study uncovered the "extraordinary power of institutional environments to influence those who passed through them" (Haney and Zimbardo 1998; also see Haney, Banks, and Zimbardo 1992). Some of the "normal" guards took on dictator roles that clearly departed from their normal behavior on the outside. Some prisoners had to be released because of extreme depression and fits of rage as a result of the mock-prison deprivations. One prisoner had to be released after he developed a psychosomatic rash. Only a few prisoners actually "survived" until the experiment ended. Throughout the duration of the six-day experiment, there was evidence of prisoners openly defying staff, bargaining, submitting, casting out fellow prisoners for personal benefit, and even organizing collective attempts to subvert guards by barricading doors with bed frames and mattresses. At one point, one of the released inmates was said to be forming a group to come break the others out of the mock prison. Again, this experiment lasted only six days. Had the experiment lasted longer, we might have received a more in-depth look at the deprivations of imprisonment, and we would have likely seen the development of a stronger prisoner subculture—including more-refined aspects of group solidarity, a unique language, prisoner values, and designated prisoner roles. While preprison characteristics are important in the development of the prisoner subculture, the deprivations of imprisonment also seem to be a powerful factor, as evidenced by the Stanford Prison Experiment (Haney and Zimbardo; Haney, Banks, and Zimbardo).

Each model may be more or less influential, depending upon the type of behavior. For example, the fact that heterosexual men engage in homosexual relations in prison may be best explained by the deprivation of normal sexual relationships. Upon release, these inmates resume a heterosexual identity. This would be an example where deprivation would possibly be the best explanation. Alternatively, when inmates engage in violence, this may be indicative of the fact that they were violent before prison, and continue to be violent in prison regardless of deprivations. In the description of prison provided by Abbott (**Box 4-2**), both models seem to be at work.

Prisons from the 1940s to the 1960s

Early studies of the prisoner subculture that began in the 1940s focused on factors such as the inmate code; prisoner slang, or argot; prisoner roles; and prisoner values. Further efforts examined relationships among prisoners, particularly homosexual relationships and rape, and the concept of prisonization (Clemmer 1958). Regardless of how prisoner subcultures develop—as a result of prison deprivations, importation of preprison characteristics, or a combination of importation and deprivation—these aspects of the prisoner subculture warranted much attention in early research on the prisoner subculture.

prisonization the socialization of a new inmate to the norms, values, and behaviors of the prison.

Box 4-2

Everyone is afraid. It is not an emotional, psychological fear. It is a practical matter. If you do not threaten someone at the very least someone will threaten you. When you walk across the yard or down the tier to your cell, you stand out like a sore thumb if you do not appear either callously unconcerned or cold and ready to kill. Many times you have to "prey" on someone, or you will be "preyed" on yourself. After so many years, *you are not bluffing*. No one is (Abbott 1981, 144).

The Inmate Code

The inmate code was a major focus of all early studies of prison subcultures. The code entailed a *sub rosa* set of rules by which prisoners "should" run their lives in the prison setting. Early descriptions of the inmate code portrayed a world where convicts "did their own time," shunned guards, and never lost their "cool" even in the face of extreme provocation. The prisoner who followed the code valued loyalty, autonomy, and strength. Some of the features of this inmate code are displayed in **Box 4-3**.

This is not to say that all inmates followed such a code. It is hard to imagine that a group of the most violent and least stable individuals in American society, who were unable to follow the simplest laws of society, would have complete allegiance to an informal inmate code. However, early researchers uncovered a prisoner subculture that, to a large degree, subscribed to these principles.

The inmate code was perceived to be part of every prisoner subculture. Few researchers attempted to look at women in prison to see whether they endorsed the code, too. Giallombardo (1966) was one of the few researchers who made any attempt to describe the female-prisoner subculture. In her study of Alderson, a federal prison for women, she found that different themes were emphasized in the female prisoners' world compared to that of men, but that a code did exist. Some examples include:

> *You can't trust other women.*
>
> *To live with other women is to live in a jungle.*
>
> *If I don't get there first, someone else will.* (Giallombardo 1966, 100)

Box 4-3

The Inmate Code

Don't interfere with inmate interests.
Never rat on a con.
Don't be nosy.
Keep off a man's back.
Don't put a guy on the spot.
Be loyal to your class.
Be cool.
Do your own time.
Don't bring heat.
Don't exploit inmates.
Don't cop out.
Be tough.
Be a man.
Never talk to a screw.
Have a connection.
Be sharp.

(Sykes and Messinger 1960)

Giallombardo found that certain elements of the male code, such as "do your own time," were rejected in the women's culture. For example, she found that female prisoners were more likely to be involved in relationships that encouraged interaction with other inmates as opposed to doing one's own time and "keeping off a man's back." Contemporary research still shows that this is the case; female prisoners have different interactions than do men, especially concerning staff relations. Female inmates are much more likely to confide in staff at length about their problems or to use guards as surrogate counselors (Pollock 2002; Girshick 1999). This was not the case in the male subculture. Barbara Owen (1998; Owen and MacKenzie 2004) found that the convict code is more potent among male than among female prisoners, but that there still is a female version of the convict code. For example, Owen's (1998) research indicated a female code that at least tacitly approved of relationships ("take care of each other"), certainly more so than what would be found in the subculture in men's prisons. Some additional features of the female code uncovered by Owen and MacKenzie are:

> *Mind your own business.*
>
> *The police are not your friend; stay out of their face.*
>
> *If asked to do something (by staff), you do not tell.*
>
> *Do not allow rat-packing—fight one on one only.*
>
> *Take care of each other. (Owen and MacKenzie 2004, 159)*

The male and female inmate codes have changed from when early researchers began writing about prisoner subcultures. One result seems to be that inmate solidarity has eroded considerably, if it ever existed to the degree suggested in early studies (Irwin 1980; Owen 1998). For example, Hassine (1999) talks about a code that has changed as a result of younger and more-violent offenders entering prisons. Such a contemporary code includes prescriptions such as, "Don't gamble, don't mess with drugs, don't mess with homosexuals, don't steal, don't borrow or lend, and you might survive" (Hassine, 42). Hassine's portrayal of prison in the 1980s paints a picture of an inmate subculture where avoiding others altogether may be a more promising avenue than subscribing to inmate solidarity and interaction via the inmate code. There is similar evidence that female prisoners also seek isolation rather than solidarity, but this trend may not be as extreme as in men's prisons (Greer 2000; Owen 1998).

Prison Slang

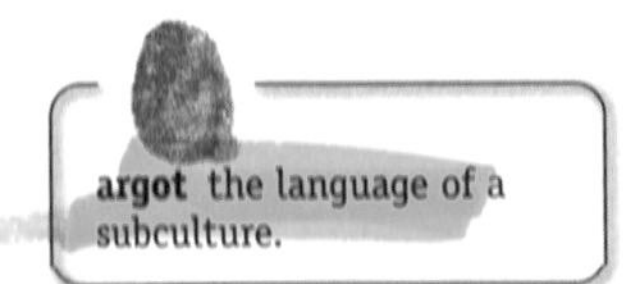

argot the language of a subculture.

Prison slang, or argot, is the language of prisoners and is thought to be an important part of the prison subculture. Just as lawyers, mechanics, and car salespeople have special terms and language native to their profession, so do prisoners. According to Hargan (1934), prison argot serves as a secret code against prison officials, and therefore acts to promote solidarity among inmates. This may or may not be correct, because guards usually know the meaning of prisoner slang as well as the prisoners do, and may use it to a significant degree (Lerner 2002; Sykes 1958). Perhaps more important than a secret language to subvert the keepers, prisoner argot serves as a symbolic expression of group loyalty, the use of which serves as a measure of integration and allegiance to the inmate subculture (Garabedian 1964; Sykes 1958).

There are thousands of examples of prison slang; terms tend to change over time and vary between institutions, and across different regions of the country. Indeed, as in any language, prison slang is a dynamic, constantly evolving entity. Some of the earliest examples of prison slang may have very different meanings today, if they are used at all. Despite these caveats, prison slang is used in all prison subcultures, with special terms designated for everything from prison food to guards, and especially the different types of inmates and roles in the prison setting.

Clemmer (1958) documented some of the earliest forms of argot and included an entire appendix of slang terms in his book *The Prison Community.* Some of the prisoner slang terms and their definitions are displayed in **Box 4-4**. Sixty-one years after Clemmer's original 1940 publication, Glenn (2001, 345–362), a former Texas prison warden, provided a long list of contemporary prison slang he had encountered in his years in the Texas prison system. His contributions are also displayed in Box 4-4.

Box 4-4
Prison Slang

1940: Ace = A dollar.
Clown = A policeman.
Fish = One newly arrived to prison.
G-Note = A $1,000 bill.
Baby Grand = A $500 dollar bill.
Hack = A night policeman.
Hand Jig = To masturbate.
Junk = Any narcotic.
Lunch Hook = A human hand.
Mouthpiece = A lawyer.

2001: Ace = A best or close friend.
Bitch Up = To cry, give in, or act like a whining woman.
Boss = An inmate term for officers; also "sorry son of a bitch" spelled backward.
Brownie Queen = A male homosexual who takes the female role.
Head Running = Unnecessary or excessive talking.
Jack Books = Pornographic reading material with which to masturbate to.
Kill = To masturbate.
Ride = To be turned out, prostituted, pimped out, or turned into an inmate whore.
Swole = To be angry.
Tree Jumper = A rapist or sexual predator, especially a child molester.
Wood = A white convict, derived from "peckerwood."

(Clemmer 1958, 330–336; Glenn 2001, 345–362)

As mentioned, prison slang tends to vary by time, place, and institution. Prison slang can also vary based on perspective—from a former warden versus a researcher or from an "old convict" versus a new "fish." For example, Lerner (2002) described argot he learned as a new prison "fish." Being a college-educated inmate, Lerner was called a "lawdog," or someone who could assist other inmates to file and read court papers. Other argot he encountered included "dawg," which was slang for a friend or associate; "toads," which was a derogatory name for white inmates; and "fishcops," who were new correctional officers. Moreover, since getting and using drugs is an important part of the prisoner subculture, especially in contemporary prisons because of the influx of drug-related offenders, there are a number of argot terms for various drugs. For example, "smack," "crap," "H," and "brown" are some of the many names for heroin. Argot for marijuana may be the most numerous, with traditional names from "Mary Jane," "grass," or "Aunt Mary," to less-recognized names such as "lubage" and "mutha." Slang for drugs in general may simply be labeled "junk." Argot terms for drugs are so numerous that every conceivable combination and potency of a drug may have several different names.

Prisoner Roles

argot roles a classification of inmates based upon activities they were involved in before prison or how they behave in prison.

Prisoner roles (or **argot roles**) serve to sort and classify inmates based on the activities in which they are involved in prison, or on the activities they were involved in before they came to prison (Irwin and Cressey 1962). In short, prisoner roles may be based on both in-prison and preprison behavior. Gresham Sykes (1958) identified a number of prisoner roles—including those of rats (informers), centermen (those aligned with correctional officers), gorillas (aggressive predators), merchants (involved in the underground, or **black market**, in prison), wolves (sexual aggressors), punks (those forced into homosexual activity against their will), fags (those who engage in homosexuality voluntarily or because they "like it"), real men (old-style convicts who do their own time), ballbusters (often involved in violent confrontations), toughs (affiliated with violent crime), and hipsters (newer, drug-involved criminals) (Sykes 1958, 84–108). Roles such as wolves and centermen describe in-prison behaviors, while roles such as toughs or hipsters concern preprison criminal pursuits.

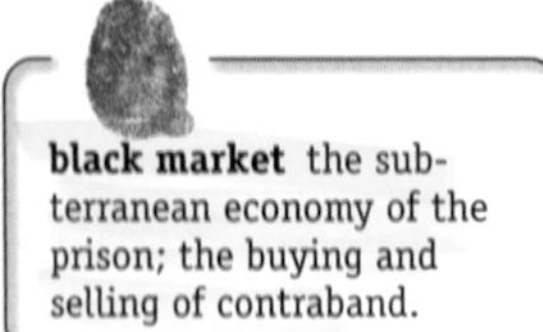

black market the subterranean economy of the prison; the buying and selling of contraband.

Schrag's (1944) roles have perhaps become the best known. He listed prisoner roles such as the square john, right guy, con politician, and outlaw. The "**square john**" was the type of inmate who did not have an extensive criminal record and identified more with prison officers than with other inmates. Other inmates did not trust him. The "right guy" was characterized as the old-style convict. He was respected for his criminal profession, usually being a professional thief, bank robber, or a member of organized crime. He was respected as a leader in the yard and was not afraid of violence, but used it sparingly and selectively. The "con politician" was a title ascribed to those inmates who held the most-formal leadership roles in prison—for example, as a representative of the inmate council—and were very visible in interaction with the prison administration. The "outlaw" inmate was characterized as one who did not follow any code of behavior, even the inmate code. Anyone was subject to being his victim, and power and violence were his tools of making it in prison.

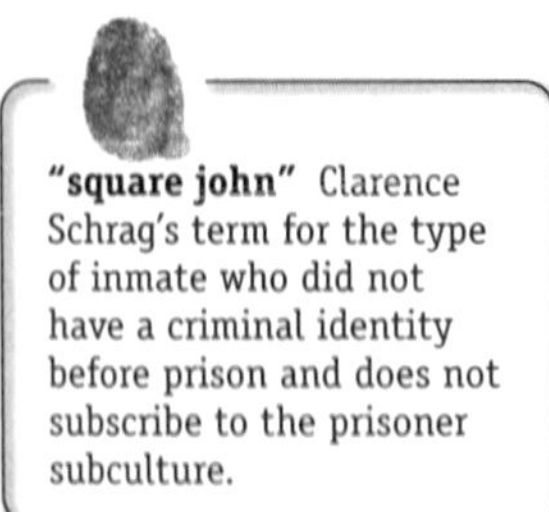

"square john" Clarence Schrag's term for the type of inmate who did not have a criminal identity before prison and does not subscribe to the prisoner subculture.

There are numerous typologies that have been developed about prisoner roles in the inmate subculture, and the foregoing discussion includes only a small sampling (see Hayner 1961; Irwin 1970). There is only a little research on female social roles, however. Although few researchers have examined women's roles, those who have done so have generally found that argot roles are different from those in the male subculture. For example, Giallombardo (1966) wrote that there was no female equivalent to the "real man" found in men's prisons, because the qualities associated with the role (doing one's own time, standing up to authority) had no place in the female world—at least at the time she examined it. In fact, women were described as spiteful, deceitful, and untrustworthy. Despite the lack of overlap, Giallombardo uncovered numerous social roles among female prisoners—including snitchers (informers), inmate cops (affiliated with correctional officers), squares (square-john equivalents), jive bitches (those who stirred up trouble), rap buddies or homies (friends), boosters (shoplifters), pinners (a trusted inmate who serves as a lookout when illicit activities were taking place), and a number of roles associated with homosexuality—penitentiary turnouts, butches, lesbians, femmes, stud broads, tricks, commissary hustlers, chippies, kick partners, cherries, punks, and turnabouts (1966, 105–123). Again, very few roles overlapped with male roles, with the exception of snitches (rats) and squares (those who identified with officers more than with inmates), and most female roles appeared to have been derived from in-prison behaviors and activities (jive bitches), though some roles spoke to behavior prior to prison (boosters).

Prisoner Values

An examination of the early inmate code uncovers a world where group loyalty, violence, strength, and resistance to the "man" were the major values of the prisoner subculture through the 1970s. In the prisoner subculture, violence and snitching are two examples of the most esteemed and vilified activities in prison. Snitching may be the most heinous prison "crime" from the perspective of inmates, and extreme violence may propel one to elevated status where "viciousness and recklessness" are to be respected and admired (Hassine 1999, 13; Santos 2004).

Violence is a predominant theme in all prisons, except for possibly minimum-security or treatment facilities. The same can be said in women's facilities, where, although violence may be more muted, inmates are still fearful and aggressors often intimidate the weak. In maximum-security institutions for men, the potential for violence is omnipresent—and, at times, lethal.

Violent episodes can make the "man" in prison (Fleisher 1989, 217; Santos 2004). Inmates not opposed to violence are viewed as "honorable men" in the prisoner subculture because they are willing to attack or defend with extreme violence in the face of great provocation. Although keeping one's "cool" is valued in the inmate code, it is viewed in a much different way in prisons as opposed to in the free world. For one to use extreme violence may be losing one's cool in the free world, but in prison, such violence does not mean an inmate is out of control. Rather, violence is rational, directive, and useful, as it indicates toughness, strength, and the ability to exact retribution on those who "interfere with an inmate's business."

Group loyalty is an important factor in the prisoner subculture. Being a snitch and shirking loyalty by "talking to the screw" and "riding leg" may be the foremost affront to the inmate prescription to be loyal.

> *When a man commits to the subculture of prison, definitions of honor, respect, integrity, and character take on entirely new meanings that are completely at odds with the world of noncriminals. To a stand-up guy, perhaps the pinnacle of the social pecking order, robbers, racketeers, extortionists, drug dealers, and even killers are good guys. They're honorable men as long as they never rat. (Inmate quoted in Santos 2004, 100)*

The values connected with group loyalty and the proscription against snitching point up to the importance of the rat in prison society. Virtually every writer has at least mentioned this theme of prison life, which is evidenced by the disproportionate number of names used to refer to this group in the prison setting—the "rat," "squealer," "snitch," "cheese-eater," and "stool pigeon," among others.

Officials and inmates alike often scorn the person identified as a snitch, yet the role of a snitch is complex. Officials indicate they could not run the prison without informers, as they provide valuable information on the presence of contraband and planned violence so that officers can preempt such activities. Despite the social and sometimes lethal sanctions against rats and the threat they pose to the inmate code, they seem to exist in abundance (Johnson 1996/2002; Pollock 2004).

Snitching is present in both female and male subcultures, but sanctions against this behavior were (and are) more heavily enforced in prisons for men. Both male and female prisoners use social isolation to sanction snitches, but gossip is used in women's prisons, and only infrequently do the sanctions for female snitches include violence. As Owen and MacKenzie (2004) found in today's female subculture:

> *It used to be, you snitched, you got cut. It was strict about stealing and ratting. Now the "ho's" can rat all day long and no one will kick their ass. Before you would get a chance if you messed with my woman, but no chance for stealing or ratting. (Owen and MacKenzie 2004, 160)*

In prisons for men, snitches may be targeted with extreme violence. Perhaps one of the most gruesome attacks on snitches occurred during the New Mexico Prison riot in 1980. Inmates took over the New Mexico Prison and eventually reached Cell Block 4, the housing area for snitches. Using acetylene torches, rioters carved their way into the snitch cell block with purpose. Morris (1980, 100–101) details the massacre:

> *In the next cell they cut a large hole in the door and come through one by one, stripping and holding the prisoner down for the last killer, who comes through with an acetylene canister. Outside, a prison official hears a whistling sound from 4, scans the block with binoculars, stops on an incredible scene through the window, and watches in horror as four or five inmates hold a man down as another burns*

his head and face with a blowtorch. When his eyes explode out of the back of his head, the inmates burn his groin, then mutilate the body with shanks, and torch him again. When they are through, one seared corpse, a man who weighed over 200 pounds, will weigh less than 50.

Snitching that results in the loss of contraband, such as some extra commissary items or extra pillows, may receive a mild response in the prison subculture. Snitching resulting in a major staff response, such as the discovery of a "shiv" or "shank" that results in segregation time and loss of privileges or good time, may mean severe repercussions for the informer.

Homosexuality and Rape

Prison rape is possibly one of the most sensational elements of the prison subculture, and perhaps the most researched of all prisoner relationships. With a few exceptions, however, homosexuality was largely ignored in the early research of male-prisoner subcultures, yet studied to the virtual exclusion of all other aspects of the subculture in prisons for women. This may be because women were seen primarily as sexual beings, or possibly because homosexuality in women's prisons seemed so different and more open than in men's prisons. Today, the situation is different—there is attention to both male and female homosexual relationships, with a special focus on forced rape in prisons for men.

Homosexuality in men's prisons was (and is) characterized by violent assaults and coercion where older and more-experienced convicts (wolves) offered protections to younger and less-inexperienced inmates (punks) for sexual favors and commissary articles (Lockwood 1980). Perhaps more so in men's than in women's prisons, sex is a status grabber because to get it, one must usually use violence, intimidation, and exploitation. The sexually violent inmate, not unlike any regularly violent inmate, gains an increase in prestige while the victim suffers a loss of status and prestige (Hanser and Trulson 2004). Such status differentials are evidenced by the roles assigned to the sexual perpetrators and victims in prisons for men. Aggressive and dominant inmates get powerful names such as "wolf" and "daddy," while submissive and weak inmates are "punks" and "fags" who are forced "to ride" to make their way in prison (Knowles 1999). Within this dynamic, only punks and fags are identified as homosexuals, and besides being the virtual slave of the wolf, the role completely destroys his male identity as he becomes, for all intents and purposes, a "wife."

Alternatively, homosexuality in women's prisons was (and is) more often consensual, and those women (femmes) who are interested in a relationship compete for the few women who have assumed the male role (butches) (Gagnon and Simon 1968). Such consensual relationships might be examined by the stigma (or lack of stigma) attached to them. For example, being labeled an inmate's "housewife" or "punk" in a male prison certainly has a more derogatory tone than being labeled an inmate's "woman" or "girl" or "femme" in a female prison. There simply may be no corresponding role of the punk in the women's prison because of the largely consensual nature of female-inmate relationships. Even a woman who takes on the role of a man in a prison for women does not suffer the same stigma attached to a man's taking on (or being forced to take on) the role of a woman in a prison for men. In the female prison, for example, being the butch female (or man) may mean the benefit of homosexual relationships without stigma, the right to the material possessions of her femme, and the ability to have multiple sexual partners. Thus, the butch role for women may be more advantageous than the punk role for men, though each involves role reversals. This may be because one role reversal is voluntary (butch) and one role reversal is forced (punk), there is less societal stigma associated with homosexuality among women than there is for men, and because the male role carries more benefits and status inside the prison (Pollock 2002, 137).

Another unique feature of homosexual relationships among women prisoners is the concept of "**pseudofamilies**." Pseudofamilies are make-believe family systems that include all the familial roles—fathers, mothers, daughters, sisters, cousins, and so on (Gagnon and Simon 1968). Such pseudofamily groupings are thought to be unique to female prisoners for several reasons, though such reasons smack of old stereotypes of the woman's "place in the home," their subservient position to men, and their domestic "nature." For example:

"pseudofamilies" make-believe families found in some women's prisons where women take on familial roles and act accordingly in an informal social grouping.

> *These family systems [in prisons for women] seem to arise from three sources. One source is a process of compensation; the majority of females in these institutions are from severely disordered homes, and the creation of the pseudofamily often compensates for this lack. A second source results from the socialization of women who, unlike males who form gangs in self-defense, tend to form families, the basic institution in the society in which they have stable and legitimate roles. Finally there is the fact that a pseudofamily operates to stabilize relationships in the institution and to establish orders of dominance and submission, the primary model which women have in family relationships with fathers, husbands, and children. . . . It is not odd that women model their experience on the institution that they know best in the outside community.* (Gagnon and Simon 1968, 27–28)

In both men's and women's prisons, it is difficult to gauge the extent of homosexual behavior, either forced or consensual, because competing definitions and difficulties in detecting all such behavior are common. For example, definitions of "homosexual relationships" have run the gamut from note writing and hugging, to propositioning and harassment, to genital sexual relations (Struckman-Johnson and Struckman-Johnson 2000). Comparing studies with different definitions varies the incidence of this behavior tremendously. Moreover, homosexual behavior is probably not something that is advertised in prisons, especially in prisons for men, and many studies have used unreliable estimates of homosexual activity, such as reports from officers or other inmates. Such reports probably underestimate such activities in men's prisons (but may overestimate these activities in women's prisons), and particularly when homosexuality comes in the form of rape. Inmate victims of homosexual rape are not likely to report such practices to researchers, and especially not to prison guards. One reason might be the lack of sympathy from guards to such activities. For example, some inmates who choose to snitch or go to guards for protection may be told that they have three options: ". . . submit, fight, or go over the fence" (Weiss and Frier

1974, 3). This indifference by guards may be based on the belief that inmates deserve such treatment if they cannot fend for themselves. Even self-reports from aggressors may not accurately represent participation rates in any form of homosexual behavior, forced or voluntary, because some inmate aggressors do not identify themselves as homosexual, despite participation in activities that are obviously of that nature.

It seems that there are varying amounts of homosexuality in prisons, but the exact incidence of the behavior is subject to much interpretation (Hanser and Trulson 2004; Knowles 1999). Although the extent of homosexuality and sexual victimization in prison may be "one of the most illusive statistics in the criminal justice field," about 15 to 20 percent of a prisoner sample reported that "minor" forms of sexual victimization (rubbing, fondling, or harassment) happened to them at least once or was the "worst incident" of sexual victimization experienced by them. Actual completed rapes are less frequent—about 1 to 7 percent of inmates reported at least one episode of forced rape (Hanser and Trulson; Struckman-Johnson and Struckman-Johnson 2000, 379). Despite the elusive nature of sexual-victimization statistics involving prison inmates, the Prison Rape Elimination Act (PREA) passed by Congress and signed into law by President Bush in 2003 has placed a renewed emphasis on such victimizations in prisons. This renewed emphasis may result in a clearer picture on the nature and extent of sexual victimizations in prison.

Adapting to Prison: Prisonization

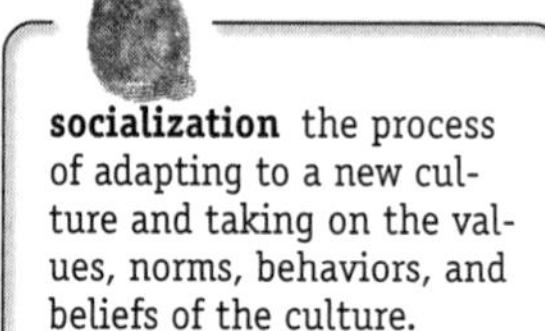

socialization the process of adapting to a new culture and taking on the values, norms, behaviors, and beliefs of the culture.

A discussion of the inmate code, slang, roles, values, and relationships suggests a prison subculture that was supposed to give most inmates a sense of solidarity—but did it? Early writings of the prisoner subculture focused on the concept of prisonization, or socialization within the prisoner subculture. Whereas socialization refers to the process whereby one adopts the values, behaviors, and norms of the subculture, *prisonization* may be referred to as the process whereby inmates become so acclimated to prison life that the free world becomes strange and the prison becomes normal. Early writers questioned, however, whether all or most inmates became totally prisonized and accepted the prison routine as "normal."

Early writers found that the degree of prisonization, and hence solidarity, varied among inmates. For example, Goffman (1961, 61–63) discussed four types of adaptations among inmates: situational withdrawal, intransigent line, colonization, and conversion. "Situational withdrawal" was characterized by a removal and alienation from prison life. Clemmer (1958) probably first recognized these inmates and labeled them "ungrouped"—those who played a solitary or semisolitary role in the prison environment. "Intransigent line" inmates were those who bucked the system and intentionally challenged the institutional regime by flagrantly refusing any and all cooperation with staff members. Perhaps the closest social-role comparison may be found in Schrag's (1944) "outlaw" or Giallombardo's (1966) "jive bitches," who followed no set of rules, formal or informal. "Colonized" inmates were described by Goffman (1961) as those who maximized the perks allowed in prison and created their whole world out of them. Schrag's "con-politicians" seem to come close to this role, as do Irwin and Cressey's (1962) "inmates," as opposed to the "felons" type. Colonized inmates may be considered the most prisonized because it appears that they have come to accept prison as a home, including most tenets of the inmate subculture. Finally, "converted" inmates are those that had completely accepted the institutional role and were described as the "perfect inmate" by staff and administrators, but not necessarily by inmates (Goffman 1961, 63). The colonized inmate might be considered a "suck ass" or a "leg rider" by other prisoners, and is probably the equivalent to a social-role comparison of Giallombardo's (1966) "inmate cops" or Sykes's (1958) "centermen."

The foregoing discussion paints a picture of prisonization as a fragmented and variable process, and early writers suggested that certain factors appear more important than others in determining the degree of prisonization. Clemmer (1958, 301) cited the following factors affecting prisonization: long sentences, an unstable personality, few positive relations with outside people, readiness to join prison primary groups, a blind acceptance to the mores of the primary group, chance placement with similar others in prison, and readiness to participate in gambling and "abnormal" sexual behavior. The bottom line is that Clemmer saw socialization as becoming stronger the longer the exposure to the prison subculture, and as being more likely among those that were more antisocial coming in, with few contacts with the outside world. Thus, to Clemmer, those who had not yet become prisonized were those that had probably just entered the prison world, whereas those who were institutionalized had been there longer.

Where Clemmer saw prisonization as a linear process becoming stronger as time spent in prison increased, Stanton Wheeler (1961) viewed prison socialization as interacting with expectation of release. He characterized prisonization as a U-shaped curve whereby adherence to pro-social norms was high at entry, then decreased over time to align with the inmate subculture until the inmate approached release, at which time adherence to pro-social norms and values increased. Goffman (1961, 61) hinted at the same when he explained that inmates may "employ different personal lines of adaptation at different phases," and that no inmate is bound to one role or the other.

The prison subculture has been described as an underground organization replete with its own code of rules, terminology, social roles, and values. One must put aside the view of a completely cohesive subculture, however, when presented with evidence that the prison described by earlier researchers consisted of individuals who isolated themselves from most others, where the strong preyed on the weak, and where snitches were prevalent and scorned. As one examines the prison society of the 1960s and beyond, this seems even more accurate as prisoners became considerably more diverse and fragmented.

Prisons from the 1960s to the 1980s

Gresham Sykes (1958, 8) noted that ". . . the prison wall is far more permeable than it appears, not in terms of escape . . . but in terms of the relationships between the prison social system and the larger society in which it rests." Sykes was suggesting that what happens outside of prisons will eventually filter into the prison walls. This may have been most accurate in roughly the 1960s to the 1980s, when the free world spilled into the prison and resulted in an era of significant change to the prison setting, and hence, the prisoner subculture.

A Changing Free World

The 1960s through the 1970s were marked by the beginning of a wave of social movements that redefined the status of "marginalized" groups in society—minorities, poor people, women, children, illegal aliens, gays, and the disabled (Jacobs 1983). It was a time of massive social change by way of legislation and judicial decisions that extended rights to a greater and more diverse proportion of the population by "recognizing the existence and legitimacy of group grievances" (Jacobs, 35). As we will see in Chapter 8, this free-world civil-rights movement permeated the prison walls, and inmates were no longer considered "slaves of the state" with no rights. Being acknowledged in such a way led to massive changes in the prisoner subculture.

A Changing Prison World

The general societal unrest of the 1960s led to students, civil libertarians, and others being sent to prison for acts of civil disobedience and forms of criminal behavior that were often justified (by them) as responses to the unfair treatment by society. Once there, these politically charged individuals sensitized prisoners to a new form of power. This power was grounded in the use of legitimate avenues for securing rights once thought inapplicable to prisoners. The period may be illustrated by saying that inmates quit "beating on each other" and started "beating on the bulls" (Brody 1974, 104). Although "beating on the bulls" was literal in some sense, it was probably more accurately characterized by the stream of lawsuits by prisoners that challenged not only the basis, but also the conditions, of confinement.

In the 1960s and 1970s, socially connected inmates were willing to fight for their rights. They were able to garner the attention of the most prolific civil-rights attorneys in the nation, representing large civil-rights organizations such as the American Civil Liberties Union (ACLU) and the National Association for the Advancement of Colored People (NAACP). With their connections, these inmates (and their lawyers) filed numerous federal lawsuits in the 1960s challenging all aspects of prison life. The result was that by the 1970s and 1980s, prisoners were granted rights in a number of areas—including court access, religious equality, due process in punishment, and equal protection by race. Inmates also received the benefits of larger cases that attacked broader "conditions of confinement" such as corporal punishment, overcrowding, health care, and sanitation (Feeley and Rubin 2000; Jacobs 1983). These issues will be covered more fully in Chapter 8.

Inmate solidarity may have reached its peak around the late 1960s and early 1970s, but it appears that this solidarity was short-lived, and, in some ways, one-sided. The Black Awareness movement of the 1960s and 1970s spurred the prisoners'-rights movement, but it also may have spurred a prison subculture that revolved around racial identity whereby the prison, although becoming officially desegregated via judicial decisions, became increasingly self-segregated along racial and ethnic lines (Carroll 1974). Chapter 2 detailed the process whereby the comparatively safe "Big House" became a violent and unstable social jungle.

Prisons from the 1980s to Today

In the 1980s, prisons started to undergo considerable change. Notwithstanding the impact of the prisoners'-rights movement in the 1960s and 1970s (Jacobs 1983), perhaps no other factor fueled these changes more than the massive prison-population explosion that began in the 1980s, carried through the 1990s, and continued into the early years of the 21st century. One product of the burgeoning prison populations was that members of minority groups came to represent the majority on the inside. This demographic change exacerbated the racial cleavages that formed during the prisoners'-rights movement of the 1960s and 1970s, and such pronounced racial divisions contributed to the formation and stronger organization of violent gangs by the 1980s. Another by-product was the inability of correctional agencies to effectively control the flow of prison contraband and the black-market trade. The black market has always created tension in prisons, but this tension soared when it was coupled with the development of racially based gangs who vied for complete control of this illegitimate trade.

On a more practical level, population growth also led to a reduction in many forms of inmate programming, to more-restricted contacts with the outside world, and also to the abandonment of furloughs and work-release programs (Johnson 1996/2002). The removal of these programs and activities led to a prison regime that served to "manage" bodies rather than "rehabilitate" people (Feeley and Simon 1992). Such an environment may be seen as more depriving, where inmates became more frustrated and willing to do whatever is necessary to lessen the deprivation.

Population increases in the 1980s and 1990s can be traced to the tough sentencing of drug offenders, and this had many implications for the prisoner subculture. These modern-day "hipsters" had little or no allegiance to a criminal subculture on the streets. It was not surprising, then, that they had little allegiance to the inmate code—these detached inmates shunned the tenets of "don't snitch" (Pollock 2004, 100). Perhaps one of the greatest change agents for the inmate subculture, however, was the influx of younger and more-violent inmates who appeared to take control of some prisons by the 1980s. Much different from the early prisoners, the growing number of "lowriders," or "state-raised youth," who came to prison had little respect for any code, inmate or otherwise, and were ready to use violence in any situation, on any inmate, at any time, for even the slightest provocation (Irwin 1980, 189).

> *Most of these early lowriders were young juvenile prison graduates . . . unskilled, lower- and working-class criminals who had low respect among older, "solid" criminals and regular convicts. But they were a constant threat to the other prisoners who were trying to maintain peace. For most of the 1950s and 1960s, other prisoners disparaged, ignored, and avoided the lowriders, whose activities were*

> *kept in check by the general consensus against them and the belief (accepted by the lowriders and most other prisoners) that if the lowriders went too far, the older prison regulars would use force, including assassination, to control them. Lowriders steadily increased in numbers. In the states . . . whose youth prison systems expanded to accommodate the increase in youth crime, the adult prisons began to receive growing numbers of tough youth prison graduates and criminally unskilled, more openly aggressive young urban toughs. They could no longer be controlled. They entered the growing racial melee and stepped up their attacks and robberies on other prisoners. When there were no successful countermoves against them, they took over the convict world. (Irwin 1980, 189)*

Changes to the Inmate Code and Subculture

Early researchers such as Sykes, Clemmer, and others might have a hard time recognizing the inmate subculture today. Indeed, they paid little attention to race, and the social groupings that they uncovered would hardly compare to the gangs that permeate prisons today. Moreover, the violence they described was not the sort of nondescript and lethal violence encountered in today's prison, erupting from the violent mix of young, violent, drug-related, and gang-related youngsters who have literally spent more time behind the walls of state institutions than in their homes (Irwin 1980).

More recently, Hassine (1999) suggests a subculture where the best way to make it is to avoid interaction altogether—stay in one's cell and avoid the yard. This is much different from the earlier-described inmate code that promoted inmate solidarity against the "bulls," replete with values of loyalty and respect. This is not to say that an inmate code has vanished from prisons. It is perhaps more accurate to say that a code operates among prisoners, but that inmate loyalties to the code have changed (Mays and Winfree 2002). Inmates today have become far more diverse and unpredictable than when early researchers entered the prison walls (Johnson 1996/2002). A select few "right guys" and "real men" no longer play a major role in dealing with inmate problems; rather, an infiltration of wolves, gorillas, and "young bucks" have taken their place in the prison hierarchy. Inmate relationships are characterized as weak and detached, and loyalty to an "inmate class" has changed to "loyalty to one's race, ethnic group, clique, or gang" (Irwin 1980; Mays and Winfree, 184).

Some of the same can be found in the contemporary female-prison subculture, though by most accounts, the fragmentation is not as pronounced as in prisons for men (Owen 1998; Pollock 2002). Women's prisons are still characterized by a greater degree of cohesiveness whereby the inmate culture serves to provide women with emotional support (Pollock 2002). There has been some erosion of female cohesiveness, however, as Owen uncovered a female-prisoner subculture where the inmate code has fragmented because younger "kids" have failed to respect any code, and where new commitments or first-timers become involved in the "**mix**"—a term used for women who continue the kind of behavior in prison that led to their imprisonment in the first place—such as involvement in drugs, fights, and other negative in-prison behaviors (Greer 2000; Owen).

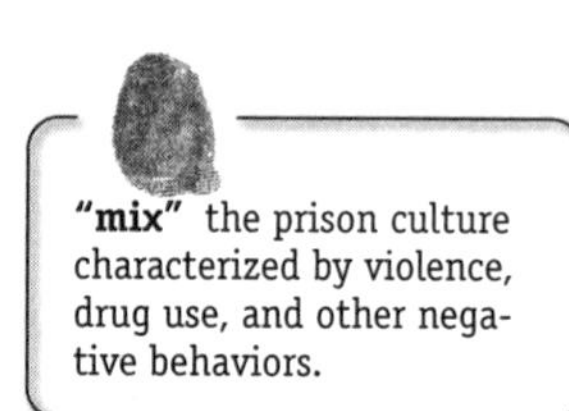

"mix" the prison culture characterized by violence, drug use, and other negative behaviors.

The inmate code in the contemporary prison increasingly takes an individual focus where self-rule may be closer to the norm than the exception. When solidarity is found, it is usually within violent cliques and racial gangs, and not the result of the entire prisoner subculture operating on a unitary set of principles. The prison world today, particularly among male prisoners, can be characterized by detached and predatory relationships where your best friend today may be your executioner tomorrow. Such is the characterization by Johnson (1996/2002, 126), who describes the prison subculture as a process of "prowling the yard," where violence aimed at fellow prisoners was once "more subtle or covert," but is now "open and bold" (Johnson 1996/2002, 137). This may be most evident among prison gang members, who seem to have transformed the inmate subculture to a gang culture (Pollock 2004).

Prison Gangs

The downfall of the white-dominated prison described by Irwin (1980), where older convicts maintained order and were the guardians of the inmate social system, can be traced to the rising ethnic identification of the prisoners'-rights movement of the 1960s and 1970s. It can also be traced to the population explosion and the disproportionate incarceration of minority inmates, and the importation of "lowriders" who had coerced other inmates into race-specific groups by the 1980s (Hawkins and Alpert 1989). The result was a changing structure in prisons filled by inmates who became loyal to racial cliques and gangs.

Gangs formed in all major prison systems in the United States. Probably the best-known gangs formed in Illinois, California, and Texas. According to Hawkins and Alpert (1989, 251), racial minorities organized the first prison gangs for control of the prison black market. They were followed by other racial groupings, particularly by whites, who organized gangs as a response for protection. By all indications, however, gangs had different ways of forming—some were imported through the incarceration of extant street-gang members, some formed solely in prison, and some were formed by a combination of these methods. James Jacobs (1977) examined the presence of gangs in Illinois, and noted that their development came generally from the importation of major street-gang members. For example, Chicago's major street gangs had branches in Stateville Penitentiary (the Devil's Disciples, Blackstone Rangers, Vice Lords, and Latin Kings). However, while African-American and Chicano gangs were imported by way of street gangs in Chicago, white gangs appeared to have emerged in prison and in response to African-American and Chicano groupings (Jacobs).

Theodore Davidson (1974) described gang formation in California's San Quentin Prison, explaining that they were formed in prison first by Chicano inmates, though eventually African-American and white inmates also developed gangs within prisons. Irwin (1980) recalls that the Chicanos formed their first in-prison gang called the Mexican Mafia or "EME." These individuals were generally connected because of their street identity (the area in which they lived in Los Angeles), so, in some ways, their formation was imported into prison. As the result of a 1967 EME-ordered stabbing death of another Chicano inmate at San Quentin, independent Chicano inmates attempted to organize and assassinate members of the Mexican Mafia. They were not overly successful in their assassination attempt, but in response, formed an organization of their own. The result was a rival Chicano gang called La Nuestra Familia (Irwin 1980, 190). These gangs battled mainly each other during the early periods of formation, but eventually victimized unaffiliated whites and African-Americans. According to Irwin, the formation of the Mexican Mafia and La Nuestra Familia resulted in solidarity among groups of white and African-American prisoners. Consequently, African-American inmates formed the Black Guerrilla Family, and white inmates formed the Aryan Brotherhood to protect themselves from attacks from Chicano gangs.

Building Tenders inmates in the Texas prison system who were chosen by officers to enforce order against other inmates.

The formation of gangs was different in Texas. Until the late 1980s, the Texas inmate subculture was absent of gangs because it was dominated by an elite group of white inmate guards, called **Building Tenders** (BTs), who were authorized by staff to control inmates by any means necessary. The BT system was highly coercive, but by most accounts was very effective in suppressing the development of rogue groups and cliques, and kept the most violent disorder at bay in the Texas prison system (Crouch and Marquart 1989). Ironically, as the result of a prisoners'-rights case, *Ruiz v. Estelle* (503 F. Supp. 1265 [1980]), the BT system was dismantled. When the BTs were removed from power, a power vacuum was created that witnessed whites, African-Americans, and Hispanics squaring off and vying for control of the prison. Guards could not be hired fast enough to replace the semistaff roles occupied by the BTs. Ralph and Marquart (1992, 47) noted the result was that inmates "killed more people in a two year span than had been killed in the twenty years prior." The vying of power eventually resulted in the formation of several race-based gangs in Texas prisons—including the Texas Syndicate (Hispanic), the Aryan Brotherhood of Texas (white), and the Mandingo Warriors (African-American).

The creation of gangs among female prisoners, although by no means clear, seems to be different from gang formation among men (Pollock 2002). Some research indicates that gangs among female prisoners are growing, whereas other researchers have found little evidence of this phenomenon (compare Mahan 1984 to Greer 2000; Pollock 1996, 2002). Perhaps the comparative lack of gang membership among female prisoners may be related to the pseudofamily groupings discussed earlier. These make-believe families may provide the same services for women—including protection, goods, and group belonging—as gang membership does for men. Greer found that women are less violent and less involved in gang activity because of these relationships (Greer; Owen 1998). However, Greer did note that relationships among female prisoners are less familylike than in the past, presumably because of a higher degree of mistrust among female prisoners of today.

Possibly the greatest factor for the comparative lack of gang membership in prisons for women is that safety is less of a concern for women; thus, there is less of a need to have gangs for protection—although this is one of the most important reasons for the formation of men's gangs in prison (Allender and Marcell 2003). Moreover, since pseudofamily groupings and homosexual relationships are more racially integrated and voluntary, this precludes the formation of gangs as a result of racial tensions, as found in men's prisons. Also, because there is less marketing in drugs and other contraband in women's prisons, there is less incentive to group together for control over this illicit trade. Whether the absence of female-prison gangs will remain is unclear.

The rise in gangs created a serious dilemma in prison settings, especially in prisons for men. It may not be too much of an overstatement to say that for a period of time, control had been lost in American prisons. Free from an inmate code, inmates became loyal to race-based cliques, and the ultimate symbol of respect and loyalty was predatory violence against other gang members. Some argue today that gang influence and power have been reduced in the prisoner subculture (Hunt, Riegel, Morales, and Waldorf 1993). This may be accurate, as gangs are becoming increasingly diverse and numerous in American prisons, are constantly changing, and appear to have more shifting allegiances and loyalties than when they were first organized in prisons (Pollock 2004). In some cases, there is more fragmentation and shifting loyalties *within* racial groups. Thus, not only are gang members fighting other races, they are also fighting within their own race. Such conflict naturally reduces solidarity.

The responses of correctional systems to gangs are worth mentioning because these practices have arguably lessened gang influence and power. For example, gang members in the past developed intricate coded communication methods and were able to order "hits" on other inmates and staff. Though this still occurs today (Fong 1990), to curb such "reach," special gang-intelligence departments have been created to investigate and preempt such attacks by gang members—or, more appropriately today—members of "security threat groups." Another practice, which is not new but has expanded, is the development of special control units or areas in prison to segregate prison gang members (Hawkins and Alpert 1989). This practice has been effective in the battle against gangs by eroding some of their power base and preventing complete organization.

Racial Conflict in Prisons

The earliest gang conflicts in prisons were described as intraracial affairs (same race) in which competing gangs began to dominate the drug black market or use extortion practices against other prisoners who had not formed any group alliance (Irwin 1980). This was not the case for long. By many accounts, when gang members victimized other racial groups (interracial), this translated into severe racial tensions and violence between racial groups (Irwin).

There is a paucity of systematic research that has examined racial violence among prisoners, particularly racial violence as the result of gang membership. Usually, research examining race relations and violence among prisoners has come from excellent qualitative accounts by researchers who have entered the prison environment or were actually (or currently) prisoners themselves. These examinations—such as those from Carroll (1974) and Jacobs (1977; 1983) or from Irwin (1980) and Hassine (1999)—provided a glimpse into race relations and conflict through various forms of participant observation. The general synopsis from these accounts was that racial conflict and violence was omnipresent in the institutions they examined. As Jacobs (1982, 120–121) explains:

> *It should hardly cause surprise to learn that relations among the racial groups of prisoners are extremely tense, predatory, and a source*

of continual conflict. Prison populations contain disproportionate numbers of the least mature, least stable, and most violent individuals in American society. If that were not sufficient to make successful race relations unlikely, prison conditions themselves generate fear, hatred, and violence. . . . Racial conflict . . . is a reality of institutional life in prisons around the country. . . . Such settings do not promote understanding among the races; on the contrary they are breeding grounds for racism.

In the late 1960s and 1970s, federal-court decisions mandated that prisons become racially integrated in all aspects, including job, program, and housing assignments (such as in-cell integration). Most academics and correctional officials predicted an explosion of interracial violence. In perhaps the only large-scale systematic examination of the outcomes of racial integration and violence to date, Trulson and Marquart (2002) found that, over a span of 10 years (1990–1999) in the Texas prison system, rates of officially recorded interracial (different race) violence among prisoners were less than the rate of violence among prisoners of the same race—even when different-race offenders were integrated in a double cell. This supports other contemporary studies where researchers have asked wardens to report on the levels of violence following integration. Most wardens believed that integration had little impact on interracial-violence levels (Henderson, Cullen, Carroll, and Feinburg 2000).

More specific to gang violence, Trulson and Marquart (2002) found that less than 3 percent of the more than 40,000 incidents they studied were gang related, and an even smaller percentage was motivated by race. Caution should be noted, however, as Trulson and Marquart found that racial violence usually occurred as the result of a few "irreducible" or "extraordinary" prisoners—such as gang members. Although extraordinary prisoners represent a very small fraction of the prison population, they account for a disproportionate share of violence. Trulson and Marquart hypothesized that correctional-system policies that proactively segregate this "extraordinary minority" before they strike or immediately after they strike served to significantly dampen gang-related racial violence over the long term, thus the finding that few racial incidents in the Texas prison system were gang related.

Although they do not deny that racial unrest continues in prisons, Trulson and Marquart (2002, 2004) do note that it is not any more likely to result in officially recorded violent incidents than other reasons—especially once the smaller number of predatory gang members or violently racist inmates are removed from the mix. However, officially recorded incidents and unofficially held tensions are two different things.

The racial imbalance in prisons has led to a situation where minorities, particularly African-Americans, have become the majority group in most prisons. Being the majority in prison translates into an increase in power in the prison setting. In many cases, this power differential is experienced firsthand by white inmates. Because they are less organized than Hispanic inmates, white inmates seem to have the hardest time with integration and seem to feel the most tension as a result. At least for some inmates, then, integration has created considerable anxiety, and in some cases has caused inmates to commit severe acts of violence against members of another race so they would qualify for a racially segregated cell or so they could be isolated in a single cell—though the consequences, including the loss of good time or free-world prosecution for prison acts, are sometimes severe (Zatukel 1995). Although most do not turn to extreme violence to avoid the "pains of integration," such tension serves to significantly split the prisoner subculture and reduce inmate solidarity.

It is probably safe to say that there is much racial tension in prisons today, but for most inmates, such tension does not result in violence—except for the few extraordinary prisoners who would likely be violent inmates in any situation, integrated or not. Even when interracial violence results, however, there is some indication that factors other than race may actually precipitate it. For "ordinary" prisoners, this means that most interracial incidents may not be "racially motivated" per se, but rather they are the result of other aggravators such as arguments over television channels in the dayroom. Some evidence suggests that this may be the case with racial gangs, and interracial tension and violence may have less to do with racial prejudice than financial interests and control of the black market (Pollock 2004).

The Black Market in Prisons

It is an unfortunate reality that prisoners are able to procure contraband items in prison. Contraband can range from simple things such as pornographic material or excessive prison commissary items to other forms of contraband such as tobacco, drugs, alcohol, and weapons. Contraband items are what provide the foundation of the black market in prison—an illicit and underground economy by which inmates "traffic and trade" commodities for power and prestige. Much as with wealth and material items in the free society, those with the contraband and other illicit commodities in prison are the "haves," while those without are the "have-nots." Such arrangements make for a violent combination when gangs or other groups compete for control of the "game."

How does one get contraband in prison? In general, contraband can be brought into prisons in every conceivable way. Visitors, contractors, inmates, and even staff have been linked to smuggling contraband into prison. Staff involvement in contraband smuggling is particularly problematic. Staff may be "turned" in several ways to become an inmate's "mule," or carrier of contraband. Some staff are tempted by the large amounts of money that can be made, and voluntarily offer their services. For most correctional-staff members $1,000 to $2,000 a month can be a powerful temptation. Staff may be coerced into carrying contraband into prison by threats of violence against family members or friends. Whatever way guards and other staff members are turned, the consequences are disastrous because officer authority is undermined when inmates are able to blackmail the officer.

Perhaps one of the most shocking examples of the availability of contraband was uncovered in 1996 at Stateville Penitentiary in Joliet, Illinois, where a film surfaced showing convicted murderer Richard Speck (who died in 1991) in his cell with a smuggled video camera, using drugs, flashing currency, and engaging openly in homosexual acts with his cell partner. Even more bizarre, Speck had developed a pair of breasts attributable to the large amount of "female enhancement" drugs he was able to obtain while in prison. At Stateville, and likely at other prisons, connected inmates were literally able to procure everything from video cameras to cell phones to drugs (*Suburban Chicago News* 2004). Increasingly shocking stories are uncovered each year. For example, in May 2004, Texas prison administrators discovered that inmates had access to cell phones in certain maximum-security prisons. At the Darrington Unit in Brazoria County, Texas, prison administrators found 47 cell phones after a contraband search. Most inmates stored the cell phones in waterproof bags and flushed them down the toilet. They were able to retrieve the cell phones by attaching the bag to an elastic string taken from their prison jumpsuits. Prison administrators found that inmates used the cell phones to operate free-world criminal enterprises in addition to calling family members and friends (Axtman 2004).

Episodes such as the "Speck Tapes" and cell phones in prison have resulted in drastic steps being taken by some correctional systems to rid their prisons of contraband used for traffic and trade by gangs and other inmates. Most prisons engage in random cell searches, some conduct random drug tests, and others have attempted to make it more difficult to bring contraband into prison and have increased the penalties for anyone who does so. Some systems are doing all of these things (Pollock 2004). The obstacles to ridding prisons of contraband are numerous. In some circumstances, guards and other correctional staff may look the other way in the case of certain forms of contraband—for example, tobacco and pornographic material. They may do this based on the belief that it keeps inmates more docile and easier to manage. Other commodities of the black market are simply unable to be regulated to any degree. For example, punks are commodities in prison because they serve as "wives" to clean cells, run errands, and provide sex, but it would be hard to completely eliminate the practice (although attempts are made by isolating potential victims in special housing areas called protective custody). For serious contraband or black-market items, such as illicit drugs and weapons, staff members are not likely to turn a blind eye (unless they have been compromised). However, in the underground prison economy, even the smallest and most inconsequential items can be the basis for black-market trade because the deprivations of prison life make things we take for granted outside of prison a valuable commodity in the prison setting.

Responses to the Violent Prison Today: Containing the "Extraordinary Minority"

Although small in numbers, violent and predatory inmates contribute a disproportionate share of problems in the prison environment. Isolating them is an attempt to salvage as much order as possible in an inherently disorderly environment. To be sure, there have always been members of the prison subculture that create a disproportionate amount of trouble. So, too, have there been areas in which these individuals have been contained separately from the rest of the prisoner population for varying amounts of time. Places such as the "hole," "solitary," and the "box" conjure images of recalcitrant inmates rotting away in dark, damp, solitary cells until their attitudes get adjusted. It seems, however, that the need for containment of these inmates has been greatest beginning in the 1980s and continuing today. Prison populations started to expand, state-raised youth dominated prison admissions, and gangs formed along racial and ethnic lines. Today, these prisoners have made the prison experience a continual source of conflict, filled with fear, anxiety, and hate for some. In the past several years, states have begun to respond in a drastic fashion to reduce the influence of these violent inmates. Such efforts have meant changes to the prisoner subculture.

Perhaps the most recognizable trend in the control of serious, chronic, and violent in-prison offenders has been the development of supermaximum, or **supermax**, prisons throughout the country (King 1999). Supersecure incapacitation is possibly the most effective tool for controlling inmates who cannot control themselves. Yet this type of effectiveness comes with a price. Notwithstanding the fact that holding an inmate in a supermax facility is double or more the cost of housing a prisoner in a regular prison, this practice is perhaps more criticized for the human price it inflicts (Pollock 2004). One psychiatrist remarked: "This isn't making the community safe. . . . You're just making them the most sickest, most impulse-ridden, most enraged, paranoid, impaired human beings. And then you're just putting them right back into the community" (Jones 1999, 3). We will discuss the supermax facility again in Chapter 10.

supermax prison that is built with high security for prisoners who are deemed too violent or troublesome to live in a regular maximum security prison; typically characterized by total isolation and no programming.

Does near-total isolation in supermax work completely to suppress violence and other disorder? Apparently not. Just as penitentiary administrators at Eastern State found that they could not totally suppress prisoners, so, too, have officials in modern day equivalents, such as Pelican Bay, a supermax prison in California. In 2001, Operation Black Widow became public. In this operation, federal prosecutors targeted incarcerated La Nuestra Familia gang leaders who had been able to run street-gang operations from the Special Housing Unit (SHU), which was the segregation block, at Pelican Bay. Operation Black Widow was the largest and most expensive investigation into a U.S. prison gang ever. The Federal Bureau of Investigation (FBI) and local authorities spent $5 million over three years to control this California gang, which had expanded into phenomenal proportions by the late 1990s (Berton 2003).

Apparently, La Nuestra Familia leaders and other inmates housed in the general population at Pelican Bay had been able to get parolees and inmate visitors to smuggle small coded notes out of prisons in their body cavities, including orders for murder, drug trafficking, and racketeering. In other cases, SHU prisoners without contact visits wrote notes in their own urine on the back of drawings that were mailed out of prison. Once the note was received, an outsider would uncover the urine-written message by holding the paper to a heat source (Berton 2003; Media Awareness Project 2004). Other inmates scratched messages on the inside of legal envelopes and mailed them to a fake lawyer's address, inmates buried notes in prison recreation yards for soon-to-be-paroled inmates to receive, and others were able to fling notes to other inmates' cells using the elastic from prison-issue jumpsuits. These notes made it to other inmates and to the streets, and in this way, La Nuestra Familia members in Pelican Bay were able to order "hits." Even under the most oppressive of prison conditions today, replete with the most sophisticated technology, inmates who are described as undereducated and unstable are still able to thwart even the most technologically enhanced systems designed to prevent their "reach."

Change and Survival in the Prisoner Subculture Today

The prisoners'-rights movement, the rising numbers of state-raised youth, the disproportionate confinement of minorities, the massive influx of prisoners in the late 1980s and 1990s, the development of prison gangs, the black market, and efforts to contain the most violent of all prisoners have all contributed to change in the prisoner subculture. By all indications, the inmate subculture today is different from that described by early writers. Solidarity to a singular inmate code is described as the exception, and racial polarization and predatory violence as the rule, in the contemporary prison (Johnson 1996/2002). Although the differences may be less pronounced in the female subculture, there is still evidence that it has changed as well (Owen 1998).

The foregoing discussion is not to suggest that the contemporary prison is a raging madhouse where killings, stabbings, and other forms of violence are daily occurrences. Nor is it to suggest that the prison today is so chaotic that inmates would find it virtually impossible to function in any normal way once released. Of course, it is inaccurate to talk about prisons and prison life as if one singular type and experience existed. However, in many institutions, the elements of violence, predation, and fear are all too real.

There are many more issues that would be worth exploring concerning the prisoner subculture today in the contemporary prison. One issue is particularly interesting, and concerns niches and sanctuaries in the prisoner subculture. It seems increasingly likely that inmates will find that withdrawing almost completely from prison life, limiting their existence, and creating their own singular world to deal with its stresses may be the best way to navigate their prison sentence and survive in the contemporary prison.

Avoiding the "Mix": Niches and Sanctuaries in the Prisoner Subculture

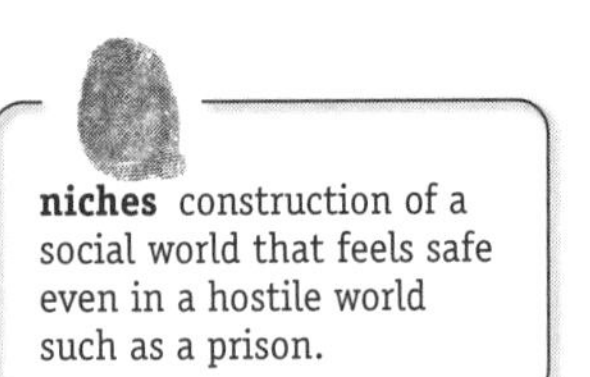

niches construction of a social world that feels safe even in a hostile world such as a prison.

Whether real or perceived, the prison world today is, for many, a dangerous and confusing place (Johnson 1996/2002). Though it is an inmate's home for the duration of their sentence, some may find that carving out certain **niches**, or sanctuaries, is the best way for them to escape the pressures of "home life" and avoid the behaviors that got them imprisoned in the first place. In essence, some inmates find it beneficial to create their own world in the larger world of the prisoner subculture. As Toch (1992, 236) noted:

> *Even where environmental conditions appear most hostile, people create . . . a fabric of life. Such adaptation also occurs in prison. To be sure, in the free world citizens can often flee stress. . . . One can quit a job, divorce a mate, develop relaxing routines. In prison there are few advertised refuges, no neighborhood bar, no supportive family, no late movie or television, few drugs, minimal alcohol, few opportunities to push the dull reality of doing time in the background. . . . The inmate, like the vulnerable person on the street, can arrange a microcosm that rarely guarantees happiness but usually guarantees survival.*

Although some find their niches by living the prison fast life with violent cliques and gangs, many inmates prefer a more structured, orderly, and predictable prison life—avoiding the yard, staying in their cells, and developing few relationships with a few other prisoners (Zamble and Porporino 1988). They may get involved in work, education, or counseling programs, staying close to guards, or may simply avoid everyone and keep to themselves. Owen (1998), for example, described how many of the women she examined preferred to avoid the yard because that is where most of the activities composing the "mix" (drugs, homosexual activity, and fighting) took place. The same is found in prisons for men, as previously mentioned, where the inmate code prescribes almost complete avoidance if one expects to avoid trouble (Hassine 1999). In the prison world described today, immersing oneself in legitimate prison activities while avoiding opportunities for trouble may be the best way for survival.

Despite the many differences between female- and male-prisoner subcultures that have been described in this chapter, and the changes that have come to both over the years, niches and sanctuaries may be the most similar part of both subcultures. It is certainly one of the most enduring components of the prisoner subculture. For example, Clemmer (1958) noted solitary or unaffiliated inmates as early as 1940. A finding of niches and sanctuaries in early and contemporary male and female subcultures suggests that the prison subculture may never have been dominated by one inmate code. There is no reason to believe that prisoners of the 1940s are much different from prisoners of today concerning this dynamic. It may be, however, that there was less of a need to seek out niches and sanctuaries in the early prisons. There may be more need for such a solitary existence today because of the nature of the contemporary prison.

Seeking out individual sanctuaries and safe places may be the only way that inmates can survive the deprivations and pains of prison life and retain some semblance of what is "normal" to them. For some, this is always going to be a problem since, for many, their life has been anything but normal. Niches and sanctuaries for these individuals are found in the fast and violent life, and normality is the "ripping and roaring" of the yard. Perhaps, like prisons of old, the contemporary prison offers niches and sanctuaries for everyone, not just for those who prefer a secluded round of life by avoiding trouble.

> *Secretly, we all like it here. This place welcomes a man who is full of rage and violence. Here he is not abnormal or perceived as different. Here rage is nothing new, and for men scarred by child abuse and violent lives, the prison is an extension of inner life.* (Masters 2001, 205)

Conclusions

There is no denying that elements of the inmate culture have changed from when researchers first entered the prison. Prisoners have more freedoms, more rights, and access to better programs. However, prisoners are more likely to be young, violent, and drug- and gang-related, and the increasing freedom means more opportunity for violence. This is the reality a prisoner must face as they enter this unique social world. It is a world of power and interchanges where efforts to make comfortable an uncomfortable existence usually benefit only the sickest, strongest, and most violent. The rest are left to submit, bargain, or dodge trouble as they attempt to "do their own time." Although the experience may be somewhat different for women—less violence, fewer gangs, and more-consensual relationships seem to characterize the women's prison—it is still an unpleasant place. The strong continue to prey on the weak in a world where "normal" is a relative term.

KEY TERMS

argot—the language of a subculture.

argot roles—a classification of inmates based upon activities they were involved in before prison or how they behave in prison.

black market—the subterranean economy of the prison; the buying and selling of contraband.

Building Tenders—inmates in the Texas prison system who were chosen by officers to enforce order against other inmates.

"checking"—tests of a new inmate's strength or courage (by inmates) or obedience (by officers).

closed system—organization where there is little or no permeability in communication with the outside world.

congregate system—type of prison like Auburn Penitentiary where inmates slept in separate cells but were released each day to work and eat together; also called New York System or Auburn System.

convict criminology—a developing sub-field comprised of researchers and professors who served time and offer their unique perspective to the study of prisons and crime.

degradation ceremonies—procedures in a total institution toward new entrants that depersonalize and strip them of identity and self esteem.

inmate code—an informal but powerful set of rules for behavior for inmates.

"kite"—prison notes that are thrown to their recipient; today a "kite" is known as any form of written communication.

"mix"—the prison culture characterized by violence, drug use, and other negative behaviors.

niches—construction of a social world that feels safe even in a hostile world such as a prison.

participant observation—form of research where the researcher lives or works with, or simply observes, the target population.

prisoner subculture—a sub rosa system of power and exchange that includes the special rules, norms, values, and behavior patterns of prisoners.

prisonization—the socialization of a new inmate to the norms, values, and behaviors of the prison.

"pseudofamilies"—make-believe families found in some women's prisons where women take on familial roles and act accordingly in an informal social grouping.

"rapping alphabet"—the type of communication inmates at Eastern Penitentiary developed by tapping on their sewer pipes as a type of early Morse code.

separate system—type of prison system, like Eastern Penitentiary, where inmates were totally isolated and spent their entire sentence in their cell and private exercise yard. Also called Pennsylvania System or Philadelphia System.

socialization—the process of adapting to a new culture and taking on the values, norms, behaviors, and beliefs of the culture.

"square john"—Clarence Schrag's term for the type of inmate who did not have a criminal identity before prison and does not subscribe to the prisoner subculture.

supermax—prison that is built with high security for prisoners who are deemed too violent or troublesome to live in a regular maximum security prison; typically characterized by total isolation and no programming.

total institution—Erving Goffman's term for an institution where the boundaries of work, play, and sleep are eliminated.

REVIEW QUESTIONS

1. Who is in prison? (Provide a complete demographic profile.)
2. What are the ways to study prisoner subcultures?
3. Compare and contrast the male- and female-prisoner subcultures.
4. List and discuss Sykes's deprivations of imprisonment.
5. Discuss "degradation ceremonies." What is the purpose of such ceremonies in prisons? Compare degradation ceremonies to other organizations that may employ similar types of ceremonies. Do the purposes of these organizations' ceremonies differ from the purpose of prison ceremonies? If so, in what way?
6. What are the problems in trying to determine the extent of sexual victimization in men's and women's prisons?
7. Discuss how the prison subculture today is different from the subculture in the 1940s through 1960s.
8. What is a pseudofamily? Why are these found only in women's prisons?
9. What are the major factors that served as change agents for the prisoner subculture from the 1960s to today? In your opinion, what is the most important factor, and why?
10. Make a list of advantages and disadvantages to racial integration in prison. One study in this chapter found that integrated inmates in cells had a lower rate of violence than inmates not integrated in cells. To what do you attribute this finding?
11. What are two methods used to deal with the most violent or "extraordinary" prisoners? If you were the director of a large prison system, which method would you use, and why?
12. What are three reasons cited in this chapter why gangs are not as prevalent in women's, as opposed to men's, prisons?

FURTHER READING

Carceral, K. C. 2004. *Behind a Convict's Eyes*. Belmont, CA: Wadsworth.

Carroll, L. 1974. *Hacks, Blacks, and Cons: Race Relations in a Maximum Security Prison*. Lexington, MA: Lexington.

Clemmer, D. 1958. *The Prison Community*. New York: Holt, Rinehart, and Winston.

Ian Ross, J., and S. Richards. 2003. *Convict Criminology*. Belmont, CA: Wadsworth.

Lerner, J. 2002. *You Got Nothing Coming: Notes from a Prison Fish*. New York: Broadway Books.

Santos, M. 2004. *About Prisons*. Belmont, CA: Wadsworth.

REFERENCES

Abbott, J. 1981. *In the Belly of the Beast*. New York: Vintage.

Allender, D., and F. Marcell. 2003. "Career Criminals, Security Threat Groups, and Prison Gangs." *FBI Law Enforcement Bulletin* (June): 8–12.

Axtman, K. 2004. "The Newest Prison Contraband: Cellphones." June 11, 2004. *Christian Science Monitor*, 1–4. Retrieved August 1, 2004, from www.csmonitor.com/2004/0611/p01s04-usju.html.

Berton, J. 2003. "Black Widow's Web." *Maxim* (November): 137–144.

Brody, S. 1974. "The Political Prisoner Syndrome." *Crime and Delinquency* 20: 94–110.

Bunker, E. 1977. *The Animal Factory*. New York: St. Martin's Press.

Bunker, E. 2000. *Education of a Felon*. New York: St. Martin's Press.

Bureau of Justice Statistics. 1997. *Correctional Populations in the United States*. Washington, DC: U.S. Department of Justice.

Bureau of Justice Statistics. 2001. *Trends in State Parole 1990–2000*. Washington, DC: U.S. Department of Justice.

Bureau of Justice Statistics. 2003. *Educational and Correctional Populations*. Washington, DC: U.S. Department of Justice.

Bureau of Justice Statistics. 2004. *Number of Persons in Custody of State Correctional Authorities by Most Serious Offense, 1980–2001*. Washington, DC: Department of Justice. Retrieved August 1, 2004, from www.ojp.usdoj.gov/bjs/glance/tables/corrtyptab.htm.

Cao, L., J. Zhao, and S. Van Dine. 1997. "Prison Disciplinary Tickets: A Test of the Deprivation and Importation Models." *Journal of Criminal Justice* 25: 103–113.

Carceral, K. C. 2004. *Behind a Convict's Eyes*. Belmont, CA: Wadsworth.

Carroll, L. 1974. *Hacks, Blacks, and Cons: Race Relations in a Maximum Security Prison*. Lexington, MA: Lexington.

Cleaver, E. 1968. *Soul on Ice*. New York: McGraw-Hill.

Clemmer, D. 1958. *The Prison Community*. New York: Holt, Rinehart, and Winston.

Conover, T. 2000. *Newjack: Guarding Sing-Sing*. New York: Random House.

Crouch, B., and J. Marquart. 1989. *An Appeal to Justice: Litigated Reform of Texas Prisons*. Austin, TX: University of Texas Press.

Davidson, T. 1974. *Chicano Prisoners, The Key to San Quentin*. New York: Rinehart and Winston.

Feeley, M., and T. Rubin. 2000. *Judicial Policymaking and the Modern State: How the Courts Reformed America's Prisons*. New York: Cambridge University Press.

Feeley, M., and J. Simon. 1992. "The New Penology: Notes on the Emerging Strategy of Corrections and Its Implications." *Criminology* 30: 449–472.

Fleisher, M. 1989. *Warehousing Violence*. Beverly Hills, CA: Sage.

Fong, R. 1990. "The Organizational Structure of Prison Gangs: A Texas Case Study." *Federal Probation* (March): 36–43.

Gagnon, J., and W. Simon. 1968. "The Social Meaning of Prison Homosexuality." *Federal Probation* 32: 23–29.

Garabedian, P. 1964. "Social Roles and Processes of Socialization in the Prison Community." *Social Problems* 11: 137–152.

Giallombardo, R. 1966. *Society of Women: A Study of a Women's Prison*. New York: John Wiley and Sons.

Girshick, L. 1999. *No Safe Haven: Stories of Women in Prison*. Boston: Northeastern University Press.

Glenn, L. 2001. *Texas Prisons: The Largest Hotel Chain in Texas*. Austin, TX: Eakin Press.

Goffman, E. 1961. *Asylums*. Garden City, NY: Doubleday.

Greer, K. 2000. "The Changing Nature of Interpersonal Relationships in a Women's Prison." *Prison Journal* 80: 442–468.

Hagan, F. 2003. *Research Methods in Criminal Justice and Criminology*, 6th ed. Boston: Allyn and Bacon.

Haney, C., C. Banks, and P. Zimbardo. 1992. "A Study of Prisoners and Guards in a Simulated Prison." In E. Aronson (Ed.), *Readings About the Social Animal*, 6th ed., pp. 52–67. New York: Freeman.

Haney, C., and P. Zimbardo. 1998. "The Past and Future of U.S. Prison Policy: Twenty-Five Years After the Stanford Prison Experiment." *American Psychologist* 53: 709–727.

Hanser, R., and C. Trulson. 2004. "Sexual Abuse of Men in Prison." In F. Reddington and B. Kreise (Eds.), *Sexual Assault: The Victims, The Perpetrators, and the Criminal Justice System*, pp. 147–159. Durham, NC: Carolina Academic Press.

Hargan, J. 1934. "The Psychology of Prison Language." *Journal of Abnormal and Social Psychology* 30: 359–361.

Harlow, C. 2003. *Education and Correctional Populations: B.J.S. Report*. Washington, DC: U.S. Department of Justice.

Harrison, P., and A. J. Beck. 2003. "Prisoners in 2002." Washington, DC: U.S. Department of Justice.

Harrison, P., and J. Karberg. 2004. "Prison and Jail Inmates at Midyear 2003." Washington, DC: U.S. Department of Justice.

Hassine, V. 1999. *Life Without Parole: Living in Prison Today*. Los Angeles: Roxbury.

Hawkins, R., and G. Alpert. 1989. *American Prison Systems: Punishment and Justice*. Englewood Cliffs, NJ: Prentice-Hall.

Hayner, N. 1961. "Characteristics of Five Offender Types." *American Sociological Review* 26: 96–128.

Henderson, M., F. Cullen, L. Carroll, and W. Feinburg. 2000. "Race, Rights, and Order in Prison: A National Survey of Wardens on the Racial Integration of Prison Cells." *Prison Journal* 80: 295–308.

Hunt, G., S. Riegel, T. Morales, and D. Waldorf. 1993. "Changes in Prison Culture: Prison Gangs and the Case of the Pepsi Generation." *Social Problems* 40: 398–409.
Ian Ross, J., and S. Richards. 2003. *Convict Criminology*. Belmont, CA: Wadsworth.
Irwin, J. 1970. *The Felon*. Englewood Cliffs, NJ: Prentice-Hall.
Irwin, J. 1980. *Prisons in Turmoil*. Boston: Little, Brown.
Irwin, J., and D. Cressey. 1962. "Thieves, Convicts, and the Inmate Culture." *Social Problems* 10 (Fall): 142–155.
Jackson, G. 1970. *Soledad Brother: The Prison Letters of George Jackson*. New York: Bantam Books.
Jacobs, J. 1977. *Stateville: The Penitentiary in Mass Society*. Chicago: University of Chicago Press.
Jacobs, J. 1982. "The Limits of Racial Integration in Prison." *Criminal Law Bulletin* 18: 117–153.
Jacobs, J. 1983. *New Perspectives on Prisons and Imprisonment*. Ithaca, NY: Cornell University Press.
Jiang, S., and M. Fisher-Giorlando. 2002. "Inmate Misconduct: A Test of the Deprivation, Importation, and Situational Models." *Prison Journal* 82: 335–358.
Johnson, R. 1996/2002. *Hard Time: Understanding and Reforming the Prison*. Belmont, CA: Wadsworth.
Johnston, N. 1994. *Eastern State Penitentiary: Crucible of Good Intentions*. Philadelphia: Philadelphia Museum of Art.
Jones, R. 1999. "High Tech Prison Designed for Toughest of the Tough." August 31, 1999. *Milwaukee Journal Sentinel*. Retrieved May 4, 2004, from www.jsonline.com/news/state/aug99/max01083199.asp.
King, R. 1999. "The Rise and Rise of Supermax: An American Solution in Search of a Problem." *Punishment and Society* 1: 163–186.
Knowles, G. 1999. "Male Prison Rape: A Search for Causation and Prevention." *Howard Journal of Criminal Justice* 38: 267–283.
Lerner, J. 2002. *You Got Nothing Coming: Notes from a Prison Fish*. New York: Broadway Books.
Lockwood, D. 1980. *Prison Sexual Violence*. New York: Elsevier.
Mahan, S. 1984. "Imposition of Despair: An Ethnography of Women in Prison." *Justice Quarterly* 1: 357–385.
Marquart, J. 1986. "Prison Guards and the Use of Physical Coercion as a Mechanism of Prisoner Control." *Criminology* 24: 347–365.
Martin, D., and P. Sussman. 1993. *Committing Journalism: The Prison Writings of Red Hog*. New York: W. W. Norton.
Masters, J. 2001. "Scars." In D. Sabor, T. Kupers, and W. London (Eds.), *Prison Masculinities*, pp. 201–206. Philadelphia: Temple University Press.
Mays, L., and T. Winfree, Jr. 2002. *Contemporary Corrections*. Belmont, CA: Wadsworth.
Media Awareness Project. 2004. "Inside Pelican Bay." Retrieved May 4, 2004, from www.mapinc.org/drugnews/v01/n707/a04.html.
Morris, R. 1980. *The Devil's Butcher Shop: The New Mexico Prison Uprising*. Albuquerque: University of New Mexico Press.
Owen, B. 1998. *"In the Mix": Struggle and Survival in a Women's Prison*. Albany: State University of New York Press.

Owen, B., and D. L. MacKenzie. 2004. "The Mix: The Culture of Imprisoned Women." In M. Stohr and C. Hemmens (Eds.), *The Inmate Prison Experience*, pp. 152–172. Upper Saddle River, NJ: Pearson.

Paterline, G., and D. Peterson. 1999. "Structural and Social Psychological Determinants of Prisonization." *Journal of Criminal Justice* 27: 427–441.

Pollock, J. 1996. A Needs Assessment of Texas Women Inmates. Paper presented at the American Society of Criminology Conference. Boston, MA.

Pollock, J. 2002. *Women, Prison, and Crime*, 2d ed. Belmont, CA: Wadsworth.

Pollock, J. 2004. *Prisons and Prison Life: Costs and Consequences*. Los Angeles: Roxbury.

Ralph, P., and J. Marquart. 1992. "Gang Violence in Texas Prisons." *Prison Journal* 72: 38–49.

Rothman, D. 1980. *Conscience and Convenience: The Asylum and Its Alternatives in Progressive America*. Boston: Little, Brown.

Rothman, D. 1990. *The Discovery of the Asylum*, rev. ed. Boston: Little, Brown.

Santos, M. 2004. *About Prisons*. Belmont, CA: Wadsworth.

Schrag, C. 1944. "Social Role Types in a Prison Community." Master's thesis, University of Washington.

Sorenson, J., R. Wrinkle, and A. Gutierrez. 1998. "Patterns of Rule-Violating Behaviors and Adjustment to Incarceration Among Murderers." *Prison Journal* 78: 222–231.

Struckman-Johnson, C., and D. Struckman-Johnson. 2000. "Sexual Coercion Rates in Seven Midwestern Prison Facilities for Men." *Prison Journal* 80: 379–390.

Suburban Chicago News. 2004. "The Speck Tapes." *Suburban Chicago News*. Retrieved August 1, 2004, from www.suburbanchicagonews.com/joliet/prisons/speck/tapes3.html.

Sykes, G. 1958. *The Society of Captives*. Princeton, NJ: Princeton University Press.

Sykes, G., and S. Messinger. 1960. "The Inmate Social System." In R. Cloward (Ed.), *Theoretical Studies in the Social Organization of the Prison*, pp. 6–10. New York: Social Science Research Council.

Toch, H. 1992. *Living in Prison: The Ecology of Survival*. Hyattsville, MD: American Psychological Association.

Trulson, C., and J. Marquart. 2002. "Racial Desegregation and Violence in the Texas Prison System." *Criminal Justice Review* 27: 233–255.

Trulson, C., and J. Marquart. July 2004. "Prison Letters on the Limits of Racial Integration in Prison." Paper presented at the Law and Society Association Annual Meeting. Miami, FL.

U.S. Census Bureau. 2004 "Quickfacts." U.S. Census Bureau. Retrieved May 4, 2004, from quickfacts.census.gov/qfd/states/00000.html.

Weiss, C., and D. J. Frier. 1974. *Terror In Prisons: Homosexual Rape and Why Society Condones It*. Indianapolis: Bobbs-Merrill.

Welch, M. 2004. *Corrections: A Critical Approach*, 2d ed. New York: McGraw-Hill.

Wheeler, S. 1961. "Role Conflict in Correctional Communities." In D. Cressey (Ed.), *The Prison: Studies in Institutional Organization and Change*, pp. 229–260. New York: Holt, Rinehart, and Winston.

Wynn, J. 2000. *Inside Riker's: Stories from the World's Largest Penal Colony*. New York: St. Martin's Press.

Zamble, E., and F. Porporino. 1988. *Coping Behavior and Adaption in Prison Inmates.* New York: Springer-Verlag.

Zatukel, D. 1995. "White Man in a Texas Prison: Forced Integration Works No Better in Prison Than it Does Anywhere Else." *American Renaissance* 6. Retrieved June 4, 2004, from commonselseclub.com/Texas.html.

CASES CITED

Ruiz v. Estelle, 503 F. Supp. 1265 (1980)

5 Industry, Agriculture, and Education

William Stone
Texas State University–San Marcos

Chapter Objectives

- Understand the history of prison labor.
- Discuss the various issues regarding prison labor in contemporary prisons.
- Be familiar with some of the problems with prison agriculture.
- Know the types of prison education and what evaluations show regarding recidivism.
- Be familiar with the differences between programs for men and women, both historically and today.

Traditionally, we are conditioned to think of forced prison labor as an integral by-product of incarceration; however, it is probably true that imprisonment is actually the by-product of the need for forced labor. In recent years, the issue of inmate idleness has caused us to reexamine how inmates spend their time while in state custody. Is the role of prison simply to educate and reform prisoners? Or, is there a broader obligation to make restitution to society as well as to improve the inmate?

History of Inmate Labor

Early systems of punishment generally were based on the principles of corporal punishment, capital punishment, or banishment. These punishments were logical and efficient in that they required few government resources to carry out sentences. There is some evidence of other considerations in early systems of punishment; for example, both the Sumerian Codes and the Hammurabic Code contain references to the use of forced labor as punishment (Allen and Simonsen

1986/1992). It is very difficult to determine how frequently these early societies used forced labor because so very few records from this time period exist.

As governments became more complex and started providing more public services such as road and sewer construction, the need for public labor obviously increased. With this increased need for labor came a corresponding need for containing the forced laborers while the public projects were completed. The **Mamertine Prison**, built under the sewers of Rome in 64 B.C., was an example of an early prison built to hold inmates condemned under the Roman penal-servitude concept (Johnston 1973). It is also well known that Roman roads, aqueducts, and sewers were built throughout the empire with the assistance of forced labor. Western civilization was not alone in its desire to capitalize on the use of convict labor. Offenders sentenced under the Chinese laws of the time provided much of the labor used to construct the Great Wall of China. While most of these early convict laborers were maintained in camps, cages, and natural sites such as mines, the need for mass labor obviously contributed to the need for mass confinement.

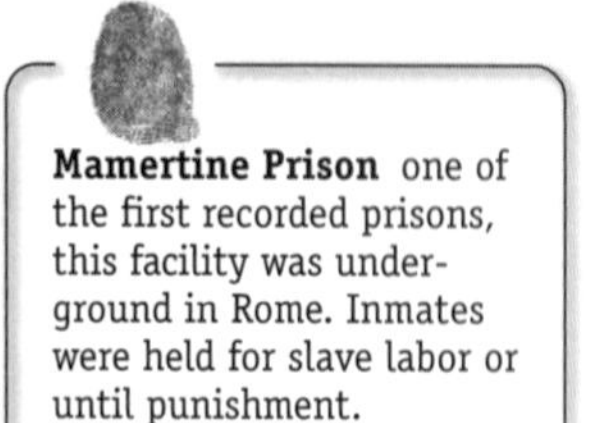

Mamertine Prison one of the first recorded prisons, this facility was underground in Rome. Inmates were held for slave labor or until punishment.

England

The link between imprisonment and labor can be traced throughout history from the earliest forms of prisons to very recent penal philosophy. The English House of Corrections built at Bridewell in 1557 was based on a strict work ethic. As noted in a previous chapter, Americans generally credit the Walnut Street Jail with being the forerunner of the modern penitentiary, while the English give this credit to the Bridewell Institution (Carney 1979). The forced-work ethic of the Bridewell would be reflected in most of the **houses of corrections** of the time, and they all came to be known as **bridewells**. The famous Hospice of San Michael, established in 1704, was also based on an ethic of hard labor and silence. Generally, the labor in these early institutions was intended to be productive in nature, such as gardening, food preparation, and tailoring. It was thought that harsh, productive labor helped support the institution, taught a trade, and instilled moral discipline. Traditionally, men and women were assigned work suitable to their social roles of the time. For example, women were assigned food preparation and "domestic" labor, and men were assigned traditional farm and manufacturing jobs.

This same work-oriented philosophy was proposed for more-formal penal institutions as well. In 1696, John Bellers, a Quaker businessman, proposed his College of Industry. Bellers's plan involved creating a communitarian industrial colony within the walls of the prison at Bristol. In the same year, John Locke proposed the creation of formal industries within penal workhouses as a means of reducing the public cost of imprisonment (Ignatieff 1978). The intent of these early reformers was to devise a system that would benefit the inmates by training them in a usable trade, and benefit the public by reducing the cost of confinement. There were times, however, when the "productive-work" philosophy would be subverted and result in such correctional extremes as the treadmill and shot drill, two classic forms of nonproductive inmate labor. **Treadmills** (or treadwheels) were originally designed as a source of industrial power, but were later converted to provide forced unproductive labor. They consisted of a wheel, equipped with treads, that a person would step on to make the wheel move. Weights or paddles

houses of corrections the general term for bridewells or poorhouses after they became associated with the custody of minor criminals such as pickpockets and thieves.

bridewells early British institutions for the itinerant poor. They were a form of workhouse or poorhouse where the poor were housed, but made to work for their room and board.

treadmills early form of punishment. Also called treadwheels, this contraption was a wheel with treads so that a person would be forced to continuously step on the treads to move the wheel until being removed.

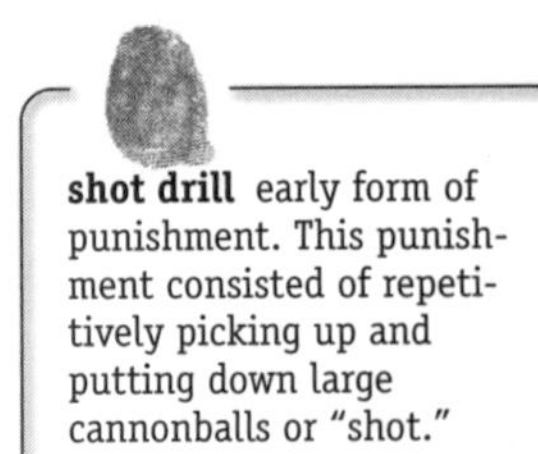

shot drill early form of punishment. This punishment consisted of repetitively picking up and putting down large cannonballs or "shot."

in water provided resistance. The **shot drill** consisted of repetitively picking up and setting down large cannonballs, or "shot." (These two means of forced labor might be considered the historical version of modern-day stair-steppers and free weights!) In general, productive labor such as the roadwork found in the Irish penal system was more typical of inmate labor (Carney 1979).

Early America

The philosophy of sentencing inmates to hard labor was also common practice in the United States. With the adoption of William Penn's "Great Law" in 1682, the concept of substituting hard labor for corporal punishment was clearly established (Clare and Kramer 1976). When the first state prisons were being established in the late 18th and early 19th centuries, provisions for inmate labor were incorporated into all the prison systems. The Pennsylvania prison system, with its philosophy of isolation, would adopt a system of laboring in individual cells at spinning wheels, small textile looms, and shoemaking (Wines and Dwight 1867). The Auburn Prison, with its congregate philosophy, would apply a more mechanized approach and operate traditional textile mills and other industries within its walls. With the ultimate dominance of the Auburn type of correctional institution and its ability to utilize industrialized production systems, it was not surprising that the "**factory prison**" evolved in the United States.

"factory prison" is a term used for early northeastern prisons that adopted the congregate care system and placed prisoners in prison factories to work during the day.

The early prison industries focused on several labor-intensive types of enterprises. The choice of work was heavily influenced by the fact that inmates generally represented an unskilled labor force. Probably the most common type of early industry was the textile mill. Textile mills could be operated by relatively unskilled labor and were a high-profit industry of the time. State legislators of this time period could see the obvious advantage of industrializing their prisons. Industrialization would reach from Maine to the frontier areas such as Texas. The focus of these early industrial efforts was clearly for the benefit of the state, not to teach inmates a trade or to improve the conditions of their confinement. The industrialization of American prisons was simply a state-based version of the evolving industrial systems in the United States that were exploiting the immigrant classes during this same period. In 1850, the director of the Texas prison was denied his request for $10,000 to finish the outer walls of the prison and provide living quarters for the inmate population. However, in 1854, the same legislators would appropriate $40,000 for the construction of the prison's first textile mill (Stone, McAdams, and Kollert 1974).

Private-Sector Interest in Prison Labor

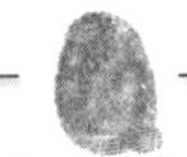

lease-labor system was a form of income for early prisons by "leasing" inmate labor to private entrepreneurs who would reap the profit of the low-cost labor.

The rise of prison industries would bring about the first efforts at privatization of prisons; the earliest form of privatization was the **lease-labor system**. An individual would bid for the temporary ownership of a state prison. The winning bidder would then become the leaseholder and gain almost total control of the prison and its labor. The leaseholder was responsible for providing the inmates with food, clothing, and other essentials, as well as being in charge of the industrial operations of the prison. A leaseholder could maximize his profits by working the inmates as hard as possible and spending as little as possible on their care. The leases were frequently very profitable. In 1855, Zebulon Ward leased the Kentucky prison for an annual rent of $6,000. At the end of his four-year lease, he had amassed a personal fortune close to $75,000 (Wines and Dwight 1867, 260).

In Texas, the legislature passed a law in 1866 permitting the prison to lease out inmates in groups no smaller than 20 inmates per lease. These leases allowed the inmates to leave the prison to work on plantations and railroad construction, with the security being provided by the lessee. While the leasing of complete prison systems was subject to some problems, these remote-site leases, such as the ones in Texas, produced the worst abuses. The escapes and deaths under the Texas lease system jumped from a total of 104 in 1873 to 376 by 1875, one of the worst years for the lease system. During the calendar year of 1875, a total of 26 percent of all the inmates in Texas either escaped or died (Stone, McAdams, and Kollert 1974).

It had become clear by the 1850s that prison industries could be run at a profit. Inmates generally represented unskilled and unmotivated workers, but it was still possible to make a profit because they could be obtained so cheaply. Inmate laborers generally were worked from "bell to bell," this representing the amount of daylight hours available. During the winter months, the workday was about 8 hours; however, with the coming of summer, the workday might extend to 12 hours. Moses C. Pilsbury, a prominent penologist in 1866, best described the profitability:

> *The case stands thus: The labor of twelve convicts will cost no more per day than that of four citizens; yet the convicts will do nine days work while the citizens will do four. Thus every dollar paid for convict labor will produce as much as two dollars and a quarter expended on citizen labor. (Wines and Dwight 1867, 256–257)*

The thought was that inmate labor was not the equal of free-world labor, yet it was possible, with effort and corporal punishment, to make inmate labor a good investment.

It is important to recognize that these early prison industries were not all run in a corrupt or exploitative fashion; neither were the legislators of the time evil men. Most states established boards of labor or prison inspectors to make sure that inmates were not treated in an inhumane manner. Some of the early leases included clauses that required the contractor to rotate the job assignments to ensure that the inmates learned all the different parts of the trade they were practicing (Wines and Dwight 1867). Prison industries suffered many scandals and problems, yet they existed in almost every prison and prospered into the first part of the 20th century.

Civil War and the Rise of Prison Agriculture

While the latter part of the 19th century was prosperous for prison industries located in the northern states, the same could not be said for the prison industries of the South. In the economic depression following the Civil War, southern prison industries could not find suitable markets for their goods, and the supply of many of the raw materials necessary for production was very erratic.

In an attempt to support the southern prisons during this period, many southern states largely abandoned traditional industries and started to focus on agriculture as a way to generate money from the available prison labor force. During the post–Civil War depression, land was very cheap, and southern states started purchasing land in sufficient quantities to create state-run plantations. In Wines and Dwight's (1867) descriptions of northern state prisons in 1867, the premises (including the surrounding grounds) ran from 0.5 acre in New Hampshire to 20 acres in Wisconsin. In contrast, Texas purchased the Harlem Farm in 1885, which contained 5,011 acres of farmland. Agricultural production seemed a logical choice for many southern prisons because it required a small investment, used unskilled labor, and promised a reasonable profit. By the early 20th century, immense farm-labor programs evolved in Texas and Oklahoma (Clare and Kramer 1976).

Women generally were exempt from farm-labor assignments because they were not perceived as suitable, except perhaps for working in "kitchen gardens." Women during this time period normally were confined separately, and were used for support activities such as garment manufacturing and repair (Pollock 2002). The total number of women in prisons was relatively low during the 19th century, and the cultural values dictated that they be treated with greater leniency (Rafter 1990).

The 20th Century and Prison Labor

At the start of the 20th century, both industrial-based and agricultural-based prison labor encountered significant problems. The development of efficient farming equipment changed the labor-intensive nature of the agricultural industry. Prison farms, even with relatively cheap labor, could not compete with mechanized private farming. By the 1930s, prison farming was largely reduced to growing food crops and raising livestock for consumption by the inmate population. One of the few exceptions to this was the Texas prison system, where cotton crops, processed in the prison textile mill, continue as a viable enterprise to this date. Prison agriculture for internal consumption still represents a major function in many prisons today. In 1991, prison agricultural production exceeded $45 million in value (American Correctional Association 1992).

State-Use Laws

Traditional manufacturing industries fared little better than agriculture in the first half of the 20th century. With the growing power of the American labor movement and the initial effects of the Depression, opposition to prison manufacturing increased. In almost every state, labor groups and manufacturers raised complaints that prison manufacturing represented unfair competition from the state. By the early part of the century, more than half of the states had adopted state-use laws (Clare and Kramer 1976). The state-use laws provided that goods manufactured with prison labor could not be sold on the open market. Under these laws, only the prison itself or other state agencies or subdivisions could use prison products. The most common exception to this rule was surplus agricultural production, such as milk or crops, that were prone to rapid spoilage. For instance, the Oklahoma statute specified that:

state-use laws laws passed in many states in the early 1900s that prohibited the sale of prison-made goods on the open market.

> *All counties, cities, districts or political or subdivisions, or any state agency thereof, may purchase the goods or services produced by the prison industries of the Department of Corrections through their properly authorized purchasing authority, or they may place a direct order without competitive bid, with the prison industries of the Oklahoma Department of Corrections. Others are prohibited from purchasing such goods and services, with the exception that all surplus agricultural products may be sold on the open market. (Oklahoma Use Law, sec. 549.1, "Purchase of Prison Industries Goods and Services")*

Hawes-Cooper Act was passed in 1929 and required that prison-produced goods shipped interstate be subject to the laws of the receiving state.

With the growing power of labor unions in the late 1920s came federal intervention in prison-produced goods. In 1929, the **Hawes-Cooper Act** was passed, which required that prison-produced goods shipped interstate were subject to the laws of the receiving state. This meant that those states that did not have state-use laws were blocked from shipping their goods to any state that did have such a law. The situation became worse for prison industries in 1935 with the passage of the **Ashurst-Sumner Act**. It required that all prison-made goods be clearly labeled with the name of the prison, and prohibited interstate shipment of almost all products to those states with laws restricting their sales. The final blow came with the 1940 amendment to the Ashurst-Sumner Act, which completely prohibited the interstate shipment of prison products except for some tractor parts and agricultural produce.

Ashurst-Sumner Act required that all prison-made goods be clearly labeled with the name of the prison and prohibited the shipment of prison goods to states that restricted their use.

Prison industries remained stagnant from the 1940s through the early 1970s. The state-use concept continued, and some states developed a surprising array of products, all consumed by the prison and other state agencies. In 1972, more than 360 prison industries were operating in the United States (Grisson and Conan 1981). The federal prison system also adopted the state-use concept. An act of Congress in 1934 established **UNICOR**, a wholly owned government corporation responsible for the operation of the industries within the federal prisons. UNICOR's operations include such activities as:

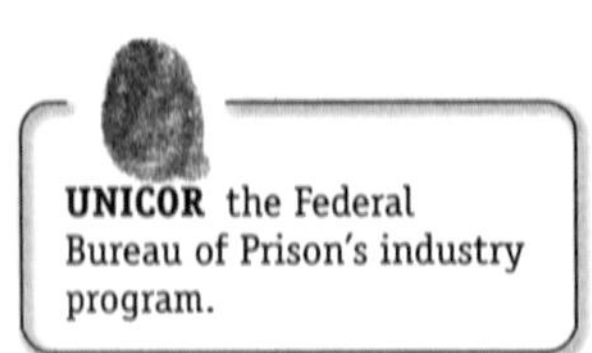

UNICOR the Federal Bureau of Prison's industry program.

1. *Providing products to federal agencies at fair market prices*
2. *Developing products that minimize competition with private sector industry and labor*
3. *Providing inmates the opportunity to earn funds*
4. *Providing inmates with job training opportunities*
5. *Reducing inmate idle time (Snarr 1992, 172)*

UNICOR became heavily involved in the production of military equipment during World War II. Several prison industries operated more than one shift per day and seven days a week during the war (Allen and Simonsen 1986/1992). The production of equipment in support of the war effort provided both economic and political support for the Federal Bureau of Prisons. UNICOR is still in operation, currently employs about 20 percent of all federal inmates, and produces more than $666 million in goods annually (UNICOR 2003). UNICOR also operates an Inmate Transition Branch (ITB) to assist inmates in job acquisition after release. During the seven years of this program's operation, the ITB has assisted in more than 350 job fairs in 100 federal prisons. More than 5,000 company recruiters and representatives from education and other community-service agencies have participated in these job fairs. In addition, ITB staff has assisted state prisons, regional jails, and federal probation services in holding both real and mock job fairs (UNICOR 2004).

Prison Industry Enhancement Certification

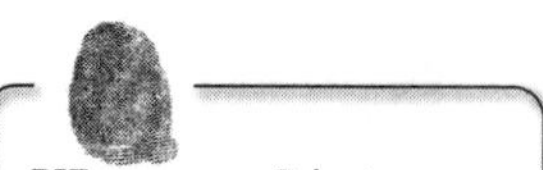

PIE program Private Sector Prison Industry Enhancement Certification Program allows state prisons to manufacture and sell goods on the open market as long as they adhere to guidelines and meet certain specifications.

In 1979, Congress removed some of the barriers to interstate shipment of prison-made goods. The passage of the Private Sector Prison Industry Enhancement Certification program, often called the PIEC or **PIE program**, permitted certified states to sell goods on the open market. States wishing to be certified had to meet all the requirements of the PIE program, which included:

1. *Pay the inmates wages comparable with similar jobs in the community*
2. *Consult with representatives of private industry and organized labor*
3. *Certify that the PIE industry does not displace employed workers in the community*
4. *Collect funds for a victim assistance program*
5. *Provide inmates with benefits in the event of injury in the course of employment*
6. *Ensure that inmate participation was voluntary*
7. *Provide a substantial role for the private sector (National Institute of Justice 1990a, 22)*

In 1984, the PIE program was revised. The original form of the PIE program provided for the certification of only 7 states to ship goods interstate. The 1984 revision of the original program provided for the certification of up to 20 states. The Bureau of Justice Assistance ultimately interpreted the Act broadly enough to permit local jails in certified states to operate open-market industries. The local jails had to meet all of the requirements of the PIE program, but did not have to seek individual certification (National Institute of Justice 1990a).

The future of prison industries was significantly brightened in 1985 with the strong advocacy of Chief Justice Warren E. Burger. Burger strongly advocated the expansion of prison industries in the United States. Burger's article called for the repeal of all laws restricting the transportation or sale of prison industrial production (Burger 1985). The Chief Justice's opinion had been heavily influenced by his experiences with international industrial prisons, such as those of the Scandinavian countries and of China. While Burger's call to action had little effect on legislation, it helped promote public acceptance of prison industries. Other prominent conservatives, such as Edwin Meese, the former U.S. attorney general and head of the Heritage Foundation, echoed this call for expansion (Stein 2003).

In the 1980s, prison industries expanded, both in the private–public partnership areas such as the PIE program and in the state-use industries. A survey conducted in 1989 found 69 prison-based industries selling goods on the open market under the PIE program. This represented a 150 percent increase over the number of similar industries operating in 1984 (National Institute of Justice

1990b). While the expansion observed between 1984 and 1989 was impressive, it did not meet the expectations that some had for the private-public-partnership concept. The private–public partnership market had made further increases by 1995, but it still employed only about 5,000 inmates. When compared to the total inmate population of U.S. prisons, this represents only about 1 percent of all prisoners. The most likely stumbling block for the PIE program was its requirement to pay inmates a wage competitive with the local community. While the requirement for competitive wages was a stumbling block in one sense, it was a blessing in another. The PIE program has a provision for withholding parts of an inmate's salary to pay victim restitution, taxes, room and board, and support for the inmate's family. Between the start of the PIE program in 1979 and June 2003, more than $264 million in inmate wages were attributable to the PIE program (Bureau of Justice Assistance 2004).

As **Figure 5-1** shows, of these wages, almost 55 percent has been withheld to reimburse victims, the correctional systems, inmates' families, and to pay state and federal taxes. Currently, 38 states have active PIE programs (Bureau of Justice Assistance 2004). The slow growth in the number of PIE programs is also partly attributable to the quality of inmate labor. The observation made by Mr. Pilsbury in 1866—that a prisoner would produce about three-fourths as much as a free-world employee—is probably as true today as it was then. When the extra costs of supervision are included, it is very difficult for private–public partnership industries in prisons to compete with traditional free-world industries.

Prison Labor Today

As we enter the 21st century, there is every reason to believe that productive prison-labor programs will gain even greater importance. In the past decades,

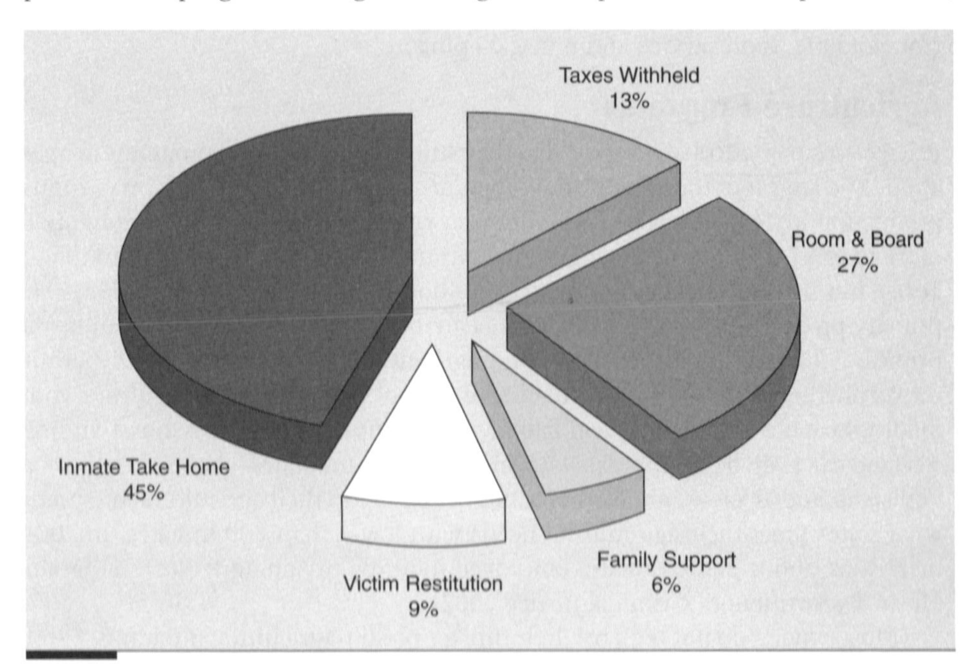

Figure 5-1 Inmate PIE Wage Withholding. ***Source:*** *Bureau of Justice Assistance 2004.*

prison populations have grown at a fantastic rate. The data for the early 2000s are showing a slower rate of increase, about 2.5 percent per year (Bureau of Justice Statistics 2003). While that percentage does not seem alarming, it represents an additional 37,000 inmates added each year.

At the same time that the prison population is increasing, politicians find themselves under increasing pressure to hold the line on taxes and spending. With the average cost of a new prison ranging from $50,000 to $80,000 per bed, and an average annual operation cost that already exceeds $20,000 per inmate, it is not difficult to see that state and federal legislators are caught in a bind. To sum up the situation, the American public wants to lock up a lot of people, but we do not want to pay a lot of money to do so. The public demands that criminals be confined, and the courts demand that we feed and care for them in a specified minimal fashion. Since we cannot reduce the overhead costs of imprisonment significantly, and since it appears that lowering the prison population significantly is not going to happen either, the only answer seems to be using the inmates to defray the cost of their confinement.

In addition to the financial argument for productive prison labor, there are other rationales as well. The United States is based on a work ethic that can be traced back to the nation's earliest history. Work is a central component of most Americans' lives, and the public generally believes that prisoners should work as well. It is also a fact of life that almost all prisoners will eventually be released, and on release, they will need jobs to support themselves. Productive prison industries, if properly operated, can reduce inmate idleness, build job skills, increase prisoner morale, and reduce the cost of confinement. Because of these advantages, all states make some productive use of inmate labor. In the past 10 years, the number of inmates employed in productive labor programs has increased significantly. The most common labor programs are in the areas of agriculture, manufacturing, public works, prison construction, and prison maintenance (for example, food service and housekeeping).

Agriculture Programs

In a recent survey, 26 states report that they still have agricultural programs in operation. The largest of the agricultural programs are in Texas and California. Annual production in Texas exceeds $33 million in value, with an outside sales figure of more than $7 million. California's total annual production is lower than that of Texas, but its outside sales exceed $10 million in value (*Anonymous* 2002). The primary products in Texas and California are beef, eggs, poultry, swine, milk, and prunes. California, as well as other large agricultural states such as Texas, permits the surplus agricultural products to be sold on the open markets. Agricultural commodities such as eggs, milk, and fruits were exempted from the Ashurst-Sumner Act, and can even be sold across state lines. While some states—such as California, Louisiana, and Texas—employ thousands of inmates in their agricultural programs, most states meet their agricultural needs with fewer than 200 inmates. In 2004, there were about 2,200 inmates employed in Texas prison-agriculture programs (Texas Department of Criminal Justice 2002).

Most experts do not see a bright future for prison-agriculture programs. While the interstate shipment of agricultural products is possible, there is little evidence that it is occurring to any significant degree. The capital-intensive, relatively low labor needs of modern agriculture make it less than ideal as a choice for a prison industry. It is also difficult to justify agricultural programs from the standpoint of job-training benefits for inmates. Currently, the job market for people trained in agricultural skills is very small. The primary function of prison-agriculture programs will most likely be limited to supplying the needs of the inmate population instead of as a source of cash income. Prison officials, even in major agricultural states such as Texas, are questioning the efficiency of the agricultural program. When compared to a commercial food-service contractor, it appears to be about 17 percent cheaper to buy the food externally than it is to grow and prepare it on-site. Recent audits by the General Accounting Office in Texas are also questioning the effectiveness of prison agriculture. They are suggesting that leasing or selling the agricultural land may be more cost-effective than cultivation (*Austin American Statesman* 2004).

Manufacturing Programs

In the most recent survey of prison industries, all but one state reported operating some form of manufacturing industry within their prisons. The product lines range from modular desks in Massachusetts to stainless commodes in Texas. Production includes items as small as eyeglasses to as large as dump-truck beds. Some of these products are sold on the open market through PIE programs, but most of them are restricted to state-use-only markets. Currently, the National Correctional Industries Association (2004) shows Internet sales links to 48 states with correctional-industries sales programs. Prison manufacturing employs about 79,000 inmates in the United States (*Corrections Compendium* 2002; UNICOR 2003). This represents about three times the number employed in prison-agriculture programs.

The Federal Bureau of Prisons has the most prisoners employed in manufacturing programs, with about 20 percent of its inmates employed in their program. The federal-prison industries (UNICOR) produce office furniture, dormitory furniture, metal storage cabinets, and general office supplies that are purchased by a wide range of federal agencies, including the Department of Defense. The operating budgets of correctional industries range from a low of $2.4 million in Delaware to a high of $166 million in California (Hill 2002).

Prison industries in California are estimated to be saving the state about $14 million per year (Hill 2002). Manufacturing programs utilizing prison labor offer probably the best opportunity for future development. If manufacturing industries are selected carefully, they offer the potential for reasonable profit and can provide good job training and experience. "How many jobs are there for license-plate makers outside of prison?" is little more than an old joke. First, prisons never used any significant number of inmates to manufacture license plates. Second, license-plate manufacturing is nothing but sheet-metal stamping, which does have a free-world job market. If barriers to the sale of prison-made goods were lifted, the potential for these industries would be excellent.

Public Works and Prison Construction

Prison labor started out with public-works projects, and we still use inmate labor in the construction of public projects today. Prisons in the United States employ inmates in a broad range of construction jobs. Inmates are currently employed in carpentry, plumbing, electrical work, bricklaying, cement work, welding, painting, surveying, and a host of other construction activities. Some prisoners are admitted to prison with the required job skills, and others are taught their skills while imprisoned.

The most common construction work is building additions to existing prisons and remodeling prisons and other public offices. Thirty-four states report using inmate labor in some form of construction, even if they don't use it to build new prisons. Florida reported saving more than $14 million on the construction of a new prison in 1990 by using inmate labor (Davis 1990). Florida does not pay its inmate construction crews, but it does award good-time credit against the maximum sentence. This means that prisoners in Florida can earn an early release through their construction labor. There are currently no accurate figures on the number of inmates employed in the construction industries. Because the number employed can change significantly with the start or completion of major projects, national figures probably would not be reliable even if they were available. One indicator of the relative importance of inmate labor in the construction and renovation of prison facilities is that in Texas, the Facilities Division has a staff of more than 400 employees responsible for supervising and coordinating inmate labor in the construction and remodeling process (Texas Department of Criminal Justice 2002).

The construction and remodeling arenas represent an employment area for inmates that is clearly larger than agricultural employment, but also smaller than traditional manufacturing employment. While it is reasonable to assume that there will be some expansion in the construction industries, most experts do not expect a dramatic expansion in these areas. The strongest argument for using inmate labor in the construction and remodeling sectors is that doing so can provide training and work experience that prisoners can translate into well-paying jobs after release.

Prison Maintenance

Prisons have historically used inmate labor for their internal-maintenance functions. The largest of these functions are normally food service, laundry service, and janitorial service. The use of inmate labor to provide these basic services can be tracked back to the early bridewells. Prisoners have cooked, cleaned, and laundered within their institutions for more than 400 years.

While there are no current surveys showing the number of inmates employed in food services, a reasonable estimate is about 7 percent of the adult prison population. This means that nationally about 150,000 inmates are employed in the food-service industries within prisons and jails. There has been some movement into externally contracted food service in recent years, but this is generally occurring in jails and smaller special-purpose prison units. In some states, the food-service functions are large enough that they run their own culinary-arts schools to train prisoners in the food-preparation areas. Experienced wardens have always recognized the importance of providing quality food service. Food is one of the few pleasures available to inmates, and poor-quality food very likely will create problems with the inmate population.

Historically, few have criticized the use of inmates in the food-service areas. There is some market in the free world for released inmates with food-service skills. However, food-service jobs are not highly desirable. The food-service industry generally represents the lower-paying end of the service industry, with many of the jobs at the minimum wage-level. There is also some question as to the type of food-service skills being taught in prisons. Inmates are being trained in a cafeteria or institutional style of food service, while today the food-service industry is dominated by "fast-food" businesses. Institutions such as public schools are the primary locations of traditional cafeteria-style food service. While the employment of parolees in school cafeterias is not impossible, it is reasonable to expect some resistance from community groups to this employment.

Laundry and janitorial-service jobs occupy about another 7 percent of the inmate population. While these services involve some training, they are considered unskilled labor. As important as these functions are to institutional operation, it is difficult to justify them as training in anything except the work ethic. In the United States, laundry and janitorial functions represent a large segment of the unskilled-service labor market. The number of public buildings, offices, and businesses that use these types of services is staggering. This market, however, is dominated by minimum-wage positions that do not represent very desirable employment. The institutional support areas such as food service, laundry service, and janitorial service represent an inmate-labor market that will probably remain unchanged into the future. Institutions will continue to provide these services internally, but there will be little opportunity for the expansion of these areas, except as inmate populations expand.

One experimental program, the Philadelphia Urban Work Cadre Program, allowed low-security federal inmates working out of a community corrections center to provide maintenance services for the Department of Defense (Byerly and Ford 1992). The Urban Work Cadre Program was implemented in the spring of 1990 at the Defense Personnel Support Center in Philadelphia. Inmates under this program mow lawns, paint, and perform general custodial duties on the Defense Department site. The transportation to and from the job site was provided by the community corrections center. While work-release programs are common in both jails and prisons, they are generally orientated toward developing job skills instead of providing a cost-savings service. This model of providing off-site custodial services could be copied in many areas where public buildings are near community correctional centers or low-security prisons.

Current Issues in Prison Labor

There are several issues that dominate the current discussions of inmate labor today. The most significant issues are:

1. Displacement of civilian jobs
2. Inmate compensation and wages
3. Inmate safety and security
4. Cost-effectiveness of prison industries

These issues were debated 100 years ago, and they still remain as the central points of debate today. Can we run humane prison industries and prison-labor programs that effectively reduce the cost of confining the rapidly increasing inmate populations?

Displacement of Civilian Jobs

The perception that prison industries compete unfairly with private enterprises is pervasive among the American public. The common belief is that "slave labor" is being used to put honest, hardworking people out of work. The evidence shows that this belief is a misconception. First and foremost is the fact that current legislation has largely blocked prison-made goods from the open market. In addition to the legislation from the first half of the century, there have been continuing efforts to restrict the open sale of prison-made goods. In 1987, federal legislation dealing with the purchase of highway signs restricted the future purchase of prison-made highway signs to no more than the purchase level of 1986 (Grieser 1989). This effectively prevented any further expansion of the prison industries involved in highway-sign manufacturing.

In those areas of state use where there is some possible conflict between prison labor and free-world labor, the real level of conflict is minimal. Prison industries generally deal in areas such as textiles and other labor-intensive manufacturing where the job market has been exported wholesale to foreign countries (Sigler and Stough 1991). Most prison industries compete with labor markets in developing nations, not U.S. manufacturing jobs. Any attempt to return a significant portion of these jobs to the United States could only benefit the U.S. economy and the public in general.

The total involvement of prison industries even in the state-use market is surprisingly small. UNICOR, the Federal Prison Industries, is the largest of all the prison-industries groups, but it currently captures only a very small percentage of the total federal market for goods and services. A similar study conducted in New York indicated that Corcraft Correctional Industries, the New York State correctional industries, was capturing only between 1 and 2 percent of the state-use market in New York. Other large prison-industries states such as California, Florida, and Texas are also expected to provide less that 2 percent of their respective state markets (Sigler and Stough 1991).

In a model proposed by Sigler and Stough (1991), a prison business would be permitted to compete in the open market when that industry could be declared an "endangered species." The criteria proposed by Sigler and Stough is that prisons could enter a production field if the United States provided less than 5 percent of the product to the world market, and if the percentage of the product produced in the United States has declined for a three- to five-year period. Following this criteria, prison industrial production would represent little threat to existing businesses and products, even if the laws restricting open-market sales were repealed.

Inmate Wages and Compensation

The second major concern involving inmate labor is the issue of wages and compensation for inmates. The conservative elements of the population are strongly opposed to paying wages to offenders. Conservatives see inmate wages as a reward for criminal activity. Under this school of thought, inmates should be made to work for the state to repay the damage they inflicted during their criminal career. To a more liberal audience, the involuntary labor of any person represents a violation of human rights. Labor unions also are opposed to paying any "subscale" wages, fearing that doing so will influence the competitiveness of free-world labor in negotiating contracts in the general labor markets (Sigler and Stough 1991). Successfully negotiating these various interest groups and still operating prison industries at a profit represent obvious challenges.

Prisons generally do not have a legal obligation to pay inmates required to work in prison or jail industries. Courts in most cases have held that prisoners do not have any constitutional right to be paid for their work. Courts also have dismissed claims that inmates are entitled to wages under federal wage laws. The laws of some states do require the payment of inmate wages, and in those states, correctional administrators must meet the restrictions that are set by their states, as seen in **Table 5-1**.

Table 5-1 Selected Court Cases on Inmate Employment and Compensation

Cases (convicted offenders)	Findings
Holt v. Sarver (1970)	Forced uncompensated labor of state convicts does not violate the 13th Amendment.
Hamilton v. Landrieu (1972)	Inmates may not be assigned to jobs where they have access to other inmates' records or to sensitive information.
Altizer v. Paderick (1978)	Inmates have no due-process rights to any particular job in an institution. No procedural due process is required to transfer an inmate from one job to another.
Turner v. Nevada State Prison Board (1985)	Nevada officials may deduct from an inmate's wages to pay for room and board and victim-compensation payments.
Hrbek v. Farrier (1986)	State-prison officials may withhold deductions from an inmate's wages to cover unpaid court costs.
Sahagian v. Dickey (1986)	State officials may withhold 15 percent of an inmate's wages in a release account that the inmate may access only on discharge.
U.S. v. Merric (1999)	Federal-prison officials may withhold deductions from an inmate's wages to cover unpaid state- or local-court costs, restitution, and fines.
U.S. v. Demko (1966)	Federal inmates may not sue for work-related injuries. Compensation is limited to the Inmate Accident Compensation Program.
Cases (unconvicted offenders)	**Findings**
Tyler v. Harris (1964)	Unconvicted offenders in federal facilities may not be subject to involuntary servitude.
Main Road v. Atych (1974)	Unsentenced state prisoners cannot be required to perform uncompensated labor.

Prison industries have developed a number of methods for dealing with the problem of inmate compensation. Inmates generally need an incentive to participate. While the opportunity to keep busy is valued by many inmates, continued motivation to work every day is not typically the pattern of their lives. Thus, nonmonetary incentives have been used since the earliest days of prison labor. The classic nonmonetary incentives were pain (through the use of corporal punishment) and hunger (through denying food to inmates who refused to work). The withholding of food and the use of corporal punishment might sound like incentives from the Dark Ages, but they were used into the second half of the 20th century. In 1951, O. B. Ellis, the director of the Texas prisons, issued a "no work, no eat" proclamation to every inmate that was found to be "malingering" in the Texas prisons. There is also evidence of corporal punishment being used as late as the 1960s by the order of George Beto, the then director of the Texas prisons (Stone, McAdams, and Kollert 1974).

Today, nonmonetary incentives are not based on the use of the lash. Modern correctional administrators have recognized that there is a broad range of benefits other than salary that can be used to reward inmates. Some of the rewards that are commonly used are:

1. Extra good time
2. Extended visiting privileges
3. Extended recreation and gym time
4. Extended hours of TV viewing
5. Short furloughs
6. Extended telephone privileges
7. Outside meals and treats

These benefits can be powerful motivators for prisoners as long as they are used properly. All incentives should be based on individual performance of inmates and not on group behavior (National Institute of Justice 1990b).

Inmate Safety and Security

tool control is the security system in prisons of frequently counting tools in order to prevent their theft and illegal use.

Any prison-industry program must consider security. Correctional industries create security hazards that the correctional administrator must address. One obvious security problem is that of tool control. When selecting a correctional industry, the administrator must carefully select the product and the manufacturing technique to fit the security needs of the institution. Heavy-metal manufacturing that involves metal-cutting tools and welding tools is obviously not a good choice for most maximum-security institutions. Tool control is important from the standpoints of inmate safety and preventing escapes. Because prisoners may use tools as weapons against other prisoners, tools must be tightly inventoried and counted on a regular basis. These security procedures represent an obvious impediment to running an efficient manufacturing industry.

Correctional industries require additional correctional officers to maintain control over the inmate population that is involved in the manufacturing process. These security officers represent staffing requirements that are above and beyond the manufacturing supervisors necessary to keep the industrial production operating. The additional security costs must be considered when comparing the cost-effectiveness of any prison business. Industries that require limited numbers of tools and limited inmate movement are ideal. One example of a minimal-security prison business is a telephone-reservation service that is operated by one state prison. After training, the inmates are assigned to a computerized phone bank that takes reservations for a number of hotels and resorts. This enterprise involves no threat from dangerous tools, and it can be located in a secure, isolated area with only minimal inmate movement. This allows the business to operate with almost no additional security costs. This is an excellent example of an industry, traditionally outsourced (moved to a foreign country), being allowed to operate in prison. While there has been some concern about credit-card fraud, there is no reliable evidence that it represents a significant issue.

Prison industries must also be selected to minimize the potential for industrial accidents. They are required to meet the same safety standards of any manufacturing industry. Prison industries receive regular inspections from the Occupational Safety and Health Administration to ensure that they are in compliance with all industrial-safety rules. The safety considerations are further complicated by the nature of the inmate population. Prisoners often have poor work habits, which can contribute to on-the-job accidents. Inmate turnover is also high, which requires a program of constant safety training and monitoring of the compliance with safety regulations.

Prison industries in the United States employ both male and female inmates in the manufacturing process. With the exception of Wyoming, Vermont, Delaware, and Rhode Island, every industrial program includes women. The latest survey shows that about 5,500 women are currently employed in correctional industries. The largest state employer of women is California, followed by Texas and Virginia (see **Table 5-2**). In 2002, women represented about 6 percent of the total inmate population, and about 7.5 percent of the inmates involved in the prison industries. In the five-year period since the previous survey, women have increased their participation in prison industries 32 percent, while men have increased only 24 percent (*Corrections Compendium* 2002). Women are used in garment manufacturing, reupholstering, data entry, microfilming, telemarketing, farming, printing, optical production, and electronic-circuit-board production (Duncan 1992).

Cost-Effectiveness of Prison Industries

Many people still question the cost-effectiveness of prison industries. Most prison industries must rely on only six to seven hours of productive labor out of every eight-hour shift due to security checks and other inmate-management problems (Grieser 1989). High inmate turnover results in higher than normal training costs. Most correctional industries also suffer from constraints in how they procure their raw materials. Almost all correctional industries have purchasing systems that must follow governmental procurement procedures. Some states such as Florida have turned their correctional industries over to a private corporation to help reduce these administrative problems.

The same governmental regulations that make purchasing difficult also make it difficult to remove inefficient employees that are involved in the industrial process. Because of these and other similar problems, the average value of pro-

Table 5-2 Females Employed in Correctional Industries 2002

Jurisdiction	No. of females	No. of males	Percentage female
Alabama[a]	108	1,425	7.58
Alaska	16	131	12.21
Arizona	357	1,090	32.75
Arkansas	14	530	2.64
California	538	6,309	8.53
Colorado	105	1,459	7.20
Connecticut	50	350	14.28
Delaware	0	194	0.00
District of Columbia[a]	29	482	6.01
Florida	107	1,926	5.55
Georgia	241	1,218	19.79
Hawaii[a]	0	54	0.00
Idaho	20	340	5.88
Illinois	122	1,291	9.45
Indiana	141	1,759	8.02
Iowa	70	450	15.55
Kansas	17	423	4.02
Kentucky	61	740	8.24
Louisiana	69	800	8.63
Maine[a]	9	126	7.14
Maryland	172	1,205	14.27
Massachusetts	17	367	4.63
Michigan	99	3,012	3.29
Minnesota	96	906	10.59
Mississippi	40	400	10.00
Missouri	113	1,268	8.91
Montana	17	340	5.00
Nebraska	48	500	9.60
Nevada	36	687	5.24
New Hampshire	20	350	5.71
New Jersey	162	1,205	13.44
New Mexico	70	361	19.39
New York[a]	107	1,387	7.71
North Carolina	69	2,125	3.24
North Dakota	20	140	14.28
Ohio	115	2,185	5.26
Oklahoma	86	1,191	7.22
Oregon[a]	30	371	8.08
Pennsylvania	46	1,633	2.81
Rhode Island	0	350	0.00

Table 5-2 Females Employed in Correctional Industries 2002, continued

Jurisdiction	No. of females	No. of males	Percentage female
South Carolina	168	1,797	9.34
Tennessee[a]	34	521	6.52
Texas	323	6,673	4.84
Utah[a]	16	328	4.87
Vermont	0	106	0.00
Virginia	192	1,296	14.81
Washington	31	1,976	1.56
West Virginia	5	266	1.87
Wisconsin	35	614	5.70
Wyoming[a]	0	129	0.00
Federal[b]	1,268	18,403	6.89
Totals	5,509	73,189	7.53

Source: Anonymous 2002. Prison Industries, *Corrections Compendium*, 27, 9 (September).

[a]Wees, G. 1997. Prison Industries. "Outside Federal System, Inmate Employees Remain an Elite Group." *Corrections Compendium* 22, 6 (June).

[b]UNICOR 2003. *FPI FY2003 Management Report and Independent Financial Audit.*

duction per inmate is estimated at only $20,000 per year. This is only about 25 percent of the anticipated production value of a free-world employee (Grieser 1989). The net impact of these constraints is that correctional industries are hard pressed to generate a profit even when selling their products on the open market at their fair value.

In spite of these concerns, most experts believe that prison industries are a valuable asset. The Escod Plant in South Carolina received an award from IBM for being one of their few feeder plants to manufacture 25,000 computer cables with zero defects—establishing that prison industries can produce high-quality goods (Sexton 1995). In addition, while the productivity of prison industries is low, so are their wages. Most states pay only a token wage to inmates participating in non-PIE prison businesses. Arkansas, Georgia, and Texas pay no wages at all for prison labor (*Corrections Compendium* 2002). Prison industries provide a structured activity that reduces inmate idleness, and they can provide valuable job experiences for inmates prior to their release.

Inmate Education

The use of productive prison labor occurred much earlier than programs designed to provide either vocational or academic education to inmates. This is easy to understand when you consider the functions of early prisons. Historically, there was little intent that the individuals sent to prison would ever rejoin society. Prior to the 16th century, prison inmates were considered to be little more than draft animals, and any efforts beyond feeding and watering were considered unnecessary. The first evidence that there might be a shift in this philosophy was the development of workhouses and *bridewells* in Europe and England.

As mentioned earlier, bridewells were somewhat like a cross between a workhouse and a prison. The first bridewell was built outside London in 1577. It was originally designed to provide vocational and moral reformation for the serfs that were migrating from the countryside to the metropolitan areas seeking employment. The economic transition from a feudal state in the 14th and 15th centuries had created large numbers of unemployed individuals who frequently were referred to as vagrants. These individuals had no marketable skills they could employ in the metropolitan centers. Without skills, many of these individuals committed minor crimes to purchase food and necessities.

The first bridewell was an attempt to take these minor offenders, debtors, homeless children, and other public charges and make them fit for city life. This attempt to improve the lot of the vagrants and other public wards represents the seeds of the first custodial-based vocational and academic education. Religious authorities delivered much of this early education, and it contained a liberal dose of religious and moral instruction that included the dominant "work ethic" of the time.

The bridewells ceased to be a historical influence on the evolution of prisons and prison education after less than 100 years. By 1640, the evolution of the British penal system had taken a new turn. Transportation to the colonies was the preferred solution for criminal offenders from this time period up until the American Revolution. There was little incentive to improve the lot of offenders when they could simply be shipped to the New World, never to be heard from again. The British made some abortive attempts to create total institutions with both vocational and secular education, but these efforts were short-lived. Influential in creating these institutions were Quakers such as John Bellers, who proposed a vocational-industrial prison model in 1690 (Ignatieff 1978). In fact, the Quakers were very influential in the entire history of prison reform. Some of the first women involved in prison reform were also Quakers. In December 1816, Elizabeth Fry led a group of Quaker wives into Newgate Prison to reform the women's and children's section of the prison. Elizabeth Fry's efforts represent what was probably the first educational and vocational programming for female inmates (Ignatieff 1978).

With the American Revolution, the British were forced to rethink their criminal-sentencing philosophy. Without the ability to transport its prisoners to the colonies, the British started to develop major penitentiaries such as the Gloucester and Pentonville facilities. In these large institutions, educational programs for inmates developed slowly. By 1860, the British provided reading, writing, and "cyphering" instruction to every convict sentenced for three months or longer (Wines and Dwight 1867).

Educational programs also evolved in the United States during the first half of the 19th century. The first documented educational program in the United States was started in 1826 (Wines and Dwight 1867). This program, at the Auburn Prison in New York, provided instruction in reading, writing, and arithmetic every Sunday after church. In 1829, Kentucky passed a legislative act requiring that

Kentucky's state prison teach all "unlearned" inmates for at least four hours every Sunday. This first statewide prison inmate educational requirement was discontinued after only five years. In 1847, the New York legislature passed a comprehensive Prison Act. This Act provided for the appointment of full-time teachers to each of the prisons in New York. These teachers operated the first full-time weekday school program designed to serve U.S. prisoners.

From the mid-1800s, prison officials increasingly saw inmate education as a logical and expected part of their function with convicts. While education had started out as a method of religious instruction, by allowing inmates to read the Bible, it ultimately evolved into a broader rehabilitative concept. It was increasingly recognized that, to prosper in the world, all people needed a minimal level of education. In 1876, the state of New York opened the Elmira Reformatory. This institution, which contained young adults ages 16 to 30, was the direct byproduct of the 1870 Prison Congress. The principles developed at the 1870 Prison Congress included a strong commitment to inmate education, which was embodied in the educational program at Elmira (Allen and Simonsen 1986/1992). Elmira would set the vocational and academic educational standard that other institutions would try to meet for the remainder of the century.

After the first survey of prison educational programs conducted in preparation for the 1870 Prison Congress, the next major survey occurred in the late 1920s. In 1927, Austin MacCormick surveyed 60 prisons in the United States, and discovered that 13 of the institutions reported not having any academic programs for prisoners. MacCormick also concluded that the programs that did exist rarely were adequate to meet the inmates' needs. Probably the most surprising discovery of the survey was the total absence of vocational educational programs for inmates. As has happened so many times in the past, the progress of one generation had been lost due to the disinterest of the next. The progress that had been made in the last part of the 19th century was sacrificed during the shifting priorities and economic hardships of the early 20th century. Many of the states would not redevelop their academic and vocational programs until after World War II (Fox 1983).

In 1945, the Correctional Education Association was formed as a result of the actions of a subcommittee of the American Prison Association. Four years later, the *Journal of Correctional Education* was founded. This journal has provided correctional educators a forum for the exchange of ideas and educational procedures throughout the nation. With these two important events, we entered the modern age of correctional education.

Contemporary Vocational-Education Programs

The development of systems designed to teach inmates a trade or vocation had made very significant progress by the 1960s. Vocational education had become available, at least theoretically, to almost all inmates imprisoned in the United States. The concept of teaching every prisoner that was capable of learning a marketable trade was accepted as a goal of corrections in almost every state. In 1965, the Manpower Development and Training Act provided federal funding to supplement the existing efforts of the states to provide vocational education. The Comprehensive Employment Training Act replaced the Manpower Development Act in 1973, thereby securing the funding of vocational programming for the remainder of the decade.

Female Inmates and Vocational Education

While the funding of vocational education was temporarily secure, there were still several problems to be faced. One of the problems to be faced in the 1970s was the changing view regarding vocational education for women. Because of the relatively small female-inmate population, and the outdated cultural belief that "a woman's place is in the home," most states were operating either no vocational-educational program for women or extremely restricted vocational programs. When the state of Texas opened a new facility specifically for vocational and academic education of female inmates at the Goree Prison in 1973, it was one of the first facilities of its kind in the United States (Stone, McAdams, and Kollert 1974). Unfortunately, this new facility, like most of the programs that preceded it, did not effectively address the needs of women in the changing society of the time.

Those states that had vocational programs for women were typically providing curricula on culinary arts (cooking), home economics (cleaning and sewing), floriculture (flower arranging), cosmetology (makeup and hair), and secretarial science. These vocational programs were suitable for some women, but they were typically preparing women to be housewives or for low-paying jobs in the "pink-collar ghetto" of the marketplace. Women who were supporting families were simply not getting vocational training in skill areas that could be relied on to provide significant salaries (Allen and Simonsen 1986/1992; Pollock 2002).

In the late 1970s, Mary Glover and a group of female inmates in the Michigan Department of Corrections filed a lawsuit with the federal courts, claiming that they were being denied equal protection under the law because of the inadequate programming available to women. The trial court found women's rehabilitative programming to be substantially inferior to that offered Michigan's male prisoners. In 1999, about 20 years after the original case, the federal courts finally released the state from federal monitoring (*Glover v. Johnson*, 198 F. 3d 557 [6th Circuit 1999]).

There were two possible solutions to the shortage of meaningful vocational-training programs for women. One solution was to begin new programs in women's institutions. A potential disadvantage of this solution was that it might move resources from institutions for men to those for women, in effect requiring a correctional administrator to deny services to 100 male inmates in order to provide a vocational curriculum to 20 female inmates. The second solution was to simply move the women to a men's institution where the programs already existed. While co-correctional facilities presented opportunities for shared resources, they also posed potential problems, such as sexual assault, pregnancy, and emotional attachments between inmates.

A survey conducted in June 1978 showed that there were 20 co-correctional institutions in operation in the United States (Smykla 1980). The Federal Bureau of Prisons at Fort Worth, Texas, opened the first of these co-correctional facilities in 1971 (Clear and Cole 1986). These co-correctional institutions solved some of the problems concerning female access to vocational programs, but they were not readily accepted in many regions of the country. Other problems existed: security concerns either dramatically increased the cost of vocational programs or were cited as the reason why men and women could not participate in the same curriculum. Women also showed little interest in traditional male vocations unless they were recruited into them. Once involved in such programs, women tended to show less enthusiasm and interest when participating along with men. By the early 1980s, the number of co-correctional facilities had dropped by almost 50 percent (Allen and Simonsen 1986/1992). From this low in 1980, the number of co-correctional institutions (not counting detention, psychiatric, and other specialized facilities) rebounded to about 30, where it remains today.

While some experts still see some potential for co-correctional institutions, others proclaim this concept a dismal failure. In their 1996 survey of research on co-correctional institutions, Smykla and Williams claimed that almost all of the studies showed that the concept was unsuccessful. It was not successful at increasing educational or vocational access for women. Instead, it had the effect of allowing men access into program resources that previously had been reserved for women.

The 1980s were a period of fiscal hardship for most state correctional systems. During this time, vocational programs for women were one of the first areas to suffer. Even the Federal Bureau of Prisons had trouble providing adequate vocational curricula for women. The Female Offender Task Force of the Federal Bureau of Prisons concluded that the vocational programs available to women were still not adequate to meet their needs in 1980 (Carlson 1981). Some states have continued to experiment with co-correctional institutions as the most effective way to deliver vocational programming to women. Illinois tried to deal with the problem of scarce funding and women's vocational needs by opening several new co-correctional facilities in the 1990s (Howell and Davis 1992). While some states are still experimenting with delivery methods, a 1998 survey of correctional institutions that contained female offenders showed that women still do not have access to adequate vocational programming (Morash, Bynum, and Koons 1998; Holtfreter and Morash 2003). In these studies, both the effectiveness and scope of current programs for women are seriously questioned.

The fiscal shortfalls of the 1980s had significant impact on the vocational programming for men's institutions as well as on the institutions for women. As most states entered the 1980s, the state treasuries were experiencing dramatic reductions in revenues. With the economic recession, available funds for prison programming became very limited. This was further complicated by the shift to a more conservative correctional philosophy.

External Contracting of Vocational Education

Some states have attempted to deal with the funding shortages by having the vocational programs delivered through local community colleges. For institutions located where they have easy access to community colleges, it is possible to get vocational programs delivered in a much more cost-effective manner. The community colleges maintain the necessary vocational-training equipment and faculty for their normal course delivery. This means that the correctional facility does not need to maintain expensive equipment inventories or full-time vocational instructors. This represents an especially cost-effective option for smaller institutions such as women's prisons. The single biggest problem with this option is the physical location of many prisons. Correctional institutions frequently are built in relatively remote rural areas where community-college access is not available.

Some of the larger state correctional systems use a very effective blend of colleges and in-house vocational instruction to deliver a broad range of vocational programs. For example, the Texas correctional system uses in-house vocational teachers to deliver 34 different vocational programs. These programs include air-conditioning, auto mechanics, business computer applications, construction trades, and dental-laboratory aide. Texas also contracts with local community colleges to deliver vocational curricula at state-prison units. The community-college vocational offerings include graphic arts, drafting, electronics, and sheet-metal manufacturing. In total, about 34,000 inmates participate in the Texas vocational-education program each year, and about 8,000 inmates are served by local colleges (Texas Department of Criminal Justice 2002).

Both New York and California run vocational programs that are very similar in size and scope to the one operated in Texas. New York and California deliver more of their total course work through community colleges, but the total course offerings are very comparable. In these three states, only between 15 and 20 percent of the inmate population may be enrolled in vocational courses each year. However, this is about twice the national average of less than 9 percent enrolled in vocational programs.

The funding for these offerings comes from a mixture of sources. Those states that have developed their own "in-prison" school district are able to tap directly into their state's general-education budgets. There is also some additional funding in most states in the form of state vocational-incentive grants, as well as some funding that is an integral part of the correctional system's budget. Educational funding is normally included as part of the programming or treatment divisions of many state correctional systems. An additional major funding source is the federal government. Inmates enrolled in vocational courses offered through community colleges were eligible for both Pell Grants and Veterans Administration grants until 1994. In 1994, the Higher Education Reauthorization Act was passed, which effectively prohibited prison inmates from receiving Pell Grants (Rose 2004). Many community-college programs were heavily dependent on federal funds; without Pell Grants, some of the programs were discontinued. Many states find that it is not politically acceptable to subsidize vocational college courses for inmates when they do not subsidize the same courses for the general population (McCollum 1994). During the late 1990s, many states tried to replace the funding lost from the Pell Grants by accessing federal programs such as the Youthful Offender Grant Program and various private agencies. However, much of the funding burden was transferred to the inmates themselves and to their families (Welsh 2002).

Pell Grants governmental grants for education.

Vocational Training and Recidivism

The ultimate goal of vocational education has been to provide the criminal offender with the skills necessary to make a successful living without resorting to crime. This assumes that if offenders are provided solid job training, they will then choose to apply those skills instead of taking the risks involved in criminal behavior. While there appears to be a logical link between crime and a lack of job skills, many experts do not believe the relationship is that simple or straightforward. The attempts to address crime through vocational training go all the way back to the bridewells of England, and obviously, we are still plagued by crime.

The evaluations of vocational training present mixed findings. There is only some evidence that seems to indicate that the presence of vocational skills reduces the chances of future criminal behavior, but many evaluations have less than adequate methodologies. The works of Taylor (1992); McCollum (1994); and Adams, Bennett, et al. (1994) are generally supportive of the link between vocational education and reduced recidivism, but most of the research that they reviewed to reach their conclusion is weak from a methodological standpoint. The research by Adams et al. (1994) was a review of 90 previous studies on recidivism that incorporated a wide range of different types of educational programs and methodologies. Of the 14 studies that were the most methodologically sound, 5 did not show a reduction in recidivism, and 9 did show a reduction. Additional research by Saylor and Gaes (1997) and Ryan and Desuta (2000), using somewhat improved methodologies, also supported the link between vocational training and a reduction in recidivism. However, probably the single best evaluation of vocational programming was conducted in the late 1980s in North Carolina. In this study, a true experimental design was used, including a randomly assigned experimental group and control group. The results of this study clearly showed that the programs were successful in transmitting the vocational skills. It also showed, however, that the vocational skills transmitted had no significant effect on recidivism over a four-year follow-up period (Lattimore, Witte, and Baker 1987).

There are several explanations for the failure of vocational-education programs to show a reduction in recidivism in the more rigorous experimental evaluations. One of the possible explanations is that while the programs can teach the skills necessary for employment, they cannot teach the attitudes necessary for successful employment. Many inmates lack the basic attitudes and social skills necessary for getting and keeping a job. Successful employment is frequently affected by such issues as punctuality, cordiality to supervisors and fellow employees, accountability, and respect for authority. Attitudes and values are transmitted in early life experiences, and attempts to change these attitudes and values at anything more than a superficial level through programs either has not been addressed or has not been very successful.

The likelihood of employment after vocational education is also affected by job discrimination and civil disabilities after release. Many employers simply do not want to hire former offenders, and in some cases, ex-convicts have difficulty getting an occupational license or surety bond necessary to employ their skills. The number of civil disabilities suffered by inmates has been dramatically reduced in recent years; however, the perceived desirability of inmates as employees has changed little. Perhaps the single greatest weakness of vocational programs is the lack of transition to outside employment. Few have job-placement services or a guarantee of suitable employment on release.

All of these issues can be further complicated for special inmate populations such as women. Female prisoners may be less motivated to participate in vocational and educational programming because of concerns for family or a misperception that the programs cannot help them. This lower motivation can translate into lower participation, less-successful participation, and lower benefits from the specific programs (Rose 2004). The special problems these female inmates experience may also complicate the work experience after release.

Contemporary Academic-Education Programs

In 1826, Judge Greshom Powers reported to the New York legislature that "fully one-eighth of the prisoners in New York's Auburn prison were either wholly unable to read, or could read only by spelling out most of the words" (Wines and Dwight 1867). An evaluation conducted 160 years after the observation of Judge Powers indicated that about one-third of inmates could not read, write, or do math at the eighth-grade level (Moke and Holloway 1986). A 1992 evaluation conducted in Louisiana indicated that recently admitted inmates were functioning at fifth- or sixth-grade level (Bates, Davis, Guin, and Long 1992). Because of this continuing problem, most states operate an Adult Basic Education program. These programs are designed to provide education to the functionally illiterate inmates that are admitted to prisons each year.

Adult Basic Education Programs

The origins of the current Adult Basic Education programs can be traced back to the compulsory-inmate-education laws passed by state legislators. In a survey conducted in 1993, every state that responded was operating an Adult Basic Education program. The five largest of these state programs are located in Ohio, Missouri, Texas, New York, and California (National Institute of Justice 1993). In a more recent federal census, all these programs were still in existence (Bureau of Justice Statistics 2003). Differences in terminology between the states makes comparisons of the programs difficult, but the percentage of inmates enrolled in basic-literacy programs has dropped in recent years to around 3 percent (Bureau of Justice Statistics). Since these programs generally have a required-attendance policy, this is hopefully a result of improvements in the education of the general public.

The Federal Bureau of Prisons established its first mandatory literacy program in 1982. In its first form, the federal program required that any inmate with less than a sixth-grade education be forced to go to school. The program was changed in 1986 to a minimum of eighth grade, and in 1991, the first mandatory high school-equivalency requirement was established (McCollum 1992).

General Educational Development Programs

The second level of educational programming that is offered in most institutions is the General Educational Development (GED) program. These programs are designed to prepare inmates to take and pass the General Educational Development test. The GED is a written test that assesses general comprehension and ability in the basic academic areas. It examines general comprehension and knowledge in reading, writing, science, and mathematics. The GED is considered to be the equivalent of a high school diploma. The GED is the minimal level of education required for entrance into many employment areas (Clear and Cole 1986).

Because of time limitations, inmates frequently choose taking the GED over completing a traditional high school program. The average inmate has a ninth-grade education at the time of his admission to prison. With the relatively short prison sentences (actual time served) of most inmates, the vast majority of inmates would simply not be in prison long enough to finish a high school degree. Because the GED is generally the most practical educational approach for most inmates, it is by far the most sought-after nonvocational educational program in American prisons. In some major prison systems, the number of GED certificates awarded in a year is 6 to 10 times the number of associate degrees awarded. In the most recent survey of prison educational programs, every state except New Hampshire reported offering the GED program to its inmates. The largest GED program in any of the state prisons is located in Texas (National Institute of Justice 1993). The Texas Correctional System, according to its most recent report, issued more than 5,000 GED certificates during the 2002 fiscal year (Texas Department of Criminal Justice 2002).

While GED certificates are the most common academic goal of the inmate population, there is a growing question as to whether the GED program actually serves the inmates' needs on release. Society is becoming more and more educated. As the educational needs of society increase, the GED may not serve the needs of the inmates when they reenter society. If the GED ceases to open the door to employment opportunities for the inmates, its utility will be largely lost. Recent evaluations of the impact of the GED on recidivism have not been promising.

A study conducted in Illinois using an experimental design that included a matched control group showed that inmates that earned their GED actually had a higher recidivism rate than those who did not earn a GED. This study followed the inmates for five years after their release, and raises some serious issues about the utility of the GED (Stevenson 1992). One of the possible explanations for the correlation between earning a GED and engaging in future criminality is the unrealistic expectations of benefits from earning the GED certificate. It appears that some of the prisoners expected the GED to open a number of employment opportunities that were simply not forthcoming. With disappointments in the job market, the inmates may have suffered negative self-image and attitude problems that contributed to recidivism. It may well be prophetic for the future of the GED-educated inmate that the U.S. military no longer accepts the GED as meeting their educational requirements.

While the best controlled research has not established a direct connection between receiving a GED and greater success after imprisonment, the jury is still out on the issue. A 2003 study using an inferior methodology did show that inmates who received their GED in prison returned to crime at a rate of 32 percent over a three-year period, while inmates who did not complete their GED relapsed into criminal behavior at a rate of 37 percent (Nuttall, Hollmen, and Staley 2003).

Prison-College Programs

Just as prisons have become increasingly dependent on local community colleges to deliver their vocational programs, the community colleges are also playing an important role in the delivery of the academic-education program. The first recorded college classes offered to inmates were in the Federal Bureau of Prisons.

In the early 1950s, a college program was started at the Fort Leavenworth, Kansas, prison by St. Mary's College in Xavier, Kansas (Carney 1979). Following this lead, several states started college programs in the early 1960s. The big growth for college programs for inmates came in the 1970s. In a survey conducted in 1993, all but 7 states reported having community-college programs available to inmates. The three largest state-based community-college programs are located in Ohio, Texas, and New York, respectively. In addition to these two-year-college programs, 17 states offer four-year-college programs, and 6 states allow inmates to pursue graduate degrees (National Institute of Justice 1993). Unfortunately, in the post–Pell Grant era, the number of states offering in-house college programs has dropped to 25 (Messemer 2003).

Evaluations of the effects of college-level education on inmate recidivism generally have produced more-favorable results than have evaluations of high school-level and high school-equivalency programs. In a study conducted in New York, inmates that completed the Community College Program were found to have a 26.4 percent recidivism rate, while a similar group of non-college graduates were found to have a recidivism rate of 44.6 percent (Clark 1991). While the research in New York is not as methodologically sound as some of the evaluations of the GED programs, it does indicate a possible rehabilitative effect from advanced education. In a study conducted in 1993, inmates in a college program were surveyed about their attitudes toward life and the future. The inmates in the college programs exhibited significantly more-positive attitudes about the future than did non-college inmates (Tootoonchi 1993). Other researchers have associated college-program participation with better institutional behavior and interpersonal-communication skills (Taylor 1992). In one of the better research studies, conducted in Massachusetts in 2001, the college-participant inmates had a 21.9 percent lower recidivism rate at the end of five years than did a comparable group of non-college inmates (Burke and Vivian 2001). Finally, Chappell (2004) conducted a meta-analysis of 10 years of evaluation studies and concluded that postsecondary education was associated with a significant reduction in recidivism.

Despite the positive research on the benefits of college education for inmates, the college programs in most states have several problems to overcome. Probably the most significant of these problems is the general lack of support from the public. Since the vast majority of citizens do not have access to free college courses, many people disagree with the idea of state-supported college programs for offenders. Expecting taxpayers to support the education of offenders when those taxpayers are having difficulty supporting the education of their own children is a difficult political hurdle. If the country becomes more politically conservative, even the Federal Bureau of Prisons may find their college programs coming under attack, despite the fact that inmates generally are paying for them. The second major stumbling block for college programs in prison is that many inmates simply cannot perform academically at the college level. Prisoners frequently have learning deficiencies that make the educational processes difficult even at the lower grades. In addition, many convicts simply do not see the need for and do not want a college degree. It is possible to force prisoners into educational classes, but the benefits of forced education are questionable at best.

Educational-Release Programs

An additional educational option for inmates is educational furlough, or **educational release**. In these programs, inmates are allowed to leave the facility to take advantage of educational opportunities. In the case of educational furloughs, prisoners may be released for significant periods of time to become traditional academic residents of colleges that may not be near their correctional facility. Educational release normally involves leaving the facility for only a few hours to attend classes, and then returning to the institution at the end of classes. The use of educational furloughs was somewhat widespread during the 1970s, but some highly visible program failures and increasing costs caused many of these opportunities to be cut back in the early 1980s (Allen and Simonsen 1986/1992). Currently, only 21 states are still operating educational-release programs. A number of other states have statutes that authorize the releases, but no inmates currently are out in the community. Since the studies conducted in the early 1990s, the participation in educational-release programs has significantly declined (Wees 1997).

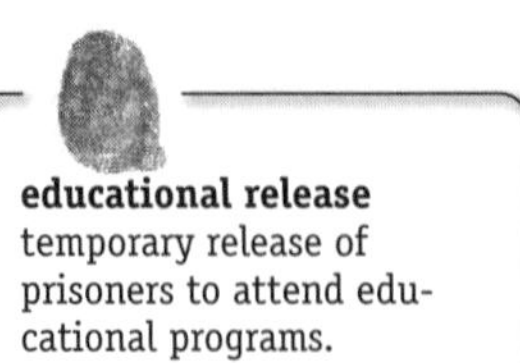

educational release temporary release of prisoners to attend educational programs.

The role of educational release is obvious. It allows inmates access to programs that are geographically remote from the correctional institution, and to curricula that are too specialized to be of interest to a significant group of inmates. Even when resources are available in a local college, it is not practical to bring in instructors to teach two or three inmates a class in anthropology. It is obviously more practical to allow the prisoners to attend classes with regular students at the local college. The drawbacks to educational release are also obvious; it involves putting the local community at some degree of risk. A single violent incident by an inmate in an educational-release program can cause a very significant political problem for correctional administrators. There have been few effective evaluations of educational-release programs because of their relatively small size and the very intense selection process that occurs before the inmates are allowed out into the community. Those evaluations that have been done show that it is very rare for a prisoner to fail to return from an educational release, and that the inmates are generally successful in their educational programs. These results do not surprise most correctional experts because educational-release inmates generally represent the best 1 percent or less of the prisoner population.

The range of educational opportunities available to prison inmates today is quite extensive. Convicts in many prisons are free to enroll in whatever type of education they choose. This does not mean that the problems of undereducated inmates—and ultimately, of prison releasees—are solved. Education programs, both academic and vocational, still have some major obstacles to overcome. In some states, sudden growth in prison populations has created a waiting list for the most desirable programs. Additionally, the poor educational background of many inmates requires that significant basic education has to be completed before more meaningful programs can be started.

Conclusions

As prison populations continue to grow, it is critical that prisons find some productive activities for increasing inmate populations. Having hundreds of thousands of prisoners in a state of forced idleness is a very dangerous situation for inmates and for correctional employees. Increases in correctional industries and increases in correctional-education programs are two of the areas of inmate programming that are most likely to receive political support in the future. In the United States, education and work are traditional American values that most people would agree should be reflected in our prisons. Educational and industrial programs occupy a good portion of inmates' time. It is important, however, that treatment programs that address problems such as substance abuse or low social skills operate in conjunction with educational opportunities. These treatment programs will be discussed in the following chapter.

KEY TERMS

Ashurst-Sumner Act—required that all prison-made goods be clearly labeled with the name of the prison and prohibited the shipment of prison goods to states that restricted their use.

bridewells—early British institutions for the itinerant poor. They were a form of workhouse or poorhouse where the poor were housed, but made to work for their room and board.

educational release—temporary release of prisoners to attend educational programs.

"factory prison"—is a term used for early northeastern prisons that adopted the congregate care system and placed prisoners in prison factories to work during the day.

Hawes-Cooper Act—was passed in 1929 and required that prison produced goods shipped interstate be subject to the laws of the receiving state.

houses of corrections—the general term for bridewells or poorhouses after they became associated with the custody of minor criminals such as pickpockets and thieves.

lease-labor system—was a form of income for early prisons by "leasing" inmate labor to private entrepreneurs who would reap the profit of the low-cost labor.

Mamertine Prison—one of the first recorded prisons, this facility was underground in Rome. Inmates were held for slave labor or until punishment.

Pell Grants—governmental grants for education.

PIE program—Private Sector Prison Industry Enhancement Certification Program allows state prisons to manufacture and sell goods on the open market as long as they adhere to guidelines and meet certain specifications.

shot drill—early form of punishment. This punishment consisted of repetitively picking up and putting down large cannonballs or "shot."

state-use laws—laws passed in many states in the early 1900s that prohibited the sale of prison made goods on the open market.

tool control—is the security system in prisons of frequently counting tools in order to prevent their theft and illegal use.

treadmills—early form of punishment. Also called treadwheels, this contraption was a wheel with treads so that a person would be forced to continuously step on the treads to move the wheel until being removed.

UNICOR—the Federal Bureau of Prison's industry program.

REVIEW QUESTIONS

1. Why would early prisons have considered unproductive labor more of a punishment than productive labor?
2. Why were most early prison programs designed for male inmates instead of female inmates?
3. Why was the state-use concept so much more successful for the federal prison system than it was for the state system, and how did World War II impact the issue?
4. How does the PIE program allow private companies to operate prison industries without destroying free-world jobs in the community?
5. Although using inmates in agricultural enterprises was probably successful in the 19th century, why is it no longer a promising way to reduce prison cost?
6. What are the issues of prison industry?
7. Although research shows that we can teach an inmate a vocational skill while they are in prison, why might that skill not be used in a job after release?
8. What do evaluations show regarding whether or not vocational programs reduce recidivism? What about educational programs?
9. Why are college courses not very common in prisons?
10. How can we run prison industries in the future to reduce the cost of imprisonment and also be sure that the inmates are not exploited?

FURTHER READING

Chappell, C. 2004. "Post-Secondary Correctional Education and Recidivism: A Meta-Analysis of Research Conducted 1990–1999." *Journal of Correctional Educatin* 55, 2: 148–169.

Ignatieff, M. 1978. *A Just Measure of Pain: The Penitentiary in the Industrial Revolution, 1750–1850.* New York: Pantheon.

Messemer, J. E. 2003. "College Programs for Inmates: The Post Pell Grant Era." *Journal of Correctional Education* 54, 1: 32–39.

REFERENCES

Adams, K., K. Bennett, T. Flanagan, J. Marquart, S. Cuvelier, V. Burton, E. Fritsch, J. Gerber, and D. Longmire. 1994. "A Large-Scale Multidimensional Test of the Effect of Prison Education Programs on Offenders Behavior." *Prison Journal* 74, 4: 433-450.

Allen, H., and C. Simonsen. 1986/1992. *Corrections in America: An Introduction*, 4th ed. New York: Macmillan.

American Correctional Association. 1992. Correctional Industries Survey Final Report. Laurel, MD: American Correctional Association.

Anonymous. 2002. "Prison Industries." *Corrections Compendium* 27, 9: 1–2, 19–20.

Austin American Statesman. 2004. "State Prison Farm System May be Put Out to Pasture." *Austin American Statesman* (October 11): B7.

Bates, P. T., T. Davis, C. Guin, and S. Long. 1992. "Assessment of Literacy Levels of Adult Prisoners." *Journal of Correctional Education* 43, 4: 172–184.

Bureau of Justice Assistance. 2004. Program Brief: Prison Industries Enhancement Certification Program. Retrieved February 12, 2005, from www.ncjrs.org/pdffiles1/bja/203483.pdf.

Bureau of Justice Statistics. 2003. Education and Correctional Populations. Retrieved on February 12, 2005, from www.ojp.usdoj.gov/bjs/pub/pdf/ecp.pdf.

Burger, W. 1985. "Prison Industries: Turning Warehouses into Factories with Fences." *Public Administration Review* 45: 754–757.

Burke, L. O., and J. Vivian. 2001. "Effects of College Programming on Recidivism Rates at the Hampden County House of Corrections: A 5-Year Study." *Journal of Correctional Education* 52, 4: 160–162.

Byerly, K., and L. Ford. 1992. "The Philadephia Story: An Innovative Work Program in the Real World." *Federal Prisons Journal* (Fall): 25–29.

Carlson, N. A. 1981. *Female Offender, 1979–1980.* Rockville, MD: National Institute of Justice Microfiche Program.

Carney, L. P. 1979. *Introduction to Correctional Science,* 2d ed. New York: McGraw-Hill.

Chappell, C. 2004. "Post-Secondary Correctional Education and Recidivism: A Meta-Analysis of Research Conducted 1990–1999." *Journal of Correctional Education* 55, 2: 148–169.

Clare, P. K, and J. Kramer. 1976. *Introduction to American Corrections.* Boston: Holbrook.

Clark, D. D. 1991. Analysis of Return Rates of the Inmate College Program Participants. Albany: New York State Department of Correctional Services.

Clear, T. R., and G. Cole. 1986. *American Corrections.* Monterey, CA: Brooks/Cole.

Davis, S. P. 1990. "Survey: Some States Save Millions Using Inmate Labor to Build Prisons." *Corrections Compendium* 15, 7: 8–17.

Duncan, D. 1992. "ACA Survey Examines Prison Industries for Women Offenders." *Corrections Today* 54: 114–115.

Fox, V. 1983. *Correctional Institutions.* Englewood Cliffs, NJ: Prentice-Hall.

Grieser, R. C. 1989. "Do Correctional Industries Adversely Impact the Private Sector?" *Federal Probation* 53: 18–24.

Grisson, G. R., and L. Conan. 1981. "The Evolution of Prison Industries." *Corrections Today* 43 (November/December): 42–48.

Hill, C. 2002. "Prison Industries." *Corrections Compendium* 27, 9: 8–19.

Holtfreter, K., and M. Morash. 2003. "Needs of Women Offenders: Implications for Correctional Programming." *Women and Criminal Justice* 14, 2/3: 137–160.

Howell, N., and S. P. Davis. 1992. "Special Problems of Female Offenders." *Corrections Compendium* 17, 9: 1, 5–20.

Ignatieff, M. 1978. *A Just Measure of Pain: The Penitentiary in the Industrial Revolution, 1750–1850.* New York: Pantheon.

Johnston, N. 1973. *The Human Cage: A Brief History of Prison Architecture*. Washington, DC: American Foundation Press.

Lattimore, P. K., A. D. Witte, and R. Baker. 1987. *Sandhills Vocational Delivery System Experiment: An Examination of Correctional Program Implementation and Effectiveness*. National Institute of Justice. Washington, DC: Government Printing Office.

McCollum, S. G. 1992. "Mandatory Literacy: Evaluating the Bureau of Prisons' Long-Standing Commitment." *Federal Prisons Journal* 3, 2: 33–36.

McCollum, S. G. 1994. "Prison College Programs." *Prison Journal* 74, 1: 51–61.

Messemer, J. E. 2003. "College Programs for Inmates: The Post Pell Grant Era." *Journal of Correctional Education* 54, 1: 32–39.

Moke, P., and M. Holloway. 1986. "Post-Secondary Correctional Education—Issues of Functional Illiteracy." *Journal of Correctional Education* 37, 1: 18–22.

Morash, M., T. Bynum, and B. Koons. 1998. *"Women Offenders: Programming Needs and Promising Approaches."* Rockville, MD: National Institute of Justice, Paper Reproduction Sales.

National Correctional Industries Association. 2004. Retrieved from www.nationalcia.org/indlinks2.html.

National Institute of Justice. 1990a. *Developing Private Sector Prison Industries: From Concept to Start Up*. Washington, DC: Government Printing Office.

National Institute of Justice. 1990b. *Operating Jail Industries: A Resource Manual*. Rockville, MD: NCJRS Microfiche Program.

National Institute of Justice. 1993. *Sourcebook of Criminal Justice Statistics, 1993*. Washington, DC: Government Printing Office.

Nuttall, J., L. Hollmen, and M. Staley. 2003. "Effects of Earning a GED on Recidivism Rates." *Journal of Correctional Education* 54, 3: 90–94.

Pollock, J. 2002. *Women, Prison, and Crime*, 2nd ed. Belmont, CA: Wadsworth.

Rafter, N. H. 1990. *Partial Justice: Women, Prisons and Social Control*. Piscataway, NJ: Transaction Publishers.

Rose, C. 2004. "Women's Participation in Prison Education: What We Know and What We Don't Know." *Journal of Correctional Education* 55, 1: 78–100.

Ryan, T. P., and J. Desuta. 2000. "Comparison of Recidivism Rates for Operation Outward Reach." *Journal of Correctional Education* 51, 4: 316–319.

Saylor W. G., and G. Gaes. 1997. "Training Inmates Through Industrial Work Participation and Vocational and Apprenticeship Instruction." *Correctional Management Quarterly* 1, 2: 32–43.

Sexton, G. E. 1995. "Work in American Prisons: Joint Ventures with the Private Sector." Rockville, MD: National Institute of Justice. Retrieved February 12, 2005, from www.ncjrs.org/pdffiles/workampr.pdf.

Sigler, R. T., and M. G. Stough. 1991. "Using Inmate Labor to Produce Products for the Open Market." *Journal of Contemporary Criminal Justice* 7, 1: 29–40.

Smykla, J. 1980. *Co-Ed Prison*. New York: Human Sciences Press.

Smykla, J., and J. Williams. 1996. "Co-Corrections in the United States of America, 1970–1990: Two Decades of Disadvantages for Women Prisoners." *Women and Criminal Justice* 8, 1: 61–76.

Snarr, R. 1992. *Introduction to Corrections*, 2d ed. Dubuque, IA: Wm. C. Brown Publishers.

Stein, N. 2003. "Business Behind Bars: Former Regan Attorney General Ed Meese Has a Way to Slow the Exodus of Jobs Overseas, Put Prisoners to Work." *Fortune* (September 3): 13, 28.

Stevenson, D. R. 1992. "Rehabilitative Effects of Earning a General Educational Development (GED) While in Total Jail Confinement as Measured by Recidivism Activity." Dissertation Abstracts International, 42. Ann Arbor, MI: University Microfilms.

Stone, W. E., C. C. McAdams, and J. Kollert. 1974. *Texas Department of Corrections: A Brief History.* Huntsville, TX: Texas Department of Corrections Printing.

Taylor, J. M. 1992. "Post-Secondary Education: An Evaluation of Effectiveness and Efficiency." *Journal of Correctional Education* 43, 3: 132–141.

Texas Department of Criminal Justice (TDCJ). 2002. FY 2002 Statistical Report. Austin: Texas Department of Criminal Justice. Retrieved February 12, 2005, from www.tdcj.state.tx.us/publications/executive/stat_report_fy02/stat_report_fy02_toc.htm.

Tootoonchi, A. 1993. "College Education in Prison: The Inmates' Perspective." *Federal Probation* 57, 4: 34–40.

UNICOR. 2003. FPI FY2003 Management Report and Independent Financial Audit. Retrieved February 12, 2005, from www.unicor.gov/customer/03Independentfinancial.pdf.

UNICOR. 2004. Inmate Transition Branch. Retrieved February 12, 2005, from www.unicor.gov/placement/ipprogram.htm.

Wees, G. 1997. "Work and Educational Release: 1996: Programs Are Numerous But Participation Is Down." *Corrections Compendium* 22, 5: 8–23.

Welsh, M. F. 2002. "Effects of the Elimination of Pell Grant Eligibility for State Prison Inmates." *Journal of Correctional Education* 53, 4: 154–158.

Wines, E., and T. W. Dwight. 1867. *Report on the Prisons and Reformatories of the United States and Canada.* Albany, NY: Van Benthuysen and Sons.

CASES CITED

Glover v. Johnson, 198 F. 3d 557 (1999), U.S. App. LEXIS 32404 (6th Cir. 1999)

6 Classification and Rehabilitation

David Spencer and Joycelyn M. Pollock
Texas State University–San Marcos

Chapter Objectives

- Understand the process of classification.
- Be able to discuss the rise and fall of the "rehabilitative ideal."
- Be able to describe the types of prison activities that inmates engage in.
- Understand the different treatment programs available in prisons.
- Be familiar with the literature on evaluating prison programs.

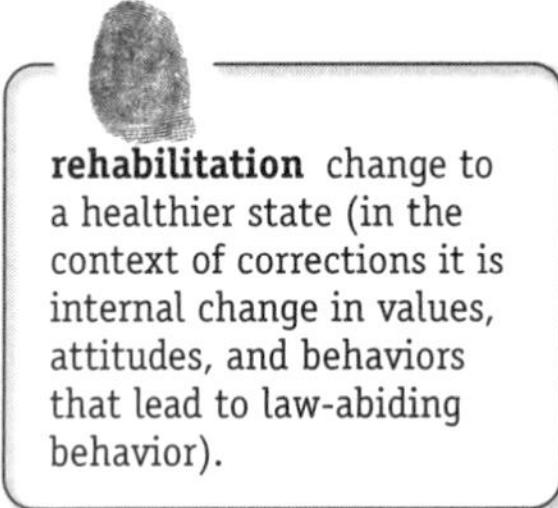

rehabilitation change to a healthier state (in the context of corrections it is internal change in values, attitudes, and behaviors that lead to law-abiding behavior).

Literature on the use of punishment in the criminal-justice system frequently cites the four major purposes to be retribution, deterrence, incapacitation, and rehabilitation. **Rehabilitation** seeks to prevent future crime by bringing about changes within the offender that make him or her less likely to reoffend. A broad array of programs have been used to bring about such changes, including education, job training, psychological and medical treatment, prison ministries, and recreation. This chapter looks at the various types of programs that are offered and their implementation.

The Development of Rehabilitation

Rehabilitation, through some means or mechanism, has long been a stated goal of corrections. The very use of the word "correction" implies an intention to bring about change in the individual. As far back as the late 1500s, a converted convent in Amsterdam was used to reform young offenders through labor and a disciplined life. As we learned in Chapter 2, the almshouses, workhouses and houses of correction used in England in the 1600s and 1700s were intended to reform the idle poor by providing work training and discipline. It was generally believed that idleness and petty crime were the result of a lack of personal discipline and inadequate moral training, and that correction through a disciplined regime was possible. In practice, however, there was frequently little difference between a correctional institution and a jail (Durham 1994).

Although moral training was a major feature of early correctional efforts, the development of the penitentiary, due mainly to the efforts of the Society of Friends and other religious groups, made religious instruction and meditation the primary element of personal reformation. Whether the inmate was in separate confinement, as in the Pennsylvania system, or in congregate confinement, as in the Auburn system, a silent regime with simple manual labor was intended to create an environment in which the offender would contemplate his life and become penitent.

In the mid-1800s, the idea of individualized treatment began to take hold. Not all offenders could be expected to respond to prison programs in the same way. Captain Maconochie's **mark system**, which tied inmate behavior to privileges and early release, and Sir Walter Crofton's **Irish system** of phased release were early attempts to encourage rehabilitation by offering incentives for good behavior. These systems also allowed room for prisoners to progress through the system at different rates, thus providing some means of treating inmates as individuals (Champion 1998).

The second half of the 19th century saw many changes in attitudes toward the purpose and practice of incarceration, especially in the United States. In 1870, the National Prison Congress (forerunner of the American Correctional Association) issued its Declaration of Principles, a broad policy statement that included recommendations for significant changes to the correctional practices that were then current (Champion 1998). Included among them were the use of indeterminate sentences and parole. **Indeterminate sentences** permitted prison officials to release inmates before the expiration of their sentences if they showed progress toward reform. Parole allowed for supervision after release and reincarceration for parolees who relapsed. By 1876, New York had opened the Elmira Reformatory, where Zebulon Brockway, its first warden, put many of the recommendations into practice.

Indeterminate sentencing fit well with the **positivist school** of philosophy and science, especially as it related to the study of crime. Cesare Lombroso and other philosophers saw the source of crime as something within the individual. Lombroso considered criminogenic factors to be biological, based on heredity. They could be identified by certain physical characteristics, or **stigmata**. However, other members of the positivist school believed that individual medical and psychological factors also played a part in crime (Alexander 2000). Regardless of the source, positivists believed that criminal behavior was determined by factors characteristic of the individual. It is a logical extension of this philosophy that these criminogenic factors could be identified and corrected, bringing about rehabilitation in the offender.

mark system Captain Alexander Maconochie's system for awarding good behavior points to prisoners on Norfolk Island prison colony off the coast of New Zealand.

Irish system Sir Walter Crofton's system in Ireland that allowed for early release of prisoners for good behavior.

indeterminate sentences are those in which the sentence length is not fixed by the court at the time of sentencing and, thus, the inmate's length of confinement is dependent upon good or bad behavior.

positivist school this is the school of criminology associated with Cesare Lombroso; adherents assumed that the cause of crime could be discovered through an empirical study of influences on the individual.

stigmata physical characteristics that Cesare Lombroso believed indicated a criminal predisposition (i.e. moles, birthmarks).

medical model under this approach to criminology crime is seen as a symptom of a pathology that can be treated.

The Medical Model

The theoretical framework of positivist criminology gave rise to the **medical model** of corrections. Positivist criminology fit well with developments in medicine and psychology. The medical model saw criminal behavior as a symptom of some pathology, or disorder, in the individual offender that could be identified and treated. Each offender was seen as the product of a unique combination of biological and environmental factors. Individualized diagnosis and treatment were seen as the appropriate way to deal with prison inmates. Indeterminate sentences permitted prison officials to grant early release to inmates who were amenable to treatment and showed satisfactory progress toward rehabilitation. The offender would then be supervised on parole, where treatment would continue through the parole officer. If the inmate relapsed, he or she could be sent back to the reformatory for further treatment (Alexander 2000).

Elmira Reformatory and similar institutions operated on these principles. Success did not come easy, and high rates of recidivism were common from the beginning of these programs (Alexander 1997). It was more acceptable to provide special programs and early release to women and juveniles than to male offenders, who were seen as a more hardened and dangerous population. The reformatory model was more frequently used for women, juveniles, and young male first offenders. In reality, however, many reformatories were run like prisons, operating on a traditional punishment model. For those institutions, rehabilitation was little more than a statement of correctional philosophy (Mitford 1973).

Regardless of these problems, the positivist philosophy was an optimistic one, and continuing advances in medicine and psychology kept the ideal of individualized treatment in the forefront of American correctional thought through the 1960s. In that decade, two social forces came into play. Although some correctional institutions provided diagnostic and treatment services, others were able to use the relative secrecy of an institution to confine inmates in conditions that were shocking to the conscience when they came to light. Chapter 8 in this text reviews the cases that established the right of prisoners to be free from extremely harsh conditions in prisons. These cases demonstrated that the stated goal of rehabilitation was offset by the brutal reality of prison life. The other important development was a growing perception among researchers that rehabilitation programs were not showing measurable success (Alexander 2000).

The Decline of the Rehabilitative Ideal

In 1966, the New York State Governor's Special Committee on Criminal Offenders began a project to investigate which offender treatment programs worked and which did not. The committee commissioned a study, apparently expecting that the results would enable them to identify and emphasize successful programs. Robert Martinson was one of the researchers. The researchers compiled information about previously published research on the success (or lack thereof) of prison-treatment programs. Their report was so pessimistic that it was not made public until after it had been obtained by subpoena in a lawsuit (Martinson 1974; also see Lipton, Martinson, and Wilks 1975). Martinson's initial conclusion, based on the report, was: *With few and isolated exceptions, the rehabilitative efforts that have been reported so far have had no appreciative effect on recidivism.* (Martinson, 1974, 25; emphasis in the original). He later modified his opinion, concluding after further research that treatment programs could be beneficial, neutral, or detrimental to recidivism, depending on the conditions under which they were provided (Martinson, 1979). For those who were opposed to rehabilitation programs, however, the original conclusion was paraphrased to a more succinct statement: "Nothing works."

Several criminal-justice researchers responded to Martinson, disagreeing with him or attempting to mitigate the impact of his work (Gendreau and Ross 1979, 1987). However, the simplicity of "nothing works" had appeal to many diverse groups. Social conservatives, who viewed the medical model as too lenient on criminals, used it as support for the idea that offenders could not be rehabilitated, so treatment efforts were useless. Fiscal conservatives saw rehabilitation programs as too expensive, with no return on the investment. Social liberals felt that indeterminate sentences were used in a discriminatory manner that violated due-process standards, and that coerced treatment was too much like brainwashing. As a result, in the late 1970s and 1980s, the federal government and many states changed their laws to move from *indeterminate sentencing* to some form of *determinate sentencing*. The new watchword of the correctional system became "just desserts"—give the offender the punishment he or she deserves, but don't expect it to change his or her behavior. In addition, funding for treatment programs was often cut since legislatures tended to see them as uneconomical. The prison-overcrowding crisis of the same time period also caused a diversion of resources from treatment to more-immediate needs, such as housing and security. While treatment was not completely abandoned during this period, it received a much lower emphasis.

In the late 1990s and through today, treatment has regained some currency (Duguid 2000). Most prisons now offer some rehabilitation programs, although both the quantity and the quality vary greatly. The material that follows gives an overview of the types of treatment programs that are available in correctional institutions. No institution has all these programs available to inmates, but most have some of them.

Classification of Inmates

An initial, and important, step in offender rehabilitation is the **classification** of the individual. Inmates are classified for a wide variety of purposes that can be included in two broad categories: prison management and offender treatment. There are areas of overlapping concern in these categories, and they are sometimes in conflict.

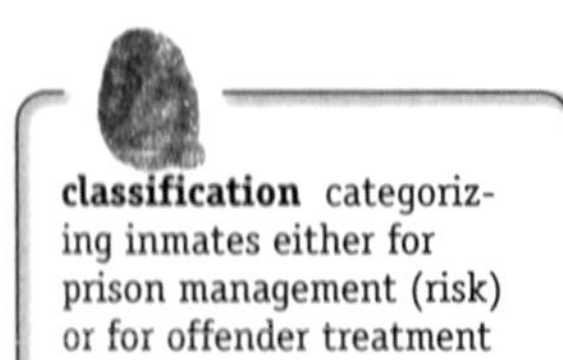

classification categorizing inmates either for prison management (risk) or for offender treatment (needs).

Classification for Management

The main function of a prison is to incarcerate individuals who have been sentenced for felonies. This means that security is of paramount importance to prison administrators. In large measure, this is an issue of public safety. The public should be protected from dangerous individuals, at least for the period of their incarceration. However, it is also important to understand that the *rule of law* dictates that those who have been sentenced to incarceration must serve their sentences in accordance with the orders of the court. Even those inmates who do not present a threat to the safety of the public must still be incarcerated as required by law. Management for security involves controlling the inmates' contact with the outside world. Prisoners must be kept inside the institution, and unauthorized people and things, such as weapons and contraband, must be kept out.

Management of inmates also involves controlling events within the prison. Movement and other actions of prisoners inside the institution must be regulated for the safety of the staff, other inmates, and the institution's property. It is, therefore, important to classify inmates according to the risks they present both to the safety of the public and to the operation of the institution.

Classification for *security* involves the first of these concerns. Security, in this sense, has to do with the degree of physical separation from the outside world that the institution imposes on the inmate. Thus, a maximum-security prison imposes the highest level of separation on its inmates. Within any given institution, regardless of the level of security, prisoners are subjected to different levels of supervision. This may be referred to as the *custody* classification. In this context, custody means something different from the security level of the institution. Even a minimum-security facility may have administrative or punitive segregation facilities that operate at a high level of security, and a maximum-security facility may have some inmates who are trusties and are allowed to work with minimum supervision. The definitions of security and custody given here are generally accepted among corrections professionals, although they may be used interchangeably (Levinson 1988).

Classification for management may be seen as an assessment of the risk posed by the inmate. At one time, this sort of assessment was done by prison staff members based on their experience with inmates. Correctional officers and counselors were expected to rate prisoners without any standard criteria. The results were **subjective-classification** systems that were, at best, idiosyncratic and could not stand up to legal challenges based on equal protection. Through the 1980s and 1990s, and into the new century, the trend has been toward the increasing use of **objective-classification** systems. These systems are based mainly on actuarial data—that is, documented information about the inmate's personal and criminal history and the current offense (Bonta, Bogue, Crowley, and Motiuk 2001). Although some institutions still use subjective classification, objective-classification criteria are generally preferred because they are easier to administer. They are also seen as more defensible if they are subjected to legal challenges based on equal protection, since all prisoners are classified by the same instrument. However, that very uniformity can be problematic if the criteria have a disparate impact on different classes of inmates. This is particularly true regarding gender. Most states that responded to a published survey reported that they used the same classification system for male and female inmates. However, the use of the same criteria may tend to classify women as greater security risks than could be otherwise justified (Burke and Adams, 1991). The use of a classification instrument that tends to overclassify women could subject states to the very liability issues that they seek to avoid.

> **subjective classification** a classification system that would depend on interviews and professionals' subjective assessment of the inmate.
>
> **objective classification** a type of classification system that uses instruments such as personality inventories or risk factor scores based on offender backgrounds to determine risk or need.

A number of standard classification instruments have been developed and are in use, but each state is free to adopt or modify an existing scale, or to develop its own. Objective classification is usually accomplished by use of a **risk scale** that rates the inmate on a number of factors. Most factors relate to criminal history and current offense, but also include demographic and social factors. Some factors commonly used in risk scales include age, gender, criminal history, prior parole, offense type, sentence length, prior time in community, history of drug abuse, employment history, marital status, and number of dependents (Bonta, Bogue, Crowley, and Motiuk 2001). Points are assigned for each of these factors, and the resulting score determines the inmate's classification. Many of these classification instruments also provide for a staff override—a method by which the classification analyst can adjust the classification if he or she decides that the score generated by the scale does not accurately reflect the risk posed by the inmate. These overrides reduce the objective nature of the classification system, but they can be useful if the staff has proper guidelines and training (Buchanan, Whitlow, and Austin 1986).

risk scale used in objective classification systems, these scales apply points based on such things as prior convictions, use of a weapon, age at first conviction, drug use history, and so on to determine the level of risk an offender presents.

Classification for Treatment

Classifying prisoners for treatment may be seen as addressing the needs of the individual for rehabilitation. This process is not entirely independent of the security- and custody-classification process, since inmates in the higher levels of management classification will have limited access to rehabilitation programs. However, identifying the rehabilitative needs of inmates can augment the management-classification system and provide direction for staff members who are responsible for programming. Needs assessment can be referred to as **internal classification** (Levinson 1988).

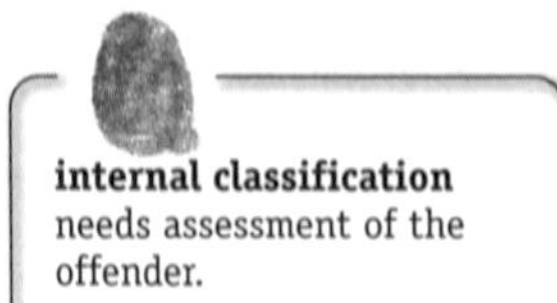

internal classification needs assessment of the offender.

Classification for treatment can be accomplished by the use of a needs-assessment instrument or by psychological-classification systems. Needs-assessment instruments are similar to the risk-assessment instruments described above, in that they are objective and based largely on historical information about the inmate. However, they tend to place more emphasis on factors such as intellectual ability, emotional problems, alcohol and drug abuse, and educational background (Van Voorhis 2000). There are three objective-classification instruments in current use that have been validated by empirical studies to address both risk and needs. One is the Level of Service Inventory-Revised (LSI-R), which is widely used throughout the English-speaking world. Another is the Community Risk/Needs Management Scale. The third is actually a family of risk/needs instruments that have been developed by various jurisdictions based on the risk/needs instruments originally developed by the state of Wisconsin (Bonta, Bogue, Crowley, and Motiuk 2001).

Psychological-classification instruments are based on various theories of personality and behavior. One type is intended to be filled out by a third person who describes the observed characteristics of the inmate. Prison psychologists, counselors, or custody staff who interact with the inmate usually complete these instruments. They frequently take the form of "checklists" that list characteristics of interest and ask whether, or to what extent, the inmate displays them.

Another type of instrument is interview based. A professional (usually a psychologist, social worker, or counselor) interviews the inmate and completes the instrument based on the responses and the interviewer's own observations.

These instruments have the advantage of gathering more-personalized information about the inmate and allowing the interviewer to ask follow-up questions in areas of particular interest. They have the disadvantage of being labor intensive, since the professional can interview only one inmate at a time and the interviews can sometimes be lengthy. In addition, the quality of the evaluation varies greatly according to the skill of the interviewer.

A third type of psychological instrument is the "paper-and-pencil" test, in which the inmate responds to questions on a written form. The responses are then compiled into standard scales that have been found by empirical research to describe certain psychological conditions. This type of test can be administered verbally to inmates who are not able to read the written form. It has the advantage of ease of administration, since many people can take the test at the same time under the supervision of a small number of corrections officers. One disadvantage is that the inmates know they are being tested and may consciously or unconsciously adjust their answers in order to serve some personal agenda.

MMPI Minnesota Multiphasic Personality Inventory is a personality inventory survey that is used to assess an individual against norms; various scales can be derived from the results, such as the depression, and paranoia scales.

A well-known example of the third type of test is the *Minnesota Multiphasic Personality Inventory* (MMPI) and its revised version, the *MMPI-2*. These are lengthy tests, consisting of a series of statements (566 for the MMPI; 567 for the MMPI-2) to which the test taker responds "true" or "false." The answers are compiled into scores on several scales. There are 10 standard clinical scales, each of which is designated by a number, name, and letter. Scales that have particular implications for prison classification are Scale 2—Depression (D), Scale 3—Hysteria (Hy), Scale 4—Psychopathic Deviate (PD), Scale 6—Paranoia (Pa), and Scale 0—Social Introversion (Si). Each of these scales has been found to be associated with a set of psychological characteristics that have implications for the individual's behavior. Researchers have developed many additional special-purpose scales over the years. Professionals who classify inmates for treatment can use one or more of these scales to assign the inmates to appropriate programs.

Three examples of psychological tests that have been developed or adapted for use in prisons are the Interpersonal Maturity Level (I-level) system, Quay's Adult Internal Management System (AIMS), and the Megargee MMPI-based Typology (Van Voorhis 2000). The I-level classification system (Warren and Community Treatment 1966; Jesness and Wedge 1983) was developed with youthful offenders, but has been adapted for use with adults (Harris 1988). The concept behind the system is that all people go through a developmental process, beginning in infancy, in which their perceptions of themselves and others become more complicated and mature. Some people go through the development process more rapidly than others do, and some become stalled at a particular stage. The level of sophistication of the person's thought processes tells the therapist something about how the person views himself and his relationship to others. Treatment is more appropriate if it is directed to the individual's current developmental state, and if it facilitates further development.

AIMS Adult Internal Management System is a type of personality scale device where a professional completes two checklists (historical and current observations) to determine the inmate's placement in the five offender personality types: asocial aggressive, immature dependent, neurotic anxious, manipulator, and situational.

The Adult Internal Management System (AIMS) (Quay 1983, 1984) is an example of a third-person evaluation based on the inmate's history and behavior. Prison staff complete two checklists. One is historical (the Checklist for the Analysis of Life History), and the other is based on current observations (the Correctional Adjustment Checklist). The checklists, taken together, locate the inmate in a taxonomy of five offender personality types; Asocial Aggressive, Immature Dependent, Neurotic Anxious, Manipulator, and Situational (Van Voorhis 2000).

The third example of a classification instrument is the Megargee MMPI-based Typology (Megargee and Bohn 1979). Although it was developed using the original MMPI, this instrument can be used with the revised MMPI-2 as well (Megargee 1994). The system uses the **profile** formed by the scores on the 10 standard scales to classify the offender into one of 10 "types." A general psychological description has been developed for each of the types. The psychological consequences of membership in a particular type have implications for custody-level assignments as well as for treatment-program assignments (Van Voorhis 2000).

profile a term associated with Megargee's MMPI-based offender typology and refers to the offender types constructed by using the scores on the 10 standard scales.

The list above is by no means exhaustive. Numerous other psychological tests are in use for treatment classification, and ongoing research will undoubtedly yield others.

Prison Programming

The terms "rehabilitation" and "treatment," while still commonly in use, are frequently subsumed in the more general concept of "programming" or "activities." Many researchers have attempted to give definitions that cover the broad spectrum of rehabilitative programs. Senese and Kalinich (1992, 223) defined rehabilitative programs as "a programmed effort to alter attitudes and behaviors of inmates which is focused on the elimination of their future criminal behavior." In the practical jargon of prison officials, "program" or "activity" can be used to refer to anything that occupies an inmate's time with an activity that does not create problems for the administration. Thus, recreation, work, and religious services can be seen as programs, as well as education, counseling, and physical therapy. Psychiatric and medical interventions that are limited to reactively treating immediate psychological or medical problems are generally not considered to be rehabilitative programs. The treatment and rehabilitation modalities that are described in this section are in general use in American prisons.

Inmate Activities

Although most inmate activities would not fall under any ordinary definition of treatment, they make imprisonment more endurable by providing some relief from boredom. They also give prisoners an opportunity to engage in cooperative efforts that may promote pro-social attitudes and assist in adjustment to prison life, and they give prison officials an opportunity to evaluate the behavior of the inmates while they participate in activities that are less structured than most inmate-staff interactions.

Recreation

Recreational programs exist in every prison. In fact, failure of an institution to provide adequate recreational opportunities is a frequent subject of litigation. Organized sports provide entertainment and an opportunity for teamwork. An inmate's ability to engage in competitive activities with a socially acceptable level of aggression is seen as an indicator of rehabilitation by many corrections professionals. Available sports activities usually include common schoolyard sports (basketball, softball, running and jogging tracks, and the like). In prisons for men, weight-lifting equipment is a standard fixture. Women's prisons are more likely to include studio activities such as dance, yoga, and Pilates. Indoor recreational activities include card games, board games, television, and movies.

As with any prison activity, security is an issue in recreational programs. Group activities give inmates an opportunity to exchange contraband. Competitive athletics can easily degenerate into violence. Many people have questioned the wisdom of allowing male inmates to "pump up" with weight-lifting equipment. However, prison administrators are generally of the opinion that recreational sports give prisoners a necessary outlet for physical tension. As one deputy warden said to the author, "I'd rather have them bench-pressing weights than bench-pressing me." The granting or withholding of recreational privileges also serves as a disciplinary tool for administrators.

Liability for injuries suffered in recreational activities is not a major issue for state-run prisons. However, privately operated prisons that do not enjoy governmental immunity for torts can be affected. The author is aware of one situation where an inmate in a private prison injured his back in the prison gymnasium. He threatened to sue on the theory that the prison had not provided proper instruction in the use of the gym equipment. The company that owned the prison was able to settle the claim without litigation, but it took the precautionary measure of contracting with a personal trainer to provide instruction to the inmates. Although this may have represented an unnecessary response to an unusual situation, liability issues are nevertheless a potential problem with inmate activities. Of course, many state systems have responded in the opposite direction and have removed all weight equipment.

Religion

The American correctional system has always regarded religion as an important element of rehabilitation. Religion is an important source of moral values for people outside of prison, and it seems intuitively obvious that worship and religious instruction might contribute to internal changes that could bring about rehabilitation. There are numerous success stories of rehabilitation through religious conversion in prison. Some authors see religion as providing "religious therapy" for inmates (Senese and Kalinich 1992), and believe that religious counseling may be combined with psychological-counseling techniques to provide spiritual as well as mental healing (Hall 2003). Counseling is usually seem as an important function of a staff chaplain, and some inmates are no doubt more comfortable talking with a chaplain than with a secular counselor. However, in recent years, legislative-funding cuts in many states have resulted in a reduction in the number of state-paid chaplains, leaving the provision of religious ministry to volunteers (Hall).

Prisons usually provide opportunities for formal worship services for most mainstream religions. These typically include Catholic, Protestant, Jewish, and Islamic services. Other religions may be allowed to hold formal services when the prison has enough inmates of that faith to warrant it. However, since groups of inmates gathering for religious services present potential security problems, prison administrators are generally reluctant to accommodate prisoners' requests for worship services when they consider the prisoners to be insincere and the religion to have been created for the purpose of manipulating the staff. An example of this is "The Church of the New Song," which originated in Texas and spread to other states. The inmates maintained that their religious practice required them to eat steak and drink Harveys Bristol Cream sherry. Not surprisingly, neither prison administrators nor the courts were willing to accommodate their demands (Pollock 2004).

Prisons often accommodate religious organizations from the outside world. These groups hold religious services and conduct other faith-based activities. Two such groups that are active in numerous prisons are Kairos Prison Ministries International, an ecumenical Christian organization, and Prison Fellowship Ministries, which was founded by former Watergate defendant and federal-prison inmate Charles Colson. Prison Fellowship Ministries operates a program called InnerChange [*sic*] Freedom Initiative in prisons in Iowa, Kansas, Minnesota, and Texas. There are several other organizations that provide faith-based activities on local, regional, and national levels.

A recent development is the creation of special prison units for faith-based programs. In 2003, the Florida Department of Corrections converted an 800-bed men's facility into a "faith-based prison." The inmates must volunteer to be in the program, must have no disciplinary record within the previous 12 months, and must be near their release date. At the time the unit was opened, the inmates represented 26 Christian and non-Christian faiths (Price 2003). Florida followed up by opening a faith-based prison for women in 2004 (Hannigan 2004). It remains to be seen whether such programs will withstand legal challenges under the First Amendment's prohibition on the establishment of religion.

Arts and Crafts

Prisoners generally have a lot of idle time, and it is better for them to occupy it in some productive activity. Working with arts and crafts has several benefits both for the inmates and for the prison administration. Inmate products may be sold in prison stores, or even in stores and galleries outside the prison. This gives the inmates a small source of income, which may be especially important in those prisons that do not pay them for labor. It also allows them to pay for their own supplies so that the prison budget does not have to support the work. The inmate general fund may get a percentage of the sales, which is used to buy welfare and recreational items that are not available in the prison budget. Some inmates become good enough that they are able to make a living with their work after release from prison. There are some security issues associated with arts-and-crafts programs, including the potential use of tools as weapons and the bartering of products in the prison black-market economy. However, prison administrators tend to regard arts-and-crafts programs as at worst harmless, and at best a contributing factor to rehabilitation.

Community Service

Performance of community-service work is a common condition of community corrections (probation and parole). Opportunities for service work in the community are obviously limited for prison inmates. However, such programs do exist in many prisons. Most obvious to the public are anticrime and antidrug activities by prisoners, such as participating in videos for use in education programs. Less well known are activities such as recording audio materials for the blind, or making and repairing Christmas toys for children of low-income families. These programs are frequently organized and facilitated by religious and community-service organizations outside the prison.

There do not appear to be any studies documenting any rehabilitative effects of these programs. However, there is anecdotal evidence that some inmates find that doing good works gives them a feeling of self-worth. In addition, these are pro-social activities that occupy the inmates' time with something worthwhile, which is a positive characteristic from the standpoint of prison management.

Treatment Programs

The difference between treatment programs and the activities above is that treatment programs exist specifically to address the criminality of the offender—either directly, or through some other problem that influences the offender's criminality.

Self-Help Groups

Few would disagree that the best-known self-help program in the United States is Alcoholics Anonymous (AA). Its success and favorable regard in the public mind have brought about the creation of several similar organizations such as Narcotics Anonymous (NA), Cocaine Anonymous (CA), Gamblers Anonymous (GA), and Sex Addicts Anonymous (SAA). Collectively, these organizations are known as 12-step groups, after the principles of the AA recovery program, which are shown in **Box 6-1**.

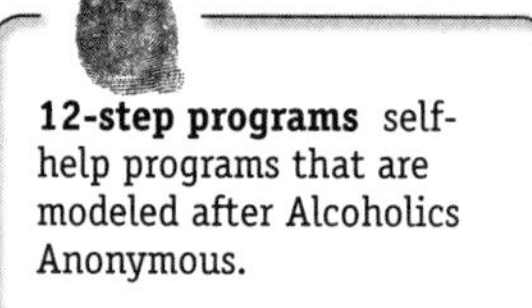

12-step programs self-help programs that are modeled after Alcoholics Anonymous.

Twelve-step programs see addiction as a disease, and addicts as sick rather than bad. An alcoholic or addict must reach a point where he or she is willing to give up the addictive behavior, referred to as "hitting bottom." To recover successfully, the person must "work the steps," beginning by admitting that he or she is powerless over alcohol (or narcotics) and cannot control his or her addictive behavior by willpower alone. Reliance on a "Higher Power" for recovery is stressed, although members are not required to profess a belief in God, and the program encourages each member to develop his or her personal understanding of a Higher Power. The only requirement for membership is that the person have a desire to stop drinking (or using drugs). The remainder of the steps set out a recovery plan that includes self-examination, an admission of wrongs, willingness to change, making amends to others, prayer and meditation, spiritual awakening, and working with other alcoholics and addicts (*Alcoholics Anonymous* 1976).

Many prison inmates have had exposure to 12-step programs before going to prison—perhaps in previous prison terms, as mandated during probation, or as an individual attempt at recovery. Groups exist in almost every prison, but the conditions of group meetings differ. Some prisons allow inmates to conduct their own groups, but this poses some obvious security issues. Other prisons will allow members from the free world to conduct meetings in the facility. Some prisons require that a staff member be able to observe, if not hear, the meeting.

Box 6-1

The 12 Steps of Alcoholics Anonymous

1. We admitted we were powerless over alcohol—that our lives had become unmanageable.
2. Came to believe that a Power greater than ourselves could restore us to sanity.
3. Made a decision to turn our will and our lives over to the care of God *as we understood Him*.
4. Made a searching and fearless moral inventory of ourselves.
5. Admitted to God, to ourselves, and to another human being the exact nature of our wrongs.
6. Were entirely ready to have God remove all these defects of character.
7. Humbly asked Him to remove our shortcomings.
8. Made a list of all persons we had harmed, and became willing to make amends to them all.
9. Made direct amends to such people wherever possible, except when to do so would injure them or others.
10. Continued to take personal inventory and when we were wrong promptly admitted it.
11. Sought through prayer and meditation to improve our conscious contact with God *as we understood Him*, praying only for knowledge of his will for us and the power to carry that out.
12. Having had a spiritual awakening as the result of these steps, we tried to carry this message to alcoholics, and to practice these principles in all our affairs.

Source: Alcoholics Anonymous 1976. Reprinted with permission.

As the name suggests, AA operates on a principle of confidentiality, and self-disclosure is an important part of group meetings. As is true with group-therapy sessions, inmates are frequently reluctant to disclose personal matters that may be passed along to staff or other inmates on the prison grapevine. The effectiveness of the program may depend on the level of trust that can be fostered in the group, as well as on the individual's willingness to participate.

One advantage of AA and similar organizations is their very pervasiveness. On release, a former inmate is almost certain to find a group that he or she can attend conveniently. Although AA, NA, and the others are separate organizations, members of one program are generally welcome at meetings of the others. Some alcoholics may look down on addicts, and some addicts may regard alcoholics as square and conventional, but people are rarely excluded from meetings unless they become disruptive.

Twelve-step groups are certainly not the only self-help organizations that seek to assist offenders. The Salvation Army, Volunteers of America, Delancey Street Foundation, and the Fortune Society are a few examples of nongovernmental organizations that provide a wide range of services to ex-offenders, including housing, education, counseling, and job training. Such groups do not usually receive governmental funding, although they may provide some contract services. They are typically supported by donations and foundation trusts. They may provide services to offenders in prison, particularly to those who are nearing release, and also to ex-inmates to help them adjust to reentry into society.

Professional Group Treatment

Self-help programs, even when augmented by members from the outside, do not involve the intervention of a trained professional counselor. Prison-treatment groups are different in that they are planned and developed by the administration. They may be mandated, or the granting of privileges or other benefits may encourage voluntary participation. They are conducted by a professional who has some training in counseling, who usually follows accepted principles of psychological treatment. Members of the prison staff may run the programs; either corrections officers who have received special training in counseling techniques, or trained counselors who have been hired as staff members. In other cases, prisons contract with licensed counselors—such as psychologists, certified drug- or alcohol-abuse counselors, or licensed professional counselors—to run groups in the prison. Sometimes the contract is with a private company that provides its own professional employees. These companies may be charitable organizations or for-profit corporations or professional associations.

Group programs run the gamut from the lower levels of "treatment"—such as self-esteem groups, life skills, or parenting skills—to more-intense psychological therapy. Groups are frequently organized along specialized lines. Groups for drug offenders and sex offenders have become common. There are also groups for survivors of abuse and incest, particularly in women's prisons. There is a growing understanding that a person's history as a victim may contribute to later criminal behavior.

Some group counselors follow a particular model of therapy. Several of these models are discussed in the following sections. However, the general trend today is for the counselor not to be locked rigidly into a particular therapy, but to provide an eclectic mix of therapeutic methods.

Bringing About Change

The medical model of treatment, referred to previously, regards crime as a symptom of a disorder, or pathology, in the offender. According to this model, if the disorder can be identified and successfully treated, the offender's criminal behavior will be reduced or eliminated. In this context, the terms "disorder" and "pathology" are used very broadly, to include any characteristic of the individual that results in criminal behavior. The pathology may be one or a combination of any number of social, psychological, or medical problems. In order to be completely successful, the treatment would have to address each of the underlying disorders. One offender might be a substance abuser, for example, and might also lack sufficient job skills to support himself even when not abusing drugs. Another offender might be undersocialized and might also be a victim of child abuse, while a third might have an antisocial-personality disorder and also be an alcoholic.

The treatment would vary for each offender in order to properly identify and address each problem. However, the procedure would be the same: identify the disorder or disorders that bring about the criminal behavior, then treat each of them in such a way as to bring about a cure, and the result will be an elimination of criminal behavior. Obviously, this model has never worked perfectly. Some critics doubt that it has ever worked at all. However, the concept has face validity, and is sufficiently attractive that it managed to retain some support even during the height of the "nothing works" period.

Important to this discussion is the question of causation. A "disorder" is generally thought to mean a condition that causes the sufferer to be dysfunctional in some major aspect of his or her life. Certainly a person who commits criminal conduct serious enough to warrant incarceration in prison can be said to be dysfunctional. The question, however, is whether the dysfunctional behavior is caused by a particular disorder. Many people have depression or low self-esteem, or lack education or job skills, but do not engage in serious criminal behavior. Treating a condition that does not cause the offending behavior may be worthwhile from a clinical standpoint, but it is not likely to change the inmate's future behavior. An aphorism common in Alcoholics Anonymous is: "If you sober up a horse thief, what you've got is a sober horse thief." In order for the medical model of treatment to work, the pathology or pathologies that bring about the criminal behavior must be identified and addressed. This is exacerbated by the fact that a diagnosis of "the problem" may be difficult and inaccurate. For example, many inmates show symptoms of drug and/or alcohol abuse, but also have histories of general antisocial behavior. A diagnosis of antisocial-personality disorder cannot be made if the client has a substance-abuse disorder because the symptoms (i.e., the client's behavior) may be much the same for both disorders (American Psychiatric Association 1994). It may be possible to make a further diagnosis of antisocial-personality disorder after the substance-abuse disorder has been successfully treated; however, treating one without treating the other may not bring about the desired change in the client's behavior.

Another issue of concern, but one that is rarely addressed, is how to measure a successful outcome. One of the criticisms that have been leveled at Martinson (1974) is his restricted definition of a successful outcome. His evaluation considered only recidivism as the criterion for a successful outcome (Alexander 2000). However, when the full report that was the source of his article was published, it examined seven different treatment outcomes: recidivism, institutional adjustment, vocational adjustment, educational achievement, drug and alcohol re-addiction, personality and attitude change, and community adjustment (Lipton, Martinson, and Wilks 1975). Recidivism is certainly important, and is the outcome measure that tends to attract public attention, but other outcomes are also important, especially within the institution. For example, an inmate's adjustment to prison has important consequences for institutional administration. If an inmate, as the result of a treatment program, completes his or her sentence with an acceptable level of cooperation and a minimal amount of disruptive behavior, then the treatment was successful from the standpoint of the institution.

The following sections give a brief overview of several treatment modalities that are found in many prisons. It must be remembered that the availability and quality of treatment programs vary greatly from state to state and from prison to prison.

Individual and Group Therapy

A standard stereotype of therapy involves the client's lying on a couch, talking about his childhood while the therapist sits out of the client's line of sight, taking notes and making occasional comments. Either psychiatrists or psychologists may practice this type of one-on-one therapy. Psychiatrists are medical doctors who have specialized in the study of the human mind. Psychologists have a nonmedical degree (usually a Ph.D. or Ed.D.) in a program that specializes in a study of human psychology. Both are qualified to conduct individual therapy, although modern practice tends to eliminate the couch in favor of a comfortable seated position, and to place the client and therapist so that they can make eye contact.

Individual therapy, regardless of the format, is not common in prison-treatment programs. It is costly, requiring the services of a highly trained professional to treat only one inmate at a time. Prison psychologists and psychiatrists typically have other duties that do not accommodate an individual-therapy schedule. They often are involved in evaluating inmates for transfer to psychiatric units, and monitoring those who have been placed under psychiatric observation. They may be called on to intervene in crisis situations without warning. Inmates who experience personal crises, such as the death of a loved one, may need emergency counseling on a one-time basis. All of these factors make individual therapy problematic.

As a result, most counseling in prison involves group therapy. Groups are more cost-effective because several inmates receive the services of the professional therapist at the same time. In addition, if the psychologist or psychiatrist is called on to perform other duties, an associate, such as a master's-level counselor or a certified drug-abuse counselor, may take over the group so that the time is not wasted. However, group meetings raise issues of security and confidentiality. Members of the group may use the meeting as an opportunity to communicate or to pass contraband. Sometimes emotions run high in group meetings, and there is a potential for violence. In addition, self-disclosure is an important element of group therapy, and a client who will not participate fully in the therapy program is less likely to benefit from it. A group member who does not trust the other members to keep his or her confidences will probably not reveal anything about his or her mental state that will be helpful to the treatment process. In this regard, the rules of the prison subculture run counter to the optimal situation for group therapy. At the other extreme are those prisoners who do not monitor their self-disclosure and may say imprudent things that can cause problems for them among other inmates. Establishing an atmosphere of trust in which the group members feel free to disclose personal matters is an essential part of creating a successful group-therapy program (Trotzer 2000).

Therapeutic Communities

Therapeutic communities may be organized inside or outside of prisons. In the free world, they are generally used for drug and alcohol treatment. In prisons, they may be used for substance abuse, sex offenders, or for mentally ill or mentally retarded inmates. There are several characteristics that identify a successful therapeutic community. First, it separates the clients from the general population. This isolates them from negative influences and distractions from the program. In prisons, therapeutic communities may be housed in a separate cell block, wing, building, or unit. Custodial-staff members are integrated into the community, and they support, rather than undercut, the benefits of the program. Inmates are carefully screened before they are placed in the program (Alexander 2000). Some communities consider voluntary inmate applications; others require a referral from the administration or from a treatment professional. Either way, the inmates are vetted for their institutional behavior and their apparent willingness to cooperate with the treatment regime. Since the conditions in therapeutic communities are frequently better than those in the general population, there are usually more than enough applicants to fill available spaces. The staff members who run the program, both administrators and counselors, should be well trained in the treatment modality and committed to the success of the program. Clients should be motivated and should have sufficient cognitive skills to understand and cooperate with the program. Treatment staff, custodial staff, and clients should all be able to work together to achieve the program goals (Singer 1996).

In a therapeutic community, all of the inmate's activities are seen as part of the treatment, and the client becomes a partner in the treatment process. Typically, new arrivals are assigned to dormitory areas with several occupants to a room and are given menial tasks, such as mopping the floors and cleaning the bathrooms. Staff and more-experienced inmates indoctrinate them into the living routine. As they demonstrate an ability to live up to their responsibilities, the new arrivals are given increased responsibility and better living conditions, moving from menial jobs to more-responsible positions, and from dormitories to smaller rooms with fewer roommates. The inmates are allowed to participate in the operation of the program, even to the extent of voting on administrative matters that do not affect the security of the prison. This empowers them and gives them a sense of responsibility for the program. All of this is seen as part of the therapy, as well as the more traditional group sessions that are conducted by the treatment staff (Alexander 2000).

Clients who have successfully completed treatment in a therapeutic community often speak highly of the experience. However, it should be remembered that this is a highly select group. First, applicants are screened for entry into the program. Within the program, there is usually a high failure rate of clients voluntarily quitting or being removed. Those who successfully complete the program, then, are those who were the most highly motivated to succeed from the beginning. When program evaluations based on randomized experimental designs are used, these programs do not appear to have much greater success than the untreated control groups (Austin and Irwin 2001).

Psychological and Psychiatric Programs

In this section, a few of the common modalities are described. Some are more commonly found than others in a prison setting, however.

Psychotherapy

Psychotherapy is defined as any form of treatment for mental illness, behavioral maladaptations, and other emotional problems by a trained person who establishes a professional relationship with a client for the purpose of removing, modifying, or retarding existing symptoms; of attenuating or reversing disturbed patterns of behavior; and of promoting positive personality growth and development (Campbell 1996). Psychotherapy was originally developed by psychiatrists (Sigmund Freud and his followers and contemporaries), but is now used by psychologists as well. It is usually thought of as an individual therapy, but the principles can be adapted to group treatment as well. It is the prototype "talk therapy," consisting mainly of verbal interaction between the therapist and the client. While there are several versions of the therapy (i.e., Freudian, Adlerian, Jungian), the primary therapeutic process is to help the client achieve self-awareness and insight into the problems that trouble them. The problems usually come from psychological trauma that the client experienced in childhood. He or she has usually tried to deal with the trauma through **defense mechanisms**, such as denial, reaction formation, blaming the incidental cause, or projection (Masters 1994). Trauma can occur as the result of a single experience, such as the sudden loss of a loved one, or can develop because of a pattern of experiences, such as dealing with parental rejection. Defense mechanisms are normal and are part of a healthy personality. However, their use can prevent a person from dealing with trauma in a healthy way, leading to the development of a neurotic personality.

defense mechanisms psychoanalytic term used to describe attempts to deal with childhood trauma, such as denial, projection, or reaction formation.

Psychotherapists generally believe that it does little lasting good to simply tell a client what the problems are; the healing process works best if the client comes to an understanding through his or her own insight. Although the therapist can guide the client by directing the content of their discussions, it is most beneficial for the client to come to a personal understanding of his or her hidden motivations. This type of spontaneous insight is sometimes called a breakthrough. Once the client has achieved this self-awareness, he or she can begin to deal with the trauma in a healthy way. Eventually, the client will no longer need to misuse the defense mechanisms to avoid dealing with the trauma.

The practice of psychotherapy developed when the science of psychology was young. The early psychiatrists were still developing their theories, and knowledge of matters of the mind was virtually nonexistent among laypeople. Psychotherapy was seen as a very slow process, often taking years to complete. This could be especially true if the client was resistant to the insight. Self-awareness can be both painful and frightening. A client might unconsciously avoid confronting the trauma by taking the content of the discussions in a safe direction, thus delaying the therapy process. The slow process of traditional psychotherapy is not conducive to dealing with prison inmates, many of whom are incarcerated for relatively short periods of time.

More recently, a variation on traditional psychotherapy has come into use. Called directive therapy, it takes advantage of the fact that people today have much more knowledge of psychological processes that they did a hundred years ago. The same psychological principles apply, but the therapist takes a more active role in directing the client's attention to problem areas. If the client wants to stray from a productive path, the therapist is likely to refocus the discussion on the important issues. If the therapist is successful, the client is likely to achieve the very breakthrough that he or she is trying to avoid, without unnecessary delay. This type of therapy frequently involves an agreement between therapist and client to work on a particular problem for a limited time. Of course, the therapist always has the option of moving more slowly if the client appears vulnerable to being retraumatized by confronting the traumatic experience too soon. Directive therapy may be more efficacious than traditional psychotherapy in the prison environment, where time is limited.

Behavior-Modification Therapy

Behavior modification is a product of the school of psychology known as **behaviorism**. As the name suggests, behaviorists focused their attention on behavior, ignoring thoughts, feelings, and attitudes. To qualify as a behavior, an action must be specific and observable. An emotional response would not qualify as a behavior, but laughing or crying would. The theory relies on the concept of reinforcement. When a person engages in a behavior, the consequence of the behavior may act to reinforce the behavior. Reinforcement may be either positive or negative. If a behavior is positively reinforced, the person is more likely to repeat it. If the behavior is negatively reinforced, it is less likely that it will be repeated. If this process is carried out as part of a therapy program, it is called conditioning.

behaviorism school of psychology that assumes that an individual's behavior is based on prior rewards; concentration is on behavior since it is believed that thoughts and emotions follow behavior.

As it is practiced today, behavior-modification therapy is a family of therapeutic techniques (Brown, Wienckowski, and Stolz 2000). The various programs may use positive reinforcement, negative reinforcement, or a combination of both. An example of a program using positive reinforcement is a **token economy** (Ayllon and Azrin 1968). Token economies are used in some special prison programs, such as therapeutic communities and honor blocks. Good behavior is rewarded with credits by which the inmates can earn special privileges. The reward positively reinforces the inmate's good behavior, making it more likely that it will be repeated. As stated above, the behavior must be specific and observable. Telling an inmate to "get your mind right" would not qualify, since the behavior is neither specific nor observable. However, telling an inmate to make up his bunk or to report for work detail on time would fall into the category of specific, observable behaviors. Modern behavior-modification programs frequently involve the use of behavioral contracts. The counselor and the client agree on behavioral goals that the client will achieve, and on the reward that the client will receive if the goals are met. These are also called contingency contracts because the reward is contingent on the client's behavior (Brown et al.).

token economy a program in an institutional setting where residents are rewarded for good behavior through the use of tokens or points which relate to privileges.

Negative reinforcement is most often used when an undesirable behavior has its own positive reinforcement. Alcohol abuse, for example, is self-reinforcing because of the immediate intoxicating effect of the alcohol. The behavior is negatively reinforced by an aversive stimulus: something that the client finds unpleasant or undesirable. The application of the aversive stimulus acts to extinguish the undesirable behavior by making it less likely to be repeated. A prescription drug called Antabuse is sometimes used as an aversive stimulus in the treatment of alcoholics. A person who drinks even a small amount of alcohol after taking Antabuse will become physically ill with a variety of symptoms. The illness is the negative reinforcement. Since inmates are not permitted to drink alcohol in prison, this might seem to have little application to institutional treatment programs. However, inmates may participate in work-release programs or be allowed privileges such as temporary furloughs that give them access to alcohol during short periods of release. Antabuse therapy may help them to avoid temptation.

A less drastic form of aversive conditioning involves the use of fines. This is often coupled with positive reinforcement in a token economy: good behavior is positively reinforced with tokens that can be used to obtain rewards, and bad behavior results in the loss of tokens, and thus of privileges. Therapeutic communities and honor blocks may employ both types of reinforcement. Both positive and negative reinforcement appear to work best when they are applied immediately so the connection to the behavior is strong.

Behavior modification has its share of critics. Books and movies such as *The Manchurian Candidate* and *A Clockwork Orange* have created a connection in the public mind with brainwashing. Professional ethics and some court cases have limited the use of this therapy, particularly the use of aversive conditioning. Electroshock treatment and drugs such as Antabuse and apomorphine may now be used only with the informed consent of the client (Wexler 1973).

Cognitive-Behavioral Therapy

Another type of treatment based on a psychological theory is cognitive-behavioral therapy, or simply cognitive therapy. While behavior-modification therapy concentrates on behavior without consideration of mental processes, cognitive therapy focuses on thoughts, beliefs, attitudes, expectations, self-statements, and other mental processes known collectively as cognitive factors, or cognitions. The underlying theory is that problem behavior arises when a person's thought processes become distorted. Everyone has basic, unstated, and unquestioned assumptions that shape the thought processes. These assumptions are known as schemata (which is the plural form; the singular is schema). They can usually be traced to things that we learned in childhood (Freeman 1983). Since the schemata are based on our perceptions of reality, and therefore are not perfect, they represent distortions of reality. For most people, the distortions are relatively minor and do not significantly impair the person's thought processes. However, if the distortions become serious enough, they can cause the person's thought processes to become dysfunctional. The schemata are so deeply embedded in the unconscious mind that we are not aware of them, but they give rise to automatic thoughts. Since these thoughts are based on unquestioned premises, they also go unquestioned. When the schemata, and therefore the automatic thoughts, are dysfunctional, they give rise to cognitive distortions that affect our conscious thoughts, and, therefore, our behavior. The following are examples of cognitive distortions.

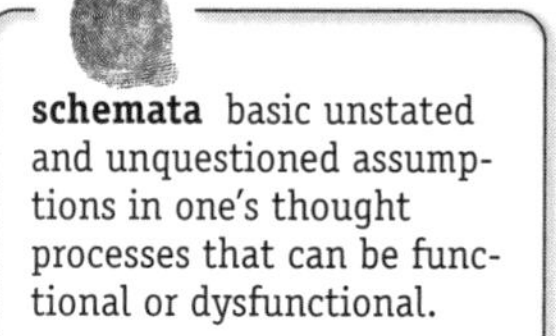

schemata basic unstated and unquestioned assumptions in one's thought processes that can be functional or dysfunctional.

All-or-nothing thinking. This is also referred to as "black-and-white thinking." A person's attitudes are polarized, with no gray area in the middle of the extremes. A person may view herself and other people as either perfect or worthless.

Overgeneralization. This distortion causes a person to think that an isolated negative experience is the norm. If the person does not succeed at some task, he might conclude that he will never get it right.

Emotional reasoning. This thinking error involves interpreting emotions as fact. A person might say, "I feel guilty, therefore I'm a bad person" (Freeman 1983).

A number of treatment strategies may be employed in cognitive therapy, but they have a common goal of breaking down the distortions by helping the client to recognize the thinking errors. Unlike traditional psychotherapy, cognitive therapy is *directive therapy*. In directive therapy, the therapist takes the role of an expert who directs the client's attention to the errors rather than waiting for the client to recognize them on his or her own (Genova 2001). This involves examining the client's irrational beliefs and confronting them with appropriate questions such as *What is the evidence for your belief? What is another way of explaining the belief?* and *If your belief is true, what is the worst that could happen?* (Alexander 2000). Questions that challenge the validity of irrational beliefs are called disputations. The main thrust of the therapy is to break down the faulty core beliefs by examination and confrontation that reveal the errors of logic that are the result (Alexander).

Cognitive therapy is highly adaptable to individuals and to special populations. Some feminist counselors have determined that women may have a particular set of irrational beliefs. Recognition of these particular beliefs allows the counselors to adapt the therapy to female clients (Fodor 1988).

One well-known version of cognitive therapy is Rational Emotive Therapy (RET), developed by Albert Ellis. The program addresses thinking errors by what are called the ABCs of cognition. *A* is the activating event, or objective fact. *B* is the individual's belief, or perception of the fact. *C* is the consequence of the belief (Ellis and Dryden 1997). If the belief represents a thinking error, the consequence is dysfunctional. Like other versions of cognitive therapy, RET involves challenging and disputing faulty thinking.

Transactional Analysis

Transactional analysis (TA) is best known because of the popular books *Games People Play* (Berne 1964) and *I'm OK—You're OK* (Harris 1969). It is a model of psychiatric therapy first set out in *Transactional Analysis in Psychotherapy* (Berne 1961). Central to this theory is the concept of the ego state. A person's current ego state is the set of feelings and attitudes and the accompanying behaviors in which the person is operating.

Everyone has three ego states: the Parent, the Adult, and the Child. The Parent is derived from the messages that the individual received from his or her actual parents or parental surrogates. Although everyone has a Parent ego state, its exact dimensions are unique to the individual because they come from the person's actual experiences with his or her parents. Much of the Parent is about rules, especially prohibitions (can'ts and don'ts) and imperatives (musts and oughts). The Parent can be disapproving and moralistic. A person acting in the Parent can be reactive and judgmental. The good side is that, like real parents, the Parent can also be loving, protective, and supportive. The Parent serves a useful function by providing ready-made rules for many situations so that the individual does not have to make numerous decisions about the routine matters of life.

The Child is the remainder of the ego state that the individual actually experienced in childhood. It is formed by the thoughts, feelings, and events that occurred in the person's childhood. The Child can be charming, fun-loving, creative, and affectionate. But like a real child, it can also be stubborn and rebellious. While the Child can be energetic and optimistic, it can also be filled with fear and guilt and have a strong need for reassurance and nurturance. It is sometimes compliant and sometimes obstinate. The Child is the ego state that wants to gratify immediate needs and wants, with little regard for the consequences.

The Adult is the ego state that is concerned with gathering and processing information and with making rational decisions. Although the name implies that it develops later in life, the Adult begins to develop as soon as an infant is able to control its movements, and therefore to make decisions about how to act (Harris 1969). The Adult makes a realistic assessment of the individual's current circumstances, and acts as a mediator between the demands of the Parent and the needs of the Child. It examines the Parent and Child to see if the information contained there is still valid or has become outdated. For example, the Parent's admonition "Look both ways before you cross the street" is a good rule throughout life, but "Don't leave the front yard" is not relevant to a grown person. The Adult is the rational, reality-testing, problem-solving ego state, but it can be somber and lack spontaneity.

transaction a communication segment between people.

game term used in Transactional Analysis to describe a pattern of communication (or transactions) between people.

Transactional analysis is based on various types of communication between people operating in their various ego states. A **transaction** is a piece of communication between people. A **game**, which takes place between two or more people, is a sequence of transactions that can be repeated. A script is a long-term pattern of transactions and games by which a person conducts his or her life (Muller and Tudor 2002).

The nature of the transaction is based on the present ego states of the people involved in the transaction. Adult-to-Adult transactions are the most useful, since both parties are trying to view the situation in a rational manner and resolve any problems. For example, one Adult may ask for information, and the other Adult may give the information or say, truthfully, "I don't know." This is the "I'm OK—You're OK" position (Harris 1969). Transactions in which neither party is acting in the Adult can easily produce conflict, since there is no mediator to resolve differences. A person who continuously operates as a rebellious Child, for example, is likely to have frequent conflict with authorities who are acting as limit-setting Parents. The police officer (Parent) may say, "Move along; you're blocking the sidewalk," and the recalcitrant offender (Child) may respond, "It's a free country; you can't make me do anything."

Transactional analysis sees games as being handed down from generation to generation. As parents raise their children, they teach them their own games, both by what they say and by what they do. People tend to select their associates (including spouses) from among others who play the same games, so children are likely to get the same games from both sides of the family (Berne 1964). All the ego states may be involved in a game, and the purpose of the game may change as the ego states shift. An example of a prison game is "How do you get out of here?" A career criminal, who sees prison as an unfortunate cost of doing business, really does want to get out. All of his ego states (Parent, Adult, and Child) are agreed on that. Therefore, he plays a version of the game called "Good Behavior," in which he makes every effort to manipulate the system in order to get the earliest release. This may include participating in rehabilitative programs and making a show of cooperation. However, some inmates play a version of the game called "Want Out." They may initially act in such a way as to get out as soon as possible, but as the actual time of release approaches, the inmate's Child becomes fearful of the prospect of dealing with the outside world. In this ego state, the inmate may sabotage the situation by engaging in misconduct sufficient to delay the release. Some other games that have been identified among offenders are "See What You Made Me Do," "Cops and Robbers," "Alcoholic," "Addict," "Now He Tells Me," "If It Weren't for You," "Little Old Me," "Courtroom," "Uproar," "Let's You and Him Fight," and "Psychiatry." (Berne 1964; Miller, Bartollas, Jennifer, Redd, and Dinitz 1974).

Scripts are the most complex of the TA structures. They involve the patterns in which a person lives his or her life. They implicate both the person's style of transactions and the games in his or her repertoire. The person who is frequently the "loser" or victim in relations with others is operating from one type of script; the successful overachiever is operating from quite another, and the bully from still another. Scripts tell us how to live our lives in relation to others.

Therapy based on the principles of TA seeks to analyze the transactions, games, and scripts to see how they are dysfunctional, and to adjust them to a more functional condition. The therapist educates the client about the ego states so that they can be recognized. The aim is to recognize how the three ego states determine the client's behavior, and to reinforce the positive aspects and alter the negative aspects. For instance, it is neither necessary nor desirable to completely suppress the Child. ("Let your child come out and play" has become a standard feature of popular psychology in modern culture.) Nor is it necessary to ignore the rules of the Parent, as long as the Adult is able to evaluate them and to decide which ones are useful and when they should be applied. The goal is to harmonize the ego states so that they create a whole, functional individual.

Transactional analysis gained a great deal of popularity during the 1960s and 1970s. Although it was based on a sophisticated psychotherapeutic model of the human mind, the popular books presented TA in such a way that it could be easily understood on a superficial level. It was well adapted to the group-counseling practices in institutions, and it could be combined with other treatment methods. However, TA may have been a victim of its own success. Its widespread popularity in the public mind caused it to be treated as little more than a parlor game. Identifying other people's "games" became a popular pastime. As a specific therapeutic method, TA fell from favor in the 1980s and 1990s, although many of its concepts are still incorporated into eclectic therapeutic models.

Reality Therapy (RT)

William Glasser (1965) developed this therapy while working with delinquent girls in California. According to the principles of reality therapy, people experience behavioral problems because they are unable to meet their essential psychological needs in a realistic way. Irrational behavior results from an attempt to fulfill these needs in a way that does not recognize the reality of the world. The dysfunctional person must learn to fulfill his or her needs and to act responsibly. Responsibility is defined as the ability to fulfill one's needs in a way that does not deprive others of the ability to fulfill their needs (Glasser, 13). There are two essential psychological needs, and both involve relations with other people. They are the need to love and be loved, and the need to feel worthwhile to others and ourselves. In order to fulfill these needs, an individual must have involvement with at least one other person that he cares about and that he is convinced cares about him (Glasser). In the therapeutic context, the other person is the therapist.

Reality therapy has three basic procedures. First, the therapist must develop an emotional involvement with the client so that the client can feel mutual caring and a sense of self-worth. This approach sets RT apart from most other therapies, in which the therapist is expected to maintain an emotional detachment from the client. Second, the therapist must reject the client's unrealistic behavior, while at the same time accept the client as a person and maintain involvement. Third, the therapist must teach the client realistic ways to fulfill his or her needs (Glasser 1965). A more specific description of the procedures follows.

1. *Achieve involvement*
2. *Understand but deemphasize personal history*
3. *Help the client understand attitude and behavior and the elements that contribute to it*
4. *Explore alternative behaviors*
5. *Get commitment to change*
6. *Monitor and evaluate; accept no excuses and enact no punishments; rewrite the commitment to ensure success (Walsh 1988, 142)*

An important aspect of RT is that it focuses on behavior rather than on thoughts or feelings. In this regard, it is aligned more with the behaviorist school of psychology than with traditional psychotherapy. The concentration is on all aspects of the client's present life and on his or her moral and ethical beliefs. Within the therapy, there is no punishment for irresponsible behavior, but the therapist analyzes the behavior to help the client understand its irrationality. The client may suffer adverse consequences for the behavior outside the therapy (i.e., punishment for infraction of institutional rules), but the therapist may use this as evidence to convince the client of his or her irresponsibility. Responsible behavior, by contrast, is praised. The therapist does not ask the client why he did something, but instead focuses on what the client did and on the consequences of the behavior.

Reality therapy takes planning and commitment from both the therapist and the client. They must analyze the entire spectrum of the client's behavior, and analyze its rational and irrational content. The ultimate goal of RT is to develop a plan by which the client can fulfill his or her wants and needs in a responsible manner. This sort of behavioral contract is also seen in various models of behavior modification and cognitive therapy. The plan must be simple so it is easily followed, specific so it is easily understood, realistic so it comports with rational behavior, and attainable so the client is set up to succeed rather than fail (Wubbolding 1988).

Reality therapy has been popular in institutional treatment because the notion that offenders are acting irresponsibly makes intuitive sense, and because the basic rules of the therapeutic process can be easily learned and understood by lay people. However, it should be remembered that Glasser and others who developed this therapy were highly trained mental-health professionals. In practice, RT is not as simple as it seems. In particular, achieving the necessary personal involvement with clients without losing professional objectivity is a task not well suited to many untrained or inexperienced prison counselors.

Evaluating the Effectiveness of Treatment

Martinson's (1974) report has become known as the death knell for the rehabilitation era, but even he softened his conclusion, later determining that some programs worked for some people (Martinson 1979). In the intervening years, there has been a steady (albeit small) set of researchers who do find success in rehabilitative efforts (Cullen 2004).

The first question to ask is: What is success? Typically, the answer to that question has been the reduction of recidivism. Measures of recidivism vary widely; recidivism may be measured as rearrest, conviction, or return to prison. If rearrest is used, figures of failure are higher because many ex-convicts are arrested even though no charges are ultimately filed. Return to prison would result in lower figures of recidivism, but some returns to prison may be for technical violations of parole rather than for new crimes. Would the technical violation of not reporting to a parole officer, although still sober and crime free, be considered a failure or a success? Interestingly, some prisoners are returned to prison after they have already been in treatment programs on the outside, because it is then that they stay in one place long enough for outstanding warrants to catch up with them. Other measures of success might be employment or attitudinal change, but these measures are used very seldom. Almost all evaluations depend on recidivism, although there may be a variety of means chosen to measure it.

Another issue is to what degree must recidivism be reduced for a program to be considered a success. Many evaluations show relatively modest reduction of 10 to 13 percent. Is it feasible to expect drops of 40 percent or higher? Are other programs that target human behavior held to these same standards? Success rates of smoking programs, weight-reduction programs, or other attempts to change behavior are fairly modest, yet we expect much more from correctional programs.

Another factor is how the program operates. If the program was designed under a certain theory of criminality, with a structure and content that was consistent with such a theory, it is important to note whether the operation of the program—how it was implemented—was faithful to the initial plan. Many times, and for many reasons, programs on paper do not resemble the programs in reality. Staff changes, poor training, or implementation problems may result in the content being different from the philosophy of the program. This is especially troublesome when one considers that certain modalities are evaluated in comparison to one another. Therefore, it would be important to know, for instance, if a reality-therapy program was actually using the assumptions, premises, and programmatic elements of reality therapy, or if these elements had somehow become attenuated or ignored. A related issue is the eclectic approach used by many programs makes replication and the ability to measure the relative efficacy of any one approach impossible.

A major problem of evaluation is high attrition. If one measures the recidivism of only those who finish the program, it may be that they also have higher motivation, which contributes to their success. What is measured may not be a treatment effect, but rather, individual differences between those who drop out and those who stick with it to the end. On the other hand, one can hardly measure a program's effectiveness by monitoring the behavior of those who dropped out.

control group a group similar in every way to the group that is subjected to a treatment, thus any differences that emerge between the two groups after treatment can be more clearly associated with the treatment itself.

random assignment is when every individual in the population has an equal chance of being placed in the control group or the treatment group; used to ensure that treatment effects are not due to some statistical bias.

Another problem of evaluation is the difficulty of obtaining an adequate **control group**. A control group is a group similar to the treatment group in all characteristics except for the treatment. Ordinarily, **random assignment** is used to ensure that the groups are the same. Random assignment is defined as every individual in the universe having an equal chance of being selected. For instance, if the universe to which you wished to project your findings were a prison population, random assignment from that population to a treatment group and a control group would be necessary. Random-number tables or other methods can be used so long as no bias is introduced.

If random assignment to the treatment group is not possible, the only control group that would be appropriate would be matching characteristics. Matching is not as effective and is very difficult to do. For instance, if one were evaluating a drug-treatment program, a control group would have to be matched to a treatment group on all potentially influential factors such as drug history, criminal background, age, family characteristics, and so on. Unfortunately, few designs are so careful, and often you see drug-treatment programs compared to general prison populations or, at best, to prisoners committed for drug crimes. Drug-addicted offenders are more likely to return to crime, so one could expect higher recidivism rates from an addict group than from a general-prison-population group. Thus, a treatment program that shows the same recidivism after treatment as a general-population control group may actually be a success since one should have expected higher recidivism figures. On the other hand, with voluntary treatment programs, program effects may be partially due to the nature of the highly motivated inmate. The only way to control for that would be to get volunteers, and then randomly assign some to a treatment group and some to a control group. This is not a popular approach in program delivery because program staff want to choose inmates.

Another important issue to address is the so-called "black box" of prison. Treatment programs are only one part of the prisoner's life. The prison experience is like a black box inside which the researcher cannot look. We can measure the outcome of the black box (higher or lower recidivism), but we can never identify which elements of the prison experience contributed to the results. It may be that treatment programs provide positive elements to a course of changing one's life, but the negative aspects of imprisonment—**prisonization** (the socialization to the prisoner subculture), violence, attacks on self-esteem, and loss of family support—may override any treatment effects.

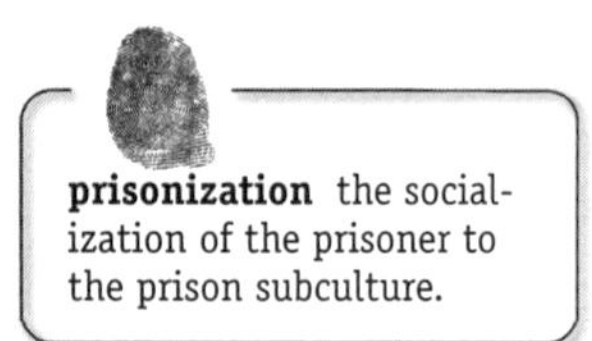

prisonization the socialization of the prisoner to the prison subculture.

Another issue is the importance of individual therapists and the related topic of individualization of treatment. Martinson (1974) observed that the favorable results that he did see may have been due to special personal gifts of the therapist rather than to the treatment itself. Another author points to research that finds that "good" counselors—those who are open, warm, accepting, and empathetic—are more similar to each other, even if they operate from very different theoretical perspectives, than any of them are to poor counselors, even within the same theoretical perspective or treatment modality (Walsh 1988, 118).

Certainly there is evidence that the strength of any program may rest more in the personnel than in any power of the modality itself. This calls into question the whole attempt to evaluate programs. Perhaps it makes more sense to evaluate the people who work in prison programs, and try to discover if their qualities of success are something that can be taught.

Many of those who continue to support treatment do so with the caveat that treatment must be individualized to the offender. Panaceas do not exist, and no treatment modality or approach will be successful with all offenders. Andrews (1995) identified three conditions under which treatment programs have been shown to be more successful: first, that rehabilitative services be delivered to high-risk offenders; second, that the criminogenic needs of offenders must be identified; and third, that treatment must be matched to client needs and styles of learning. Perhaps one of the greatest failings of therapy in the "rehabilitative era" of the 1970s was the failure to individualize treatment. Except for a few programs, there was a one-size-fits-all approach to treatment programs such as TA, group therapy, or reality therapy. No treatment approach is necessarily appropriate for all offenders. The goal should be to successfully match the type of program to each offender's needs. **Box 6-2** on page 184 lists the elements of a successful treatment program as outlined by Coulson and Nutbrown (1992).

Evaluation Studies

As mentioned before, there has been a steady thread of evaluation studies that show success. Ted Palmer (1972, 1975, 1978, 1991, 1992, 1994) has provided the longest and most consistent rebuttal to the "nothing works" rhetoric. He initially reanalyzed 82 studies that Martinson had used in his original research, and found that 39 of the evaluations (48 percent) showed some measure of success; thus, using Martinson's own data, the conclusion that "nothing works" seemed to be, at the very least, overstated (Palmer 1978). Palmer (1978, 1992) suggested that evaluations should disaggregate variables such as offender's age and treatment setting to more fully understand the effects of the programs.

Gendreau and Ross (1979, 1987) also rebutted the "nothing works" rationale. They argued that some treatment programs may be poorly implemented or there may be programs with no theoretical basis, but that did not mean that effective programs did not exist. Andrews and Bonta (2003) moved the evaluation research from rebutting the "nothing works" rationale to finding what does work. Their work is often cited for their findings that cognitive-based programming seems to provide the most significant change in individuals. Further, their findings reinforce the commonsense notion that treatment should be individualized to the offender. They found that programs that included the elements they identified as essential in success were able to show up to 30 percent reductions in recidivism (Andrews, Zinger, Hoge, Bonta, Gendreau, and Cullen 1990; Andrews and Bonta 2003).

Box 6-2
Elements of a Successful Treatment Program

1. **It provides specific guidelines for the use of positive reinforcement.** Behavioral programs have been found to be more successful than nondirective or "talking" programs.
2. **It draws from a variety of sources.** An eclectic approach may be the most effective mode of service delivery, in that some elements may work for different types of offenders.
3. **It is heavily scripted.** The value of this element is that it reduces the chance of counselor bias or diminishing program content through counselor apathy or lack of training.
4. **It is based on evaluated results.**
5. **It requires structured activity of the learner.** This principle is consistent with other learning theory that supports the notion that we learn by doing, not by listening or watching.
6. **It requires transfer of training to everyday life.** Programs that have little applicability to the offender's life will be forgotten as soon as the offender is released.
7. **It includes a method of teacher monitoring.** This is to reduce the possibility that the program is made less effective or is changed by the individual counselor.
8. **It contains an outcome evaluation.**
9. **It contains a technique and rationale for client selection.** This is consistent with several studies that indicate that certain types of programs work better for certain types of offenders.
10. **It is repetitive and integrated.**
11. **It requires active participation from the teacher.**
12. **It is constructed for a specific purpose and for a specific client**

Source: Adapted from Coulson G. and Nutbrown V. 1992. "Properties of One Ideal Rehabilitative Program for High Need Offenders." *International Journal of Offender Therapy and Comparative Criminology* 36, 3:203–208.

meta-analysis when a researcher takes the raw data of large numbers of individual studies and collapses them into a giant data set and then reexamines them.

Lipsey and his colleagues (1999; Lipsey and Wilson 1993, 1998; Lipsey, Chapman, and Landenberger 2001) have conducted meta-analyses using more than 400 treatment programs. Meta-analysis is a process whereby the researcher takes the raw data of large numbers of individual studies, collapses the data into one giant data set, and then reexamines the findings. The process they used was more refined than the simple "counting" method of Martinson and his colleagues.

Again, the findings indicated that treatment programs do show significant effects, but also that some programs work better than others.

Cullen and Gilbert (1982) provided an early summary of evaluation studies that showed rehabilitation programs were not uniformly unsuccessful. Cullen (2004) and his collaborators also illustrated the consistent findings that the public has been, and continues to be, generally supportive of rehabilitative efforts. In his recent address to the American Society of Criminology, he identifies the small group of researchers who have consistently shown that there is good evidence to support rehabilitation efforts.

What Does Work?

No general categories of programs—only individual programs—have been proved successful. Palmer (1994) believes that continuing the modest progress seen thus far depends on replicating those successful programs or creating program variations that combine the most successful elements. Several researchers have discovered that behavioral, cognitive-oriented, life-skills programs show some success; and group counseling, individual counseling, and confrontational programs show the least success (Palmer 1994).

Louis and Sparger (1990) reviewed several studies and concluded that counseling procedures that depended primarily on open communication, and friendship models that were nondirectional or involved self-help groups, resulted in negligible effects. Behavior-modification programs, on the other hand, either showed impressive successes or were dismal failures. Others report that individual and small-group counseling that is directive and focuses on sources of a subject's criminality seems to reduce recidivism in incarcerated offenders who seek it voluntarily, but not in those who are coerced to participate (Glaser 1994). Lipsey (1999) and his colleagues found that programs that were punitive or punishment based were least effective (Lipsey, Chapman, and Landenberger 2001; Lipsey and Wilson 1993; Lipsey and Wilson 1998).

What Works for Women?

The vast majority of treatment evaluations do not distinguish programs for men from those for women, nor do they use gender as a variable to understand the effect of the program on the offender. This is unfortunate since it would be logical to assume that programs that work for men may not work as well for women. While the general principles seem to apply—for example, Gendreau's (1996) principles in **Box 6-3** on page 186 make sense for all programs—women seem to have special issues that must be addressed in programming. These special issues include the fact that a large number of female prisoners are survivors of incest and sexual abuse as children. The percentage of female prisoners who share this background is larger than the percentage of women in the general population, and it is larger than the percentage of male prisoners who suffered such abuse. Parenting also poses special problems for women, as does drug abuse, since a larger percentage of women than men in prison seem to have serious drug problems (see Pollock 2002 for a review of these issues).

Only a few studies have directly looked at programs for women, or have looked at female-offender participants separately from male-offender participants. Austin,

Box 6-3

Gendreau's Principles for Effective Treatment

1. Services that are intensive and behavioral in nature
2. Behavioral programs that address the criminogenic needs of high-risk offenders
3. Programs in which contingencies and behavioral strategies are enforced in a firm but fair manner
4. Relationships between therapists and offenders that are interpersonally responsive and constructive
5. Program structure and activities that promote pro-social behavior
6. Relapse-prevention strategies provided in the community to the extent possible
7. Advocacy and brokering services between offenders and the community that are attempted whenever community agencies offer appropriate services

Source: Gendreau 1996. "The Principles of Effective Intervention with Offenders." In A. Harland (Ed.), *Choosing Correctional Options That Work*, p. 129. Thousand Oaks, CA: Sage.

Bloom, and Donahue (1992) looked at community corrections programs for women and identified some promising elements. Koons, Burrow, Morash, and Bynum (1997; also see Morash and Bynum 1995) surveyed correctional professionals to discover what they thought were the most effective programs, but this was not a systematic evaluation of the effect of such programs on recidivism. Bloom (2000) and others have coined the term "gender-responsive programming" to identify the approach whereby treatment programs must move away from a male model to one that meets women's specific needs. They include the following principles:

1. *Focus on the realities of women's lives*
2. *Address social and cultural factors as well as therapeutic interventions*
3. *Provide a strength-based approach to treatment and skill building*
4. *Incorporate a theory of addiction, trauma, and women's psychological development*
5. *Provide a collaborative, multidisciplinary approach.*
6. *Offer a continuity of care.*

However, there has yet to be an evaluation that clearly and comprehensively utilizes gender as a variable in an evaluation of treatment programming.

Conclusions

Partially as a result of the "nothing works" rationale, rehabilitative efforts get a very small share of the correctional budget (most estimate it at about 10 percent) (Senese and Kalinich 1992). Interestingly, there has always been public support for such efforts—even at the height of the "get tough" era, a substantial percentage of the public believed that treatment was happening in prisons, and more importantly, believed it was an appropriate goal of corrections and one that they approved (Durham 1994; Cullen 2004). Arguing that the only appropriate goal of prison is to punish is a stand that can be made, but is not supported by a large portion of the public, nor is it supported by the "nothing works" rationale, since the weight of evidence seems to be in favor of a conclusion that some programs do work for some offenders.

Clearly, prisons are not working to reduce crime. Recidivism hovers at around 67 percent. Furthermore, there are now about 670,000 people leaving prisons every year. Many of those individuals entered prison in the first place because of drug or alcohol problems, or chronic violence. We do have good information regarding what elements are necessary for programs to effectively change an offender's behavior. What is needed is the political will to implement them.

KEY TERMS

AIMS—Adult Internal Management System is a type of personality scale device where a professional completes two checklists (historical and current observations) to determine the inmate's placement in the five offender personality types: asocial aggressive, immature dependent, neurotic anxious, manipulator, and situational.

behaviorism—school of psychology that assumes that an individual's behavior is based on prior rewards; concentration is on behavior since it is believed that thoughts and emotions follow behavior.

classification—categorizing inmates either for prison management (risk) or for offender treatment (needs).

control group—a group similar in every way to the group that is subjected to a treatment, thus any differences that emerge between the two groups after treatment can be more clearly associated with the treatment itself.

defense mechanisms—psychoanalytic term used to describe attempts to deal with childhood trauma, such as denial, projection, or reaction formation.

game—term used in Transactional Analysis to describe a pattern of communication (or transactions) between people.

indeterminate sentences—are those in which the sentence length is not fixed by the court at the time of sentencing and, thus, the inmate's length of confinement is dependent upon good or bad behavior.

internal classification—needs assessment of the offender.

Irish system—Sir Walter Crofton's system in Ireland that allowed for early release of prisoners for good behavior.

mark system—Captain Alexander Maconochie's system for awarding good behavior points to prisoners on Norfolk Island prison colony off the coast of New Zealand.

medical model—under this approach to criminology crime is seen as a symptom of a pathology that can be treated.

meta-analysis—when a researcher takes the raw data of large numbers of individual studies and collapses them into a giant data set and then re-examines them.

MMPI—Minnesota Multiphasic Personality Inventory is a personality inventory survey that is used to assess an individual against norms; various scales can be derived from the results, such as the depression and paranoia scales.

objective classification—a type of classification system that uses instruments such as personality inventories or risk factor scores based on offender backgrounds to determine risk or need.

positivist school—this is the school of criminology associated with Cesare Lombroso; adherents assumed that the cause of crime could be discovered through an empirical study of influences on the individual.

prisonization—the socialization of the prisoner to the prison subculture.

profile—a term associated with Megargee's MMPI-based offender typology and refers to the offender types constructed by using the scores on the 10 standard scales.

random assignment—is when every individual in the population has an equal chance of being placed in the control group or the treatment group; used to ensure that treatment effects are not due to some statistical bias.

rehabilitation—change to a healthier state (in the context of corrections it is internal change in values, attitudes, and behaviors that lead to law-abiding behavior).

risk scale—used in objective classification systems, these scales apply points based on such things as prior convictions, use of a weapon, age at first conviction, drug use history, and so on to determine the level of risk an offender presents.

schemata—basic unstated and unquestioned assumptions in one's thought processes that can be functional or dysfunctional.

stigmata—physical characteristics that Cesare Lombroso believed indicated a criminal predisposition (i.e. moles, birthmarks).

subjective classification—a classification system that would depend on interviews and professionals' subjective assessment of the inmate.

token economy—a program in an institutional setting where residents are rewarded for good behavior through the use of tokens or points which relate to privileges.

transaction—a communication segment between people.

12-step programs—self-help programs that are modeled after Alcoholics Anonymous.

REVIEW QUESTIONS

1. Explain the medical model (make sure you include the history of this concept).
2. Discuss the rise and fall of the rehabilitative ideal. What factors led to its demise?
3. Discuss the many objectives and kinds of classification.
4. Distinguish inmate "activities" from "treatment programs," and provide examples for each.
5. What are "self-help groups," and how are they distinguished from other prison programs?
6. What are therapeutic communities? What are token economies? Distinguish them.
7. Explain the differences between directive therapy and traditional therapy.
8. What are cognitive-behavioral programs?

Chapter Resources

9. What are the issues involved in evaluating treatment programs?
10. What do evaluations show regarding the effectiveness of correctional-treatment programs?

FURTHER READING

Andrews, D., and J. Bonta. 2003. *The Psychology of Criminal Conduct,* 3d ed. Cincinnati, OH: Anderson.
Berne, E. 1964. *Games People Play.* New York: Grove.
Cullen, F., and K. Gilbert. 1992. *Reaffirming Rehabilitation.* Cincinnati, OH: Anderson.
Duguid, S. 2000. *Can Prisons Work? The Prisoner As Subject and Object in Modern Corrections.* Toronto, Canada: University of Toronto Press.
Palmer, T. 1994. *A Profile of Correctional Effectiveness and New Directions for Research.* Albany: State University of New York Press.

REFERENCES

Alcoholics Anonymous. 1976. New York: Alcoholics Anonymous World Services, Inc.
Alexander, R., Jr. 1997. "Juvenile Delinquency and Social Work Practice." In C. A. McNeese and A. R. Roberts (Eds.), *Policy and Practice in the Justice System,* pp. 181–197. Chicago: Nelson-Hall.
Alexander, R., Jr. 2000. *Counseling, Treatment, and Intervention Methods with Juvenile and Adult Offenders.* Belmont, CA: Wadsworth/Thompson Learning.
American Psychiatric Association (APA). 1994. *Diagnostic and Statistical Manual of Mental Disorders,* 4th ed. Washington, DC: American Psychiatric Association.
Andrews, D. 1995. "The Psychology of Criminal Conduct and Effective Treatment." In J. McGuire (Ed.), *What Works: Reducing Reoffending—Guidelines from Research and Practice.* New York: John Wiley.
Andrews, D., and J. Bonta. 2003. *The Psychology of Criminal Conduct,* 3d ed. Cincinnati, OH: Anderson.
Andrews, D., I. Zinger, R. Hoge, J. Bonta, P. Gendreau, and F. Cullen. 1990. "Does Correctional Treatment Work? A Clinically-Relevant and Psychologically-Informed Meta-Analysis." *Criminology* 28: 369–404.
Austin, J., B. Bloom, and T. Donahue. 1992. *Female Offenders in the Community: An Analysis of Innovative Strategies and Programs.* San Francisco: National Council on Crime and Delinquency.
Austin, J., and J. Irwin. 2001. *It's About Time: America's Imprisonment Binge.* Belmont, CA: Wadsworth.
Ayllon, T., and N. Azrin. 1968. *The Token Economy.* New York: Appleton-Century-Crofts.
Berne, E. 1961. *Transactional Analysis in Psychotherapy.* New York: Grove.
Berne, E. 1964. *Games People Play.* New York: Grove.

Bloom, B. 2000. "Gender Responsive Programs and Services." Paper presented at the American Correctional Association meeting, San Antonio, Texas.

Bonta, J., B. Bogue, M. Crowley, and L. Motiuk. 2001. "Implementing Offender Classification Systems: Lessons Learned." In G. Bernfeld, D. Farrington, and A. Leschied, (Eds.), *Offender Rehabilitation in Practice*, pp. 33–63. Chichester, England: John Wiley and Sons.

Brown, B., L. Wienckowski, and S. Stolz. 2000. "Behavior Modification: Perspective on a Current Issue." In P. Kratkoski (Ed.), *Correctional Counseling and Treatment*, 4th ed. Long Grove, IL: Waveland.

Buchanan, R., K. Whitlow, and J. Austin. 1986. "National Evaluation of Objective Prison Classification Systems: The Current State of the Art." *Crime and Delinquency* 32, 3: 272–290.

Burke, P., and L. Adams. 1991. *Classification of Women Offenders in State Correctional Facilities: A Handbook for Practitioners*. Washington, DC: National Institute of Corrections.

Campbell, R. 1996. *Psychiatric Dictionary*, 7th ed. New York: Oxford University Press.

Champion, D. 1998. *Corrections in the United States: A Contemporary Perspective*, 2d ed. Upper Saddle River, NJ: Prentice -Hall.

Coulson, G., and V. Nutbrown. 1992. "Properties of an Ideal Rehabilitative Program for High Need Offenders." *International Journal of Offender Therapy and Comparative Criminology* 36, 3: 203–208.

Cullen, F. 2004. *The Twelve People Who Saved Rehabilitation: How the Science of Criminology Made a Difference*. Paper presented at the American Society of Criminology, 2004 Presidential Address, Nashville, Tennessee.

Cullen, F., and K. Gilbert. 1982. *Reaffirming Rehabilitation*. Cincinnati, OH: Anderson.

Duguid, S. 2000. *Can Prisons Work? The Prisoner As Subject and Object in Modern Corrections*. Toronto, Canada: University of Toronto Press.

Durham, A. 1994. *Crisis and Reform: Current Issues in American Punishment*. Boston: Little, Brown.

Ellis, A., and W. Dryden. 1997. *The Practice of Rational Emotive Behavior Therapy*, (2d ed.). New York: Springer.

Fodor, I. 1988. "Cognitive Behavior Therapy: Evaluation of Theory and Practice for Addressing Womes's Issues." In M. Dutton-Douglas and L. Walker (Eds.), *Feminist Psychotherapies: Integration of Therapeutic and Feminist Systems*. Norwood, NJ: Ablex.

Freeman, A. 1983. "Cognitive Therapy: An Overview." In A. Freeman (Ed.), *Cognitive Therapy with Couples and Groups*, pp. 1–9. New York: Plenum Press.

Gendreau, P. 1996. "The Principles of Effective Intervention with Offenders," In A. Harland (Ed.), *Choosing Correctional Options That Work*, pp. 117–130. Thousand Oaks, CA: Sage.

Gendreau, P., and R. Ross. 1979. "Effective Correctional Treatment: Bibliotherapy for Cynics." *Crime and Delinquency* 25: 463–489.

Gendreau, P., and R. Ross. 1987. "Revivification of Rehabilitation: Evidence for the 1980s." *Justice Quarterly* 4, 349–407.

Genova, P. 2001. "There Are Only Three Kinds of Psychotherapy." *Psychiatric Times* 18, 11. Retrieved on March 1, 2005 from http://www.psychiatrictimes.com/p011140.html.

Glaser, D. 1994. "What Works, and Why It Is Important: A Response to Logan and Gaes." *Justice Quarterly* 11, 4: 711–723.

Glasser, W. 1965. *Reality Therapy: A New Approach to Psychiatry.* New York: Harper and Row.

Hall, S. 2003. "Faith-Based Cognitive Programs in Corrections." *Corrections Today* 65, 7: 108–116.

Hannigan, J. 2004. "Gov. Bush Announces First Women's Faith-Based Prison." *Florida Baptist Witness,* 122, 5 (April 29).

Harris, P. 1988. "The Interpersonal Maturity Level Classification System: I-level." *Criminal Justice and Behavior* 15, 1: 58–77.

Harris, T. 1969. *I'm OK—You're OK.* New York: Avon.

Jesness, C., and R. Wedge. 1983. *Classifying Offenders: The Jesness Inventory Classification System.* Sacramento: California Youth Authority.

Koons, B., J. Burrow, M. Morash, and T. Bynum. 1997. "Expert and Offender Perceptions of Program Elements Linked to Successful Outcomes for Incarcerated Women." *Crime and Delinquency* 43, 4: 512–532.

Levinson, R. 1988. "Development in the Classification Process." *Criminal Justice and Behavior* 15: 24–38.

Lipsey, M. 1999. "Can Intervention Rehabilitate Serious Delinquents?" *Annals of the American Academy of Political and Social Science* 564 (July): 142–166.

Lipsey, M., G. Chapman, and N. Landenberger. 2001. "Cognitive Behavioral Programs for Offenders." *Annals of the American Academy of Political and Social Science* 578 (November): 144–157.

Lipsey, M., and D. Wilson. 1993. "The Efficacy of Psychological, Educational, and Behavioral Treatment: Confirmation from Meta-Analysis." *American Psychologist* 48: 1181–1209.

Lipsey, M., and D. Wilson. 1998. "Effective Intervention for Serious Juvenile Offenders: A Synthesis of Research." In R. Loeber and D. Farrington (Eds.), *Serious and Violent Juvenile Offenders: Risk Factors and Successful Interventions,* pp. 3–19. Thousand Oaks, CA: Sage.

Lipton, D., R. Martinson, and J. Wilks. 1975. *The Effectiveness of Correctional Treatment: A Survey of Treatment Evaluation Studies.* New York: Praeger.

Louis, T., and J. Sparger. 1990. "Treatment Modalities Within Prison." In J. Murphy and J. Dison (Eds.), *Are Prisons Any Better? Twenty Years of Correctional Reform,* pp. 147–161. Newbury Park, CA: Sage.

Martinson, R. 1974. "What Works?—Questions and Answers About Prison Report." *Public Interest* 35: 22–54.

Martinson, R. 1979. "New Findings, New Views: A Note of Caution Regarding Sentencing Reform." *Hofstra Law Review* 7, 2: 243–258.

Masters, R. 1994. *Counseling Criminal Justice Offenders.* Thousand Oaks, CA: Sage.

Megargee, M. 1994. "Using the Megargee MMPI-Based Classification System with MMPI-2s of Male Prison Inmates." *Psychological Assessment* 6, 4: 337–344.

Megargee, E., and M. Bohn. 1979. *Classifying Criminal Offenders: A New System Based on the MMPI.* Beverly Hills, CA: Sage, American Psychiatric Association.

Miller, S., C. Bartollas, D. Jennifer, E. Redd, and S. Dinitz. 1974. "Games Inmates Play: Notes on Staff Victimization." In I. Drapkin and E. Viano (Eds.), *Victimology: A New Focus; Volume V Exploiters and Exploited: The Dynamics of Victimization*, pp. 143–155. Lexington, MA: Lexington Books.

Mitford, J. 1973. *Kind and Usual Punishment: The Prison Business*. New York: Alfred A. Knopf.

Morash, M., and T. Bynum. 1995. *Findings from the National Study of Innovative and Promising Programs for Women Offenders*. Washington, DC: U.S. Department of Justice.

Muller, U., and K. Tudor. 2002. "Transactional Analysis As Brief Therapy." In K. Tudor (Ed.), *Transactional Analysis Approaches to Brief Therapy*, pp. 19–44. London: Sage.

Palmer, T. 1972. "The Youth Authority's Community Treatment Project." *Federal Probation* 38, 1: 3–14.

Palmer, T. 1975. "Martinson Revisited." *Journal of Research in Crime and Delinquency* 12: 133–152.

Palmer, T. 1978. *Correctional Intervention and Research: Current Issues and Future Prospects*. Lexington, MA: Lexington Books.

Palmer, T. 1991. "The Effectiveness of Intervention: Recent Trends and Issues." *Crime and Delinquency* 37: 330–346.

Palmer, T. 1992. *The Re-Emergence of Correctional Intervention*. Newbury Park, CA: Sage.

Palmer, T. 1994. *A Profile of Correctional Effectiveness and New Directions for Research*. Albany: State University of New York Press.

Pollock, J. 2002. *Women, Prison and Crime*. Belmont, CA: Wadsworth/ITP.

Pollock, J. 2004. *Prisons and Prison Life: Costs and Consequences*. Los Angeles: Roxbury.

Price, J. 2003. "Where Punishment Must Fit the Faith." *Washington Times* (December 25, 2003): B5.

Quay, H. 1983. *Technical Manual for the Behavioral Classification System for Adult Offenders*. Washington, DC: U.S. Department of Justice.

Quay, H. 1984. *Managing Adult Inmates: Classification for Housing and Program Assignments*. College Park, MD: American Correctional Association.

Senese, J., and D. Kalinich. 1992. "Activities and Rehabilitation Programs for Offenders." In S. Stojkovic and R. Lovell (Eds.), *Corrections: An Introduction*, pp. 213–244. Cincinnati, OH: Anderson.

Singer, S. 1996. "Essential Elements of the Effective Therapeutic Community in the Correctional Institution." In K. Early (Ed.), *Drug Treatment Behind Bars: Prison Based Strategies for Change*, pp. 75–88. Chicago: Praeger.

Trotzer, J. 2000. "The Process of Group Counseling." In P. Kratkoski (Ed.), *Correctional Counseling and Treatment*, 4th ed, pp. 489–515. Long Grove, IL: Waveland.

Van Voorhis, P. 2000. "An Overview of Offender Classification Systems." In P. Van Voorhis, M. Braswell, and D. Lester (Eds.), *Correctional Counseling and Rehabilitation*, 4th ed., pp. 81–108. Cincinnati, OH: Anderson.

Walsh, A. 1988. *Understanding, Assessing, and Counseling the Criminal Justice Client*. Pacific Grove, CA: Brooks/Cole.

Warren, M., and the Staff of the Community Treatment Project. 1966. *Interpersonal Maturity Level Classification: Diagnosis and Treatment of Low, Middle, and High Maturity Delinquents.* Sacramento: California Youth Authority.

Wexler, D. 1973. "Token and Taboo: Behavior Modification, Token Economies, and the Law." *California Law Review* 61: 81–109.

Wubbolding, R. 1988. *Using Reality Therapy.* New York: Harper and Row.

Prison Authority and Prisoner Rights

III

Master of the Messhall

He strides imperiously to the serving line,
His serving line, a baton in his ham-sized
Hand, tapping it lightly against his leg,
Twisting it slightly, with a hint of menace.

. . .

"Hand back that snack, Jack"
he says to a con
who growls
lifts a leg
moves on.

"Easy on the steak, Jake"
another server busted
bold as a brass monkey
adjusting his plastic hair net
shamelessly, like nothing
happened.

"Put down the juice, Bruce"
You've got too much meat to eat
Too much garden in your salad
Too much starch for your march

He's got it goin' on now, talkin'
Smooth as butter, moving
Slow as molasses
Takin' names and
Kickin' asses.

. . .

Only he knows
How thin the veneer
How fragile the façade
That gets him through each day
Petty ain't pretty
But order matters

Let down your guard
And some cons will
Eat you alive.

Robert Johnson

7

Correctional Staff and Management

Kelly Cheeseman
Sam Houston State University

Chapter Objectives

- Be aware of the general demographics of correctional staff hiring practices.
- Be able to articulate the issues regarding correctional training.
- Understand the elements of correctional officer stress and coping strategies.
- Distinguish between program staff and custodial staff.
- Discuss the management styles of corrections and how they can be improved.

Staff members in a correctional institution are organized in a hierarchy— the warden or superintendent sits at the top, and assistant wardens down through sergeants compose the middle rungs, with correctional officers on the lowest tier. While each prison differs slightly, prison staff most often includes the following:

1. *Administration—wardens, superintendents, assistant wardens, and other managers who run the prison*
2. *Programming—medical doctors, nurses, counselors, psychologists, caseworkers, and ministers (this category could also include contract employees)*
3. *Maintenance/facilities—physical-plant supervisors, work-crew supervisors, food-service managers, industry supervisors*
4. *Custodial supervisory staff—captains, lieutenants, sergeants*
5. *Line staff—correctional officers*
6. *Volunteers—prisons ministry, Alcoholics Anonymous / Narcotics Anonymous sponsors*

This chapter will provide an examination of the individuals who work within the prison environment.

The Correctional Officer

The correctional officer, more so than any other employee in the penal environment, is in constant contact with offenders. The correctional officer's primary function is custody and control of inmates. More simply, officers serve to protect the public by keeping offenders secure and controlled. Many nicknames have been used to describe custody staff—including "hacks," "screws," "turnkeys," "keepers," "guards," and "Bossman" or "Bosslady." The American Correctional Association (ACA) has approved a resolution to discontinue the use of the term "prison guard," utilizing the "correctional officer" title instead (American Correctional Association 1993). This coincides with the shift in correctional ideologies. The "old philosophy" of dealing with inmates became obsolete as they began to win lawsuits and were granted rights through judicial intervention.

Under the old philosophy, any regulation issued by the prison system was deemed valid. As we will see in the next chapter, prisoners had little if any rights and were virtually "slaves of the state." Correctional officers had broad discretion and handled inmates as they sought fit, often employing a trip to the "**hole**" (segregation) as a means to gain compliance. Chapter 8 will describe the court holdings through the 1960s and 1980s that identified prisoners' rights and created safeguards against inhumane treatment. The prison guard of the past either adapted to the changes or made way for the new and "improved" correctional officer of today (Philliber 1987).

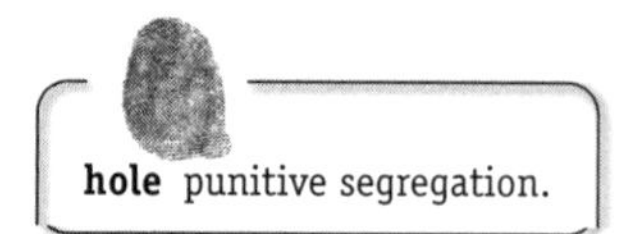
hole punitive segregation.

Training: The Transformation from Citizen to CO

Correctional training academies are a relatively recent phenomenon. Historically, officers were just given a set of keys and told to guard a cell block (Levinson 1982). In 1967, the The Law Enforcement Assistance Administration (LEAA) concluded that correctional officers were generally undereducated, untrained, and unversed in the goals of corrections (Levinson). In the late 1970s, the ACA's Commission for Accreditation for Corrections detailed the first set of training standards for correctional officer training (Josi and Sechrest 1998). The ACA specified the number of hours necessary for completion of an academy, mandating that 120 hours be spent on a "para-military" academy (Bales 1997). The total hours of academy training courses have been steadily increasing. Camp and Camp (1998) found that correctional agencies in the late 1990s were spending an average of 229 hours training at preservice academies.

Training should serve three main purposes. First, officers who have received proper training are often better prepared to act decisively when encountering a broad range of situations. Second, training in any organization leads to increased effectiveness and productivity. Third, a good training program will foster unity and cooperation (Josi and Sechrest 1998). Correctional agencies should and are making an effort to professionalize their workforce and offer training that will help correctional officers become effective in the prison environment. **Table 7-1** reveals the 10 most common areas of correctional officer preservice training.

Cheeseman (2004) examined and divided trainees into categories using the dramaturgical approach utilized by Goffman (1959). In this typology, the focus is on the "image" presented by the individual. Based on her observations, three general groups of people who attend correctional academies were identified:

1. ***"Curiosity crowd/end of the liners."*** These trainees had either no other job options or came to the academy because they were between jobs and or had time and wanted to gain "insider" information into the prison system. Typically, the "end of the liners" are people who have been laid off from other jobs or, due to a change in life circumstances (for example, divorce or death), must now work.
2. ***"In the blood."*** This category of individuals seem to be instinctively made for prison work. Even though they share similarities with those in the "curiosity crowd," they differentiate themselves by their natural aptitude and attitude.
3. ***"Former law enforcement."*** This group consists of those who had previously held jobs in corrections or law enforcement. A large portion of former military personnel also fit into this category. These individuals had "real" experience and "did their time" at the academy, although often complaining about the impracticality of following policies in the training manuals.

States vary slightly on their requirements to become a correctional officer. Requirements for employment most generally include a high school diploma or GED, and that the applicant be at least 18 years old, with no felony convictions, drug charges, or domestic violence convictions. Additionally, employees must be able to speak, hear, climb stairs, sit, stand, and walk.

On-the-Job Training (OJT)

While few would argue that the formal training is unnecessary, or is not a key component in becoming a correctional officer, the real test involves the trainees becoming immersed in the prison society. The prison environment is far removed

Table 7-1 10 Most Common Preservice Training Courses

Class	Percentage
Firearms	97
Housing and Body Searches	93
Searching for Contraband	92
Report Writing	88
Facility Rules and Regulations	85
Self-Defense	84
Key/Tool Control	78
General Safety	68
Riot Control	67
CPR	66

Source: Champion, D. 2005. *Corrections in the United States: A Contemporary Perspective.* Englewood Cliffs, NJ: Prentice Hall.

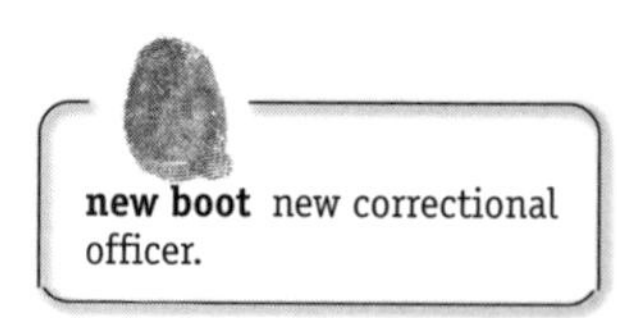
new boot new correctional officer.

from the norms and values that operate in the "free world." The "**new boot**" must adjust to the sounds and smells that are unique to the institution. As noted by Crouch and Marquart (1990, 273):

> *The ghetto-like atmosphere of a maximum-security prison is quite overpowering to the uninitiated. Perhaps as many as 2000 men live, eat, work, urinate, sleep, and recreate in a very limited concrete steel building. This concentration of life presents the new guard with an unfamiliar and at the very least distracting sensory experience as simultaneously he hears doors clanging, inmates talking or shouting, radios and televisions playing, and food trays banging; he smells an institutional blend of food, urine, paint, disinfectant and sweat. What he sees is a vast array of inmate personalities portrayed by evident behavior styles.*

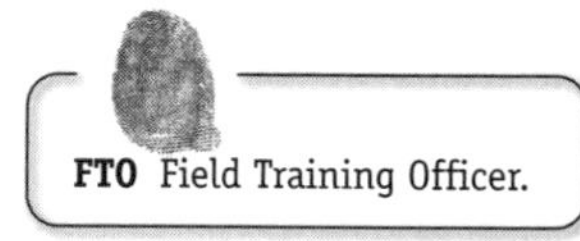
FTO Field Training Officer.

Many states are now assigning field-training officers or mentors to help new correctional officers with the transition. Field-training officers (**FTOs**) guide new officers and help them integrate the classroom instruction they received in the academy into more-practical and more-usable techniques of managing offenders. FTOs are sometimes assigned by demographics—for example, a female FTO is assigned to a female trainee, and a male FTO to a male trainee. This practice can be very important to the new female officer working at an institution for male offenders, where female officers are often faced with challenges and issues that their male counterparts are not.

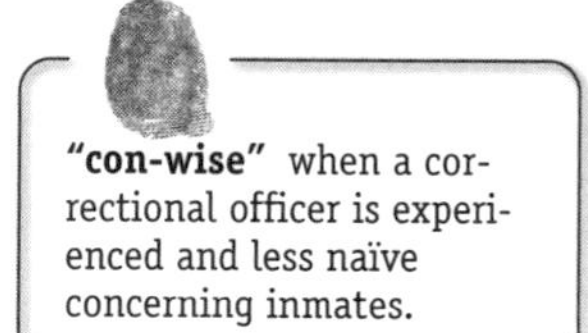
"con-wise" when a correctional officer is experienced and less naïve concerning inmates.

Traditionally, both offenders and officers test the rookie to figure out the type of officer he or she might be. Game playing is not uncommon, and inmates may drop large objects as an officer walks past on the tier. Inmates might also masturbate in view of officers to "shock" or elicit a response from the officer. Senior correctional staff might also send new recruits on missions to pick up nonexistent building reports or take a running count of all sewer drains. Through this process, officers are expected to become "**con-wise**." The officer should gain an understanding of the culture and expectations of the inmates, and develop a means of interacting with the offenders that is consistent with other officers (Webb and Morris 2002).

The Base of Correctional Officer Power

Sykes (1958) believed that the power of the correctional officer could be corrupted, and noted the "defects of total power." Sykes contended that it was impossible to control large numbers of inmates and meet the demands of the officer's supervisors without being co-opted by the inmates. An officer might overlook small or minor rule infractions to gain compliance in the other, more serious areas, such as control of dangerous contraband or fighting. Sykes (1958) argued that the officer had little or no power other than that which the convicts granted to him.

Hepburn (1985) identified five bases of correctional officer power:

1. **Legitimate power** (the formal authority to take charge over inmates). Hepburn noted (1985, 146), that ". . . the prison guard

has the right to exercise control over prisoners by virtue of the structural relationship between the position of the guard and the position of the prisoner."

2. **Coercive power** (the ability to punish). Coercion is at the heart of prison environment and is sometimes used as a method of control by correctional officers.
3. **Reward power** (using rewards to gain compliance). Correctional officers control work assignments, housing locations, and access to recreation and commissary. While the aforementioned are considered to be formal rewards, correctional officers may also reward prisoners by overlooking minor rule infractions or granting a prisoner a "favor."
4. **Expert power** (special abilities, skills, or expertise). An example of expert power could be an industry foreman who is skilled in carpentry or plumbing. When the foreman tells an inmate to complete a task, the inmate may do so because he believes that the foreman is more knowledgeable than he is and the inmate respects the foreman's expertise.
5. **Referent power** (respect). Hepburn explained this as "persuasive diplomacy," a leadership style that gains compliance because the prisoners respect and admire the CO.

Hepburn (1985) believed that of the five bases of power, officers relied most heavily upon legitimate and expert power to gain compliance from inmates. Coercive and reward power were found to be the least effective means of control. Rewards are often given out unfairly, dependent upon a particular officer's biases and dislikes. A prisoner might face harsh punishments for their action from one officer, while the same action carried out with another correctional officer might gain a reward.

"The Job": Correctional Officer Duty Positions

In order for a correctional institution to operate successfully, correctional officers work in a variety of duty posts and jobs. Lombardo (1989) divided the job assignments of officers into seven categories:

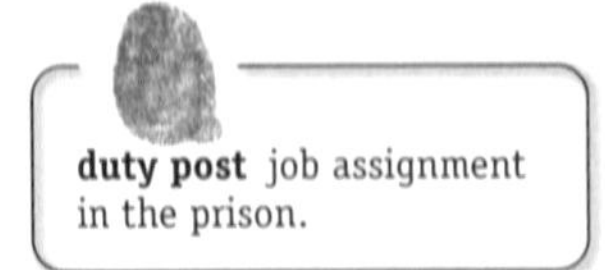

duty post job assignment in the prison.

1. Cell-block officers
2. Work-detail supervisors
3. Industrial-shop/school officers
4. Yard officer
5. Administrative-building assignments
6. Perimeter security
7. Relief officers

An overview of these job posts and job descriptions can be found in **Table 7-2** on page 202.

Some correctional agencies, such as the Federal Bureau of Prisons (BOP), rotate officers every three months through a process of bidding for job posts by

Table 7-2 Duty Assignments and Job Descriptions

Duty Assignment	Job Description
Cell-Block Officer (dormitories, cell blocks, and administrative segregation)	■ Supervise inmates in housing areas ■ Conduct counts of all inmates ■ Ensure orderly movement of prisoners ■ Search inmates and cell for contraband ■ Handle issues and confrontations in the cell-block area
Work-Detail Supervisor	■ Oversee workers in and outside institution ■ Maintain control of tools and supplies ■ Maintain running count of all inmates on detail
Industrial Shop/School Officers	■ Oversee inmates learning trades or attending classes ■ Take attendance and provide protection to instructors ■ Maintain order in shop or classroom area
Yard Officer	■ Supervise inmates on the prison yard ■ Supervise inmates moving from one area to another ■ Assist in feeding of inmates ■ Coordinate and observe inmate recreation
Administrative-Building Assignments	■ Control keys ■ Maintain facility armory and weapons ■ Supervise inmate visitation ■ Work in positions that have high levels of contact with the public
Perimeter-Security Officers (picket or wall-post officers)	■ Assigned to security towers and perimeter patrols ■ Main responsibility is to prevent escapes ■ Search for contraband dropped off around prison unit
Relief Officers	■ Replace officer who are sick or on vacation ■ Usually able to work any or all duty positions

Source: Cheeseman, K. 2004.

seniority. Federal correctional officers must also work one three-month rotation of "sick and annual," in which their schedule rotates weekly based upon vacation and sick leave. The Texas Department of Criminal Justice–Corrections Institutions (TDCJ-CI), in contrast, rotates its officers daily as duty posts are assigned by the shift lieutenant or captain. Officers may request training to work in "specialty" positions such as perimeter-security team and control picket. Many job positions are highly sought after and are selected through an application process (for example, education officer and craft-shop supervisor). Both the California Department of Corrections and the New York State Department of Correctional Services have a **bid system** that is based on seniority. Officers must post a bid on a duty post. Once having obtained the desired job, the officer may stay in that post until he or she retires or bids for a new duty assignment. It has been suggested that a system of rotation, be it daily or quarterly, assists in preventing officers, from becoming

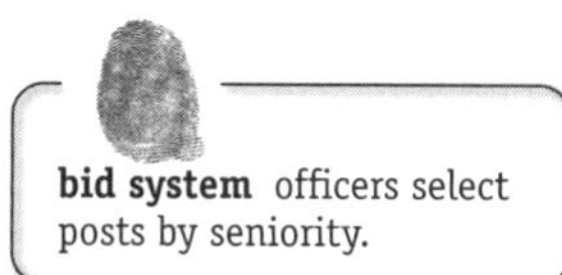

bid system officers select posts by seniority.

too friendly with offenders, and can also break up the boredom and monotony that have often been attributed to correc-tional work.

Conversely, however, shift work can be a nightmare for people who have families or for single mothers with small children. Rotating shifts often affect entire families. Work and child-care schedules may be in a state of flux, particularly for officers with little or no seniority. Shift work may also have negative effects on the body because of sleep deprivation. Grosswald (2003) listed six items she considered to be symptoms of **shift lag**:

shift lag physiological effects of shift work, which include impaired performance.

1. Impaired performance
2. Irritability
3. Gastrointestinal dysfunction
4. Depression and apathy
5. Sleepiness/sleeping at work
6. Sleep disruption during daytime sleep

Grosswald (2003) concludes that women are particularly susceptible to health problems due to shift work—including spontaneous abortion, cardiovascular morbidity, low birth weight, and preterm birth.

The Correctional Officer Subculture

There has been a great deal of research on the subculture of the police officer and its effect on supervisors, police officers, and the public. Research has also been conducted to see if there is a similar subculture amongst the "keepers." Kauffman (1988) concluded that correctional officers held a similar set of beliefs that made them distinct and unique from treatment staff and prison administrators. Kauffman (1988, 86) identified nine norms that constitute the correctional officer code:

1. Always go to the aid of an officer in distress
2. Don't "lug" drugs (bring them into the institution for an inmate to use)
3. Don't rat on another officer
4. Never make a fellow officer look bad in front of the inmates
5. Always support an officer in a dispute with an inmate
6. Always support officer sanctions against inmates
7. Don't be a "white hat" or a "Goody Two-shoes"
8. Maintain office solidarity versus all outside groups
9. Show positive concern for fellow officers

A correctional officer subculture is important to the socialization of officer trainees. The recruits observe and imitate senior officers. Crouch and Marquart (1990) assert that the subculture influences a new officer in the following areas:

1. **How to perceive inmates**—Traditionally, prisoners are viewed as the "enemy" or as "nonhuman." The correctional officer's main job is to ensure security and enforce rules and regulations. While officers have different styles of dealing with offenders, most officers strongly dislike inmates as a whole, and may often look for opportunities to "screw over" an inmate. Prisoners, regardless of their

actions, are not to be trusted. Most new officers learn these attitudes from experienced correctional officers.

2. How to anticipate trouble—Trouble can come from any prisoner at any time, but more often than not, there are signs that problems are developing. A change in noise in the cell-block area, whether it be extremely quiet when normally loud or very loud when usually quiet. An inmate who refuses to be searched could also be a sign of "trouble." The senior officer conveys this type of information to new officers.

3. How to manage inmates—As noted above, officers handle prisoners differently. Women may often use more psychological pressure or techniques to gain compliance, whereas men might use physical force. The FTO may offer strategies for dealing with offenders. Offenders are not carbon copies of each other, and senior officers often help new officers understand how to deal with mental-health inmates as well as with inmates of other races. The subculture as a whole encourages as little interaction as possible. For instance, an FTO might encourage female officers to abide by the 30-second rule. If offenders cannot convey their request in 30 seconds or less, their discussion could possibly be an attempt to engage in staff manipulation (Cheeseman 2004). The most frequently heard advice given by correctional officers to trainees is to tell an inmate "NO." If you are not certain of the answer, always say no, and then find out the answer later.

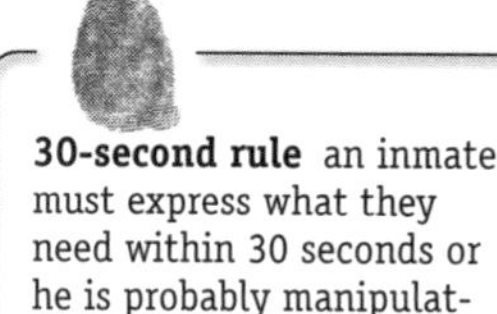

30-second rule an inmate must express what they need within 30 seconds or he is probably manipulating the officer.

One part of the correctional officer subculture is the use of slang words that describe many aspects of the job. **Table 7-3** provides an overview of correctional officer slang terms. These terms are indicative of southern correctional systems, and might not be utilized in other regions of the country.

Table 7-3 Correctional Officer Slang

Correctional Officer Slang Term	Definition
Act a fool	When an inmate is acting crazy or is out of line from their usual behavior
Aggie	A tool used in field work, commonly called a hoe
BOSS Bossman/Bosslady	A term used extensively by inmates to refer to officers working as guards (correctional officers), this began in the early years of penitentiaries as "Sorry son of a bitch" backward. Most of the inmates that are now incarcerated are not aware of this, and the term is accepted by officers
Jiggers	A term used by officers or inmates to announce the presence of a supervisor
Cell Warrior	Someone who loudly runs his mouth while safely locked in his cell, but is a coward once the cell door pops open
Bean Slot	An opening in the cell door of administrative segregation where an officer can deliver food or handcuff the inmate prior to opening the cell
Catch Out	Any person, whether it be inmate or officer, that could not handle the pressure of any area, and who left for this reason
Chain	A term meaning that an inmate is leaving (catching the chain); or the bus used to carry inmates (chain bus)
Cut Your Eyes	Meaning that a person has looked at another person or at the items they have through the sides of the viewer's eyes. Normally thought of as intent to steal the items or start a fight

Table 7-3 Correctional Officer Slang, continued

Correctional Officer Slang Term	Definition
Dropped	When an officer forcibly wrestles an inmate to the ground to be restrained
Eyeball	When someone is staring at you or at your things, they are said to be eyeballing you
Fishing Line	Made from torn sheets or string, having a weighted object tied to one end, and used to throw down the run to inmates in other cells to pass items
Free World	Life outside the prison ("When I was in the free world"); also called the world
Got Down	When one person has said or done something to another to make the other person look foolish, whether the statement is true or not, the person making the statement is said to have got down
Hog	To hog is to forcibly take property from somebody without that person's consent. Normally, there is not a fight involved, and the person allows the other to take the property out of fear. Also used as a way to describe getting one over on someone (I sure "hogged" him)
House	How most inmates refer to their living quarters: cell, dorm, cubicle, etc.
Jack book	Magazines or books depicting naked women: *Playboy, Penthouse,* etc.
Johnnie	A sack lunch
Kill	To masturbate
Killer	An inmate who is constantly masturbating in public
Kite	A letter sent from one inmate to another by using a line or fishing pole
Lay-In	When an inmate has an appointment to see somebody—for example, correctional counselor or doctor—and for this reason is excused from work for that day or period of time. Also when a doctor has given an inmate a pass to stay out of work for a period of time
Lay It Down	When an inmate quits working without authorization and refuses to go back to work
Look Out	Yelled out in a group of people to get all of them to look when you are only trying to get the attention of one of them whose name is unknown
On the Cool	When something is done quietly so that others do not become aware of what is going on, it is being done on the cool
Rack	An inmate's bunk
Shakedown	A search: cell search, pat search, or strip search for contraband
Shank	A homemade stabbing weapon made from any hard object that can be sharpened—derived from the fact that these weapons used to be made from the metal part inside the sole of a shoe, known as a shank
Slammed	When an inmate has to be forcibly wrestled to the ground and restrained by correctional officers
Sniper	An inmate who hides in corners and masturbates while looking at female employees or other inmates
Snitch	An inmate or officer who gives to the administration information about other inmates concerning drugs, contraband, or other rule infractions in hopes of receiving preferential treatment as a result
Spread	When an inmate or several inmates pool together to buy several items of food from the prison store, combine the foods into a large bowl, and make sandwiches from the mixture

continued

Table 7-3 Correctional Officer Slang, continued

Correctional Officer Slang Term	Definition
Stuck Out	To be too late to attend a specific function—showers, meals, recreation, work—to the point that the inmate will not be allowed by correctional staff to catch up to them, and must stay in his living area
Swole	When an inmate is angry, he is said to be swole
Tank	A dormitory-style inmate living quarters, where inmates can move about freely until rack time
Turn Out	Shift meeting for officers, and when a squad of inmates prepares for work

Source: Cheeseman, K. 2004.

The correctional officer subculture has both positive and negative effects. The subculture can help other correctional officers find support in a job and environment full of stress and pressure. Few people understand the world of the correctional officer the way other correctional officers do. Unfortunately, the subculture also promotes negative behavior. The subculture may encourage officers to act in ways that violate their personal beliefs, thus causing inner conflict and strife. The subculture might also encourage negative coping mechanisms, such as excessive drinking (Cheeseman 2004).

It could also be argued that the correctional officer subculture has changed as the corrections world itself has changed. With a workforce that is more diversified, including the integration of women and minorities, there are more ideas and conceptions of what a "good" correctional officer might be. Much as other subcultures adapt and change as the sentiments and values of larger society have evolved, so, too, must the correctional officer subculture adapt and change.

Minority and Female Officers

Traditionally, correctional officers have been white men. In the late 1970s, inmates became increasingly hostile to an all-white correctional officer force that supervised a prison population that was disproportionately black or other minority. **Figure 7-1** illustrates the percentage of female and male officers in prisons.

The prison riot at Attica involved inmates that asked, as one of their demands, to be supervised by a staff that included black and minority officers. Johnson (1995, 97) stated that "a diverse correctional staff can enhance communication between staff and inmates, add to the cultural awareness of staff, decrease racial tension, provide positive role models and enhance a department's public image." There are now more minorities and women working in corrections. Camp, Saylor, and Wright (2001) found that white male officers and minority male officers have different attitudes toward their employment with the Federal Bureau of Prisons. More research is needed to determine how minority officers perceive their role and whether they have any unique stressors compared to white officers.

Women were originally hired to work in men's prisons because they could search female visitors, among other reasons. Eventually, with court challenges, women were allowed to work at all duty posts in all prisons. In 2004, there were

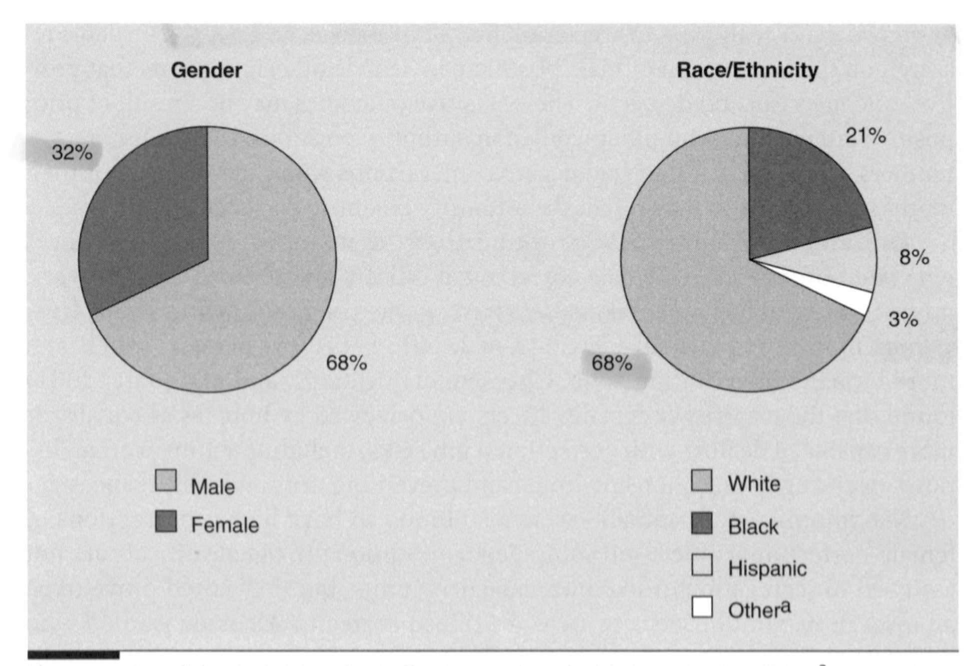

Figure 7-1 Custodial and Administrative Staff at State and Local Adult Correctional Facilities. [a]Asian, Native American, Eskimo, Aleut. ***Source:*** *American Correctional Association 2000.*

9,387 female officers among the total 24,134 officers in the Texas correctional system (about 39 percent). While few states have as many officers, many have roughly the same percentage of female officers of the total (Texas Department of Criminal Justice 2005).

Research has shown that women have a calming effect on male inmates (Morton 1981). In the early phases of confrontations, as tension rises, the equilibrating effect women bring to prison interactions may diffuse potentially violent situations. When female correctional officers are given appropriate and adequate training and placed in regular custodial assignments, they not only carry out their duties as well as male officers, but add ingredients of female skills, interest, and concern that have markedly improved the general atmosphere of the prison environment (Potter 1980). One additional study conducted in the early 1980s noted that the presence of women had also lowered the level of aggression and had improved hygiene and language within the institution (Breed 1981, 40).

In order for female correctional staff to succeed, they need to analyze their reasons for wanting to take on this type of work, and they need to be certain of who they are as individuals. Ethridge, Hale, and Hambrick (1984) offered specific coping strategies and techniques for female correctional staff in overwhelmingly male settings. Some of their suggestions included that women interact with inmates in a straightforward manner with consistency, dress appropriately, maintain a professional distance from inmates, deal with conflict decisively, and build positive relationships with male coworkers.

Female staff also have the opportunity to offer male and female inmates a positive image of women. Having to interact with women in positions of authority may induce male offenders to interact with women in a more prosocial manner. Instead of considering female correctional officers as dependent, dumb, or sex objects, male prisoners may see them as examples of integrity, intellect, and ability. Available research on the supervision of male prisoners by female officers suggests that positive outcomes can and do occur. These positive outcomes may be a result of prior positive interactions with other women in authority positions, such as teachers or mothers. Conversely, if they had negative interactions with their teachers or their mother, male inmates may resent the authority of female correctional officers.

Research by Zimmer (1986) found that the majority of male prisoners who were supervised by female correctional officers felt no invasion of privacy, no resentment at having to take orders from women, and little or no sexual frustration. Inmates may also like having female officers because of their "softer" and more humane intervention style. Cheeseman, Mullings, and Marquart (2001) found that the majority of female officers are perceived by inmates as equally or more capable of dealing with correctional job tasks, including settling verbal disputes, working in male housing units, and preventing riots and disturbances.

Maximum-custody inmates were also found to have higher perceptions of female-correctional officer job competency. Traditionally, female officers are not assigned to segregation or maximum-security units, but they could prove to be an asset in maximum-security areas if utilized correctly. A female warden who once oversaw San Quentin and Soledad came to believe that a maximum-security prison was safest when women were working some duty posts (Hunter 1993). The female warden noticed a more relaxed demeanor among the inmates, and noted that female officers were less likely to be physically or verbally aggressive when dealing with prisoners. Many authors have noted that women tend to "normalize" the prison environment and make the prison experience more tolerable. Of paramount importance is that female officers receive adequate training. Effective training and good administration can counteract many problems.

Female correctional officers can be an integral and viable part of the correctional workforce. Unfortunately, recent research has indicated that some women may be manipulated by male inmates into having inappropriate sexual relationships, which creates a threat to prison security and officer safety (Worley, Marquart, and Mullings 2003). Although this does occur, it should not be considered a compelling reason for refusing to hire female correctional officers.

The flip side of having female officers in prisons for men is, of course, having men guard women in prison. Seventy percent of correctional officers who guard female inmates are men, which can create a highly sexualized environment (Calhoun and Coleman 2002). A simple solution is to end **cross-supervision** for both sexes—allowing only male correctional officers to work with male inmates, and female correctional officers with female inmates. However, this solution infringes upon the rights of both sexes to have the equal opportunity to work in any of the state's prisons, and punishes the vast majority of officers who do their job without violating the rules. In recent years, female prisoners have been able to restrict supervision by male correctional officers by showing a history of abuse or psychological harm, illustrated by the recent case *Everson v. Michigan Department of Corrections* (222 F. Supp. 2d 864 [2002]).

> **cross-supervision** when male officers supervise female inmates or female officers supervise male inmates.

The National Institute of Corrections (NIC) offers a 36-hour course specifically dealing with issues that relate to the supervision of female offenders, including cross-gender supervision. The Arizona Department of Corrections has a

4-hour training block that covers issues of cross-gender supervision. States are not mandated to provide cross-supervision training, although it is an essential element in creating a safe and effective prison environment. With the passage of the Prison Rape Elimination Act of 2003, one of the recommendations to the standards committee could be the inclusion of cross-supervision training in all correctional officer-academy courses. The issue of cross-sex supervision will be addressed again in Chapters 8 and 10.

Stress and the Correctional Officer

stress tension that can result from physical, chemical, or emotional factors.

Stress is tension in a person's mind or body that can result from physical, chemical, or emotional factors. While stress is present in every individual's life, it is extremely apparent in the life of the correctional officer. Correctional officers may try to mask stress and anxiety by adopting a "tough-guy" image or ignoring their symptoms. Authors have cited many reasons why correctional officers might become stressed on the job. The next sections will discuss the stressors that correctional officers must deal with daily.

Role conflict

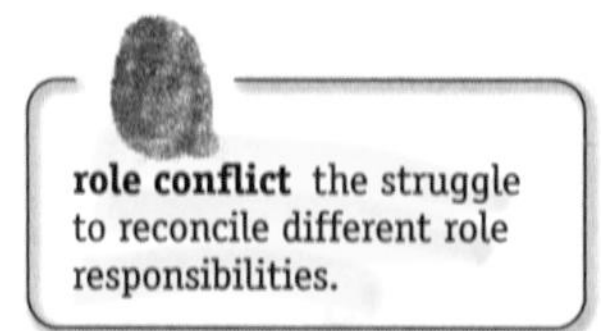

role conflict the struggle to reconcile different role responsibilities.

According to Grossi, Keil, and Vito (1996), role conflict is the predominant source of both stress and job dissatisfaction for correctional officers. **Role conflict** could be defined as the struggle of officers to reconcile custodial responsibilities (which could include maintaining security through preventing escapes and inmate violence) with their treatment functions (rehabilitation of offenders). Sykes (1958) discussed how role conflict is an inherent characteristic of the prison environment and is perpetuated by administrators. Members of the administration ask officers to enforce rules and punish those who disobey them, and also ask these same officers to assist in rehabilitative efforts. Sykes argued that this caused conflict within the officer and ultimately led to disillusionment and frustration. Stress in the correctional setting, left unsolved, may lead to high turnover and absenteeism. Cheek and Miller (1983) noted that role conflict was a source of stress that was created by prison officials who constantly changed prison goals, policies, and procedures. Grossi and Berg (1991, 75) note that many stressors involved in corrections are similar to that of police officer stress, especially those in which there are "unreasonable demands and expectations on their role as correctional officers."

Lombardo (1989) concluded that role conflict arises from experiences in which an officer is restricted from doing what he feels necessary by virtue of the rules or by adverse supervisory decisions. Research in the area of role conflict among correctional personnel has also broken down role conflict into three distinct types:

1. The officer finds that treatment and custody are incompatible
2. There is frustration when an officer tries to carry out treatment and custodial duties simultaneously
3. Problems are generated between officers who adhere to treatment orientation and those who maintain a strict custodial orientation (Crouch 1980)

Whitehead and Lindquist (1986) found that this role ambiguity and conflict could lead to significant levels of burnout, emotional exhaustion, and depersonalization. Some researchers have concluded that role conflict is higher in minimum-security institutions, where there is more likely to be mixed treatment and custody roles, than in high-security institutions. Poole and Regoli (1983) found that pursuit of both custody and treatment goals in correctional institutions raise the probability of role behavior's having to meet diverse criteria. Conflict, therefore, is likely to develop between officers who are asked to be disciplinarian and rule enforcer as well as counselor and problem solver. With these multiple roles come multiple demands. Officers are asked to quickly jump from one role to the next depending upon the situation.

Examination of the literature on role conflict points out some concerns. First, much of the literature on role conflict among correctional officers has not been collected utilizing a systematic approach. In fact, most authors are not even studying the same thing. It should also be noted that the research does not differentiate between the role conflict felt by officers because of the different roles they have to perform and the conflict felt between officers and others because they each have different tasks in corrections.

Dangerousness

Prisons are violent. No one likes to be held against his or her will, and this alone can lead to many volatile situations. Finn (1998) concluded that interviewees identified threats of inmate violence and actual inmate violence as a source of stress. Prisons are dangerous, and the correctional officer constantly faces the reality of being attacked, raped, or taken hostage. In his study of the Illinois guard force, Jacobs (1978) found that 49 percent of the sample defined "danger" as the "main disadvantage" of the job. Lombardo (1989) also found that 50 percent of the respondents identified physical strain and mental strain as a dissatisfying feature of their work. Feelings of danger may derive less from actual assaults, but more from the realization that officers face the never-ending possibility of victimization, not certain of the time, if ever, when they might be attacked. Jacobs and Grear (1977) found that of those officers that had resigned, 52 percent said that a lack of safety had influenced their decision.

Other researchers have also concluded that safety concerns contribute to work-related stress in the correctional environment. Correctional officers are concerned about their on-the-job safety and often comment upon it: "Employee concerns about safety were high enough to elicit comments in an open-ended section of the survey" (Triplett, Mullings, and Scarborough, 1996, 303). Cornelius (1994) identifies many stressors in corrections, including those related to danger and safety in institutions:

1. Inmate demands
2. Inmate arguments
3. Exposure to bodily fluids or feces
4. Cramped working conditions
5. Excessive noise and unpleasant odors
6. On-the-job injuries

7. Aggressive, violent inmates or clients
8. Inmates who are suffering from mental disorders
9. Inmates under the influence of alcohol or drugs
10. Escapes

Correctional officers' fear is universal and is not just an exclusive feeling of the younger, inexperienced officers. Those officers who are deemed most capable by their peers and supervisors are also afraid if they have frequent contact with inmates. One inmate reported that, ". . . they are all scared to death. I didn't think at any point that there was anybody (who wasn't) scared, because I saw in situations where their faces would turn white . . ." (Kauffman 1988, 215).

Low Pay

Many officers feel they are perceived—and come to perceive themselves—as occupying the lowest rung on the law-enforcement pecking order (Brodsky 1982). Many officers cite low pay as a source of stress, not only due to their "lower status," but also because they may feel they are unable to provide for their families or loved ones. Correctional officer's salaries range from as little as under $22,000 to over $52,000. The median salary was $32,670 in 2002 (United States Bureau of Labor 2005).

It should be noted, however, that Jacobs and Grear (1977) found that only 26 percent of those correctional officers who resigned left because of low pay. Triplett, Mullings, and Scarborough (1996) found that an increase in pay will relieve stress only if the amount of money is good enough to overcome other problems.

As research in the area of correctional officer stress would indicate, there are many causes of stress that staff face daily. What coping strategies are officers using to combat their high levels of stress? In order to deal with the stress and anxiety inherent within correctional environments, coping mechanisms become crucial to the officer. One of the sources that can either increase or decrease occupational stress is the amount of support given to the individual by a social-support system. There are three main aspects of social support that seem to be particularly pertinent to the correctional officer: administrative support, peer support, and family and societal support.

Administrative Support

Research conducted by Whitehead and Lindquist (1986) found that administrative support reduced job stress and **burnout** among correctional officers, while lack of support from administrators only perpetuated these problems. Some researchers conclude that the organization is vital in alleviating stressors:

> *The most useful point of intervention is the job and work setting. . . . Obviously, the problem cannot be completely eliminated until individuals and the society in which they live are changed, but much can be done before this occurs simply by changing the structure of roles, power, and norms in human service organizations (Cherniss 1980, 158).*

burnout employee has no interest in job; lethargy and minimal performance indicate burnout.

Brodsky (1982, 81) included a list of conditions that gave rise to long-term correctional employee stress, including three organizational factors:

1. Pressure designed to force them to resign or transfer
2. No backing when attacked or goaded by inmates
3. No support in dealing with problems with visitors, protestors, press

Triplett, Mullings, and Scarborough (1996) found that a majority of the stressors identified by correctional officers are those in which the officer has little or no control. Organizational responses to these might benefit officers' health, efficiency, and job satisfaction. Cornelius (1994, 61) points out three basic steps that supervisors can do to assist correctional officers' stress reduction:

1. Control their own stress
2. Recognize and help stressed-out workers cope with their stress
3. Improve physical conditions as well as the mental outlook of workers

Conflict between line staff and management is not uncommon. Managers and line employees have different backgrounds, education levels, and job responsibilities. This often leads them to view the social organization and the organization setting in contradictory ways (Fok, Hartman, Patti, and Razek 2004). Discord between staff and supervisors is found not only in correctional institutions, but also in other components of the criminal-justice system (Stojkovic and Farkas 2003). In literature on police-management interactions, there is a mistrust of supervisors on the part of "street cops," who see management as creating obstacles they must circumvent if "real" police work is to be achieved (Rothmiller and Goldman 1992). The same conflict can be found in correctional institutions.

This rivalry between management and line staff is particularly common in organizations that have a paramilitary, pyramid design—for example, prisons (Stojkovic and Farkas 2003). It could be argued that the correctional administrator is cynical by nature and views his employees suspiciously. Cynical managers may view employees as selfish, and feel it is necessary to get them before they get you (Kotter 1985). Correctional supervisors often believe that employees call in sick to protest undesirable duty posts or the denial of vacation time. Supervisors then punish the officers by doing those very things—deny days off or assign them to unfavorable duty posts.

Not surprisingly, when correctional officers see wardens as proinmate and antagonistic toward correctional officer values, they show higher levels of stress (Webb and Morris 2002). Lombardo (1989) reported that correctional employees find inmates to be the least of their problems, and hold that it is the prison administration that creates work difficulties. One officer noted that, "for the guard the work is simple, if only the administration would let him do it" (Lombardo, 164). Correctional administrators can and do take measures to alleviate correctional officer stress—which is only fair, since they also often create it.

Peer Support

The second type of social support identified by Shamir and Drory (1982) is peer support, or support from fellow officers. Peer support is known to be an important variable in occupational stress, and tends to be even more important in jobs where there is danger. Some research indicated that favorable relations with fellow officers diminishes feelings of alienation and cynicism (Poole and Regoli 1983). Some departments utilize officers as peer counselors who assist in the event of prison or personal catastrophes.

Interestingly, much of the literature on coworker support is ambiguous or shows that coworkers have a negative impact on job satisfaction. Jurik and Halemba (1984) found that those officers who reported positive attitudes toward their coworkers had negative levels of job satisfaction. Lombardo (1989, 148) noted in his research of New York officers that there was an inclination for "officers not to derive satisfaction from associations with members of their work group." Lombardo also found that, at times, correctional officers worked against one another instead of offering assistance. Finn (1998) found that 20 percent of the officers surveyed viewed "other staff" as their highest cause of stress. As noted earlier, obtaining peer support may cause correctional officers to compromise their personal integrity, values, or sense of right and wrong. In order to enhance their acceptance, the minority may remain silent but discontent. This could be an explanation as to why there are often high levels of job dissatisfaction paired with high levels of peer attachment (Grossi and Berg, 1991).

Family and Community Support

The third type of social support identified by Shamir and Drory (1982) is that of family and community support. Researchers have concluded that there is very little support from the community toward correctional officers, and that this is an additional source of stress for the officer. Members of the community may misunderstand correctional officers. Family support may vary, as the correctional lifestyle is difficult. Officers may take out their work frustrations at home, and feel as though their family members do not understand them or the pressure they are under.

The literature suggests that correctional officers experience family-related problems due to stress. Cheek and Miller (1983) note that although correctional officers did not report that they were experiencing problems at home, the rate of divorce for correctional personnel was two times that of other blue-collar workers. Cheek and Miller also found that correctional officers frequently reported letting out tensions in the wrong places (at home), tightening discipline at home and spending less time at home on their days off. Black (1982) found that correctional officers experiencing stress may damage their family relationships by displacing their frustration onto their spouses and children. Finn (1998) also suggests that shift work, long hours, and overtime make it difficult for officers to attend important family functions, further weakening their ties to a family-support system.

There are also additional sociological factors that have been identified by Cullen and Link (1985) that may facilitate coping in the correctional setting. Two of these factors are education level and correctional experience.

Education Level

Cullen and Link (1985) note that there are several reasons why education could be linked to coping. Criminal-justice reformers have suggested that education enhances the professionalization of a workforce. This would ideally allow officers to have more-positive interactions with supervisors, coworkers, and inmates. This would also facilitate officers' abilities to deal more effectively with ambiguities and complexities inherent in their correctional environment. Grossi and Berg (1991) found that education was positively linked to job satisfaction. Five years later, however, further research by Grossi, Keil, and Vito (1996) found that officers with higher levels of education were more likely to have high levels of job dissatisfaction.

Other studies have found that correctional personnel who are educated feel more emotional and physical exhaustion than do other personnel (Gerstein, Topp, and Correll 1987). Triplett, Mullings, and Scarborough (1996) found no evidence to suggest that those with higher education levels had greater or lesser amounts of stress than other correctional officers. VanVoorhis, Cullen, Link, and Wolfe (1991) found that the more educated an officer is, the more likely he or she is to experience job dissatisfaction, and concluded that this could be a lack of social integration or due to frustrated career expectations. Educated officers have higher expectations placed upon them, and therefore are more likely to experience work-related stress and job dissatisfaction. Research has also pointed to higher turnover rates among more-educated correctional officers due to more employment alternatives (Jurik and Winn 1987). Poole and Regoli (1983) made the observation that the current move toward professionalism among guards may bring cynicism because this attempt at professionalization through education may possibly create standards few correctional officers can reach. The literature on education level suggests that there is no conclusive link to education and stress level in correctional employment.

Correctional Experience

There are two differing opinions as to whether years in correctional employment affect job stress and satisfaction. One position notes that as correctional officers perform "dirty work" on a regular basis, they eventually will suffer from 'burnout' (Cheek and Miller 1983). The contrasting viewpoint suggests that job experience may function as a resource that helps officers cope effectively with the complexity and risks of prison life.

Research conducted by Toch and Klofas (1982) supports the opinion that more correctional experience often leads to higher stress levels. Those officers who worked under the more "conservative" custodial regimes are likely to experience frustration and job dissatisfaction due to more-"liberal" treatment ideologies being applied. These "older officers" would be more likely to experience role conflict, feel a lack of support from supervisors, and have higher levels of work-related illnesses. VanVoorhis, Cullen, Link, and Wolfe (1991) found that the number of years on the job was positively related to work stress and negatively related to rehabilitative focus. Triplett, Mullings, and Scarborough (1996) found that those officers who had been on the job longest had reported more work-related stress than did newer officers. It has also been argued that years of experience had a positive direct effect on role stress. The longer the officers had been employed, the more likely they were to experience role stress.

Some researchers have found conflicting results in terms of length of employment. Jurik and Halemba (1984) suggest that experience serves as a valuable resource that helps the officer to cope and mitigates stressful circumstances. They found that the number of months employed was significantly related to high levels of job satisfaction. Senior staff may be better able to develop a comfortable level of social distance and relatedness with inmates. Grossi and Berg (1991) found that the greater the amount of correctional officer experience, the greater the level of job satisfaction. The current research on correctional officer experience and its relationship to job stress is not conclusive.

Administrators and Managers

Traditionally, the task of running an institution was not subject to scrutiny by the media or the public. Prison officials did whatever they desired, rarely having to answer to anyone. Several major trends have affected the administrator of today, including greater media access to prisons and prisoners, public access to information, and legal changes brought on by the civil-rights movement. Correctional administrators must now be savvy in business as well as knowledgeable about correctional practices. Correctional managers direct and control others, take charge, and coordinate resources, all in an effort to accomplish organizational goals. The highest management level of a correctional institution is the warden or superintendent.

Management Functions

Executive managers (directors of correction or wardens in a prison) are focused on the "big picture," whereas middle and first-line managers are more concerned with daily operations and policy application. Correctional managers function like most managers in that their jobs include five basic elements: planning, organizing, staffing, leading, and controlling.

Planning

In order to create a plan of action, a manager must establish goals and objectives. Goals are defined as desired outcomes, while an **objective** is a specific, measurable way to reach a goal, which is usually accomplished under a particular timeline (Carlson, Hess, and Orthmann 1999). Planning also includes policy development and implementation. Management and staff are better able to handle daily operations if an agency has well-thought-out and well-defined policies. A lack of effective policy may create **crisis-centered management**, which is inefficient and often dangerous. This type of management spends an inordinate amount of time responding rather than proactively setting a course, and is "putting out fires" rather than spending the time to establish policies and procedures that would resolve problems before they start.

objective specific, measurable way to meet a goal.

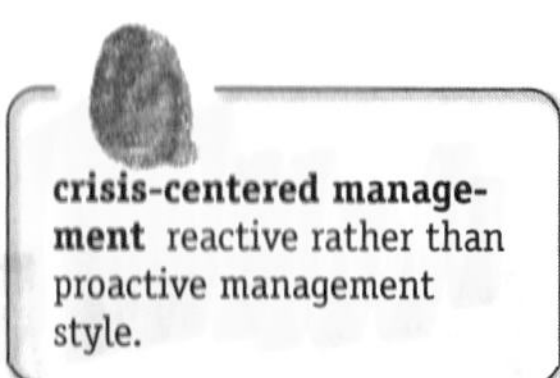

crisis-centered management reactive rather than proactive management style.

Typically, policy creation is done at the executive level, leaving lower-level managers to implement policies. This can cause potential problems when policies are set in the central state agencies without input from institutional leaders. Prisons are separate entities with issues and concerns that are unique to their own architecture and staff. At times, the central-agency personnel are not familiar with these unique issues of each prison or how central policies may impact certain prisons.

Organizing

Organizing is a broad function that encompasses the other functions of management. In order for a correctional system to run efficiently, management must divide the operations into departments and units that are able to communicate effectively with one another. These units must coordinate their efforts so as not to overlap or conflict with one another. If a system is successfully organized, it will help to reduce problems in staffing, controlling, and planning.

Staffing

Staffing is an essential skill of managers. In order for a correctional agency to do its job properly, the right people need to be doing the right things (Carlson, Hess, and Orthmann 1999). Not only must a warden concern himself or herself with having correctional officers to man the duty posts, but he or she must also be sure there are the appropriate number of food-service personnel, treatment and programming staff, and maintenance employees. The Federal Bureau of Prisons allows its wardens control over the assignment of employees at all levels, while wardens employed by the Texas Department of Criminal Justice have virtually no input on which correctional officers are assigned to their institutions. Wardens are also responsible for retaining employees. Staff retention can be optimized by providing stress- and anger-management programs for staff as well as by implementing an award system for excellent performances by staff members.

Leading

Stojkovic and Farkas (2003) view leaders as helping to develop an organizational culture. Public agencies are unique in that they have more than just one set of managers. The warden is subject to the policies of the agency itself, as well as to those who are in control of correctional budgets (namely, legislators). Thus, wardens today have less power than they might have had in the past.

The warden and his or her immediate staff (assistant wardens) are the leaders of the institution, but there is some evidence that leadership could be improved upon in many prisons. Mintzberg (1979) noted that managers spend most of their time in conversations and meetings, having coffee or lunch. While these elements are essential in disseminating information, management may need to do more to effectively lead people. Leaders must be able to transform individuals and be able to elevate the interests of their employees to generate acceptance and awareness of organizational goals. They should also be able to motivate employees to look past their own interests and work for the good of the agency or organization (Bass 1990). Moshavi, Brown, and Dodd (2003) noted that **transformational leaders** were able to give employees the following:

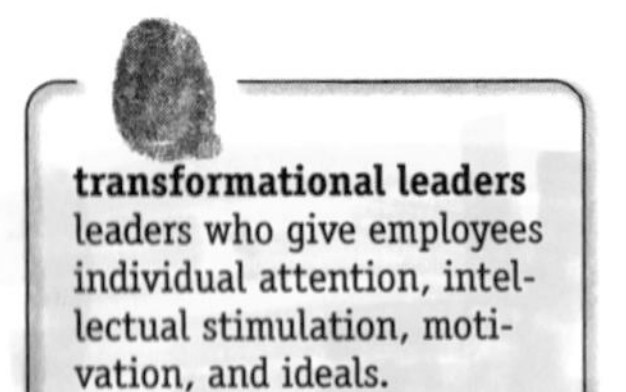

transformational leaders leaders who give employees individual attention, intellectual stimulation, motivation, and ideals.

1. Individualized attention
2. Intellectual stimulation
3. Motivation
4. Ideals

These employees also have higher job satisfaction. Correctional managers are not always correctional leaders, and there are many distinctions between the two. **Table 7-4** on page 217 presents an overview of these distinctions.

Table 7-4 Differences Between Correctional Managers and Correctional Leaders

Correctional Managers	Correctional Leaders
Operate within a structure	Deal with people in a structure
Are concerned with control	Inspire trust and involvement among employees
Live in the short term	Adopt a long-term perspective
Ask how and when	Ask what and why
Keep their eyes on the bottom line	Keep their eyes on the horizon
Imitate	Originate
Accept the status quo	Challenge the status quo
Are good soldiers	Are their own people
Do the thing right	Do the right thing
Look for things done wrong	Look for things done right
Concerned with programs	Concerned with people
Develop programs	Develop people
Driven by constraints	Driven by goals
Responsible	Responsive
Referee	Cheerleader
Direct	Coach
Preserve life	Have a passion for life

Source: Bennis. 1989. "Why Leaders Can't Lead." *Training and Development Journal* 43, 4: 35–40.

Many correctional administrators are managers and not leaders. Correctional agencies tend to follow the basics tenets of Weber's (1958/1964) **bureaucratic model**, in which a hierarchal, impersonal system or rules are implemented by management. It is difficult to be a leader in a structure that is, by design, rigid and authoritarian. The following characteristics would describe an agency that is operating under the bureaucratic model:

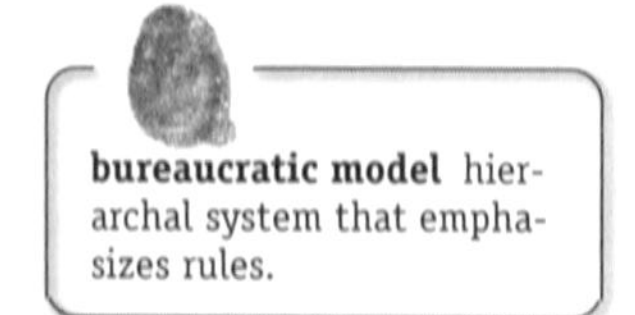
bureaucratic model hierarchal system that emphasizes rules.

- A hierarchy of power with a downward flow of power
- A division of labor with each position clearly defined and no overlap between positions
- A reliance on formal rules and procedures to guide the actions of employees (standard operating procedures or SOPs)
- An impersonal climate among and between superiors and subordinates, and a separation between professional and personal affairs (to keep one's personal life from interfering with performance at work)
- Employment and promotional decisions based on merit, with career tracks clearly defined

It has been suggested that a leadership organizational approach may be necessary for the everyday operations of corrections and to reduce job dissatisfaction and stress among correctional personnel (Stojkovic and Farkas 2003). Leadership should involve treating staff fairly and helping them to grow. Morale

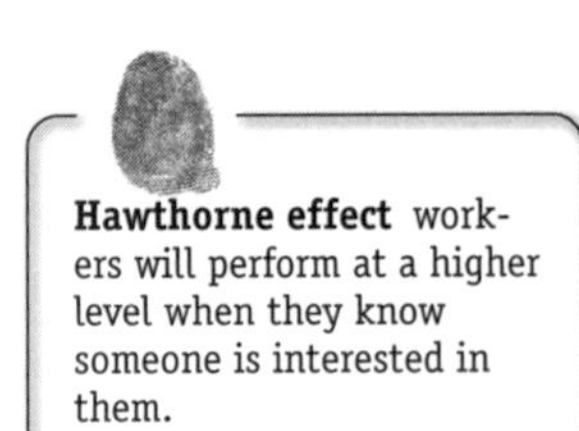

Hawthorne effect workers will perform at a higher level when they know someone is interested in them.

can and should be bolstered when managers see and recognize their subordinates' efforts. The **Hawthorne effect** refers to a research study that discovered that female assembly workers improved their production if the lighting was increased, but they also increased production when the lighting was decreased. The conclusion was that the workers were responding to the attention of the researchers. Now this phrase stands for the premise that subjects (or workers) will perform at a higher level if they know that researchers (or supervisors) care about them.

Controlling

Controlling requires the manager to consider a variety of internal and external forces, and how they affect the operation of the correctional institution. Internal control involves the assessment of departmental goals, and the analysis as to whether the institution is on track with the agency's standards. Correctional administrators are placed in an unusual predicament in that mangers must "control" inmates, and make decisions regarding which programs are working. External forces that influence daily operation within correctional facilities include state legislatures and politics, media, and public support and/or sentiment.

Carlson, Hess, and Orthmann (1999) note that correctional managers have the authority to control, but may not have the power to control. Authority is the ability to command or enforce laws, rules, and policy by virtue of rank or position. Power, however, is the ability to achieve action and results regardless of the legal right to do so. Authority may necessitate the use of force, while power involves the use of persuasion.

Management Styles

The way in which administrators have run their correctional institutions has evolved, just as the attitudes and beliefs of legislatures and citizens have also changed. The four managerial styles that have been found in correctional facilities are autocratic/authoritarian management, participatory management, shared-powers management, and inmate-control management. The age of the administrator, the age of the actual facility, the expectations of the public, and the geographic region directly influence use of these management styles.

Autocratic/Authoritarian Management

The autocratic warden was the most prevalent style of management in American penology up until fairly recently. Wardens during this time period had almost absolute control of their respective institutions—considered to be "sovereigns" over their prison kingdom. Wardens had total control over staffing and operating decisions. Authoritarian wardens did not seek input from employees. While this form of management is most often seen as outdated, some believe it is the preferred way to do things—stating that the best facilities are run "by the book," and that paramilitary, bureaucratic management is the best and most efficient way to run a prison (DiIulio 1987).

Participatory Management

As innovations in corporate management occurred during the 1970s, correctional managers turned to participatory management. This managerial style gets employees involved in decision making. It is believed that participation improves employee commitment to organizational goals. Participatory managers establish informal relationships with employees and solicit their views and opinions. Research conducted by Wright, Saylor, Gilman, and Camp (1997) found that correctional staff that had participatory managers had higher job satisfaction, greater effectiveness in working with offenders, and deeper commitment to the institutional goals. This was in direct contradiction to DiIulio's (1987) contention that prison management is best when it is impersonal. Participatory management does not give every employee a vote or equal say in decision making, but does allow for input from staff. This philosophy remains popular among prison administrators

Shared-Powers Management

Another idea that emerged with the rehabilitative era of the 1970s was the shared-powers philosophy—not only did employees get input into the running of the institution, but inmates were also included in the decision making. The shared-powers model frequently caused problems between correctional officers and inmates, as COs believed that inmates were gaining too much influence. Correctional officers responded by creating unions or correctional officer associations. With the demise of the rehabilitative ideal came the demise of the shared-powers style of management. While inmates may still have some input into basic daily operations, most administrators have abandoned the idea, preferring to focus on staff and agency needs.

Inmate-Control Management

When gang activity and inmate groups become so powerful as to dictate management decisions, some prisons could be considered to have an inmate-control management system. Many correctional agencies came perilously close to this situation in the 1980s as gang activity became widespread. Obviously, this style is more of a lack of management, and can never be considered a legitimate management style for a correctional facility.

Conclusions

Most correctional managers operate under some variation of the above management styles, as each facility and institution has unique concerns. Managers and correctional line staff are the key to running a good prison. It is an unfortunate reality that management science has not been applied in corrections in the same manner that one sees in private business. Good management can avert many of the problems found in today's prisons, and good management requires hiring and keeping good employees. Instead, what happens all too often is that administrators and employees are in conflict.

While there is no one perfect way to close the chasm between administrators and line personnel, there are potential remedies. First, one could utilize what Likert (1976) coined the "employee-centered" approach: supervisors that focused more on people and relationships had higher levels of productivity than did those managers who made decisions themselves and dictated orders to subordinates.

Human-resource approaches to management suggest that if you take care of your people, they will take care of you.

Human-resource strategies might not work in every situation, but one could increase autonomy and participation by creating some sort of work groups in which employees at each level of the organization could have input on how to deal with correctional policy that affects the institution. This would not only serve as a motivator, but would also lessen the division between supervisors and correctional officers. Webb and Morris (2002) noted that if prisons are to become better places for inmates to live and correctional officers to work, prison administrators will have to treat officers and their ideas with more respect. Most correctional agencies—indeed, most bureaucracies—avoid and resist change. In the corrections field, change occurs very slowly (Stojkovic and Farkas 2003).

KEY TERMS

bid system—officers select posts by seniority.

bureaucratic model—hierarchal system that emphasizes rules.

burnout—employee has no interest in job; lethargy and minimal performance indicate burnout.

"con-wise"—when a correctional officer is experienced and less naïve concerning inmates.

crisis-centered management—reactive rather than proactive management style.

cross-supervision—when male officers supervise female inmates or female officers supervise male inmates.

duty post—job assignment in the prison.

FTO—Field Training Officer.

Hawthorne effect—workers will perform at a higher level when they know someone is interested in them.

hole—punitive segregation.

new boot—new correctional officer.

objective—specific, measurable way to meet a goal.

role conflict—the struggle to reconcile different role responsibilities.

shift lag—physiological effects of shift work, which include impaired performance.

stress—tension that can result from physical, chemical, or emotional factors.

30-second rule—an inmate must express what they need within 30 seconds or he is probably manipulating the officer.

transformational leaders—leaders who give employees individual attention, intellectual stimulation, motivation, and ideals.

REVIEW QUESTIONS

1. What are the types of trainees, and what do they study in an academy?
2. What are the "defects of total power"? What are the five bases of CO power?
3. Discuss the CO subculture. What are the elements of it? Why does it exist?
4. Discuss the findings regarding the effectiveness of female COs in prisons for men.
5. What are the problems of cross-sex supervision?
6. What are the stressors for COs?
7. What are some coping factors for the stress that COs experience?
8. How can administrators reduce stress for COs?

9. What are the five basic elements of managing?
10. What are the four management styles? Describe them.

FURTHER READING

Crouch, B., and J. Marquart. 1990. "On Becoming a Prison Guard." In S. Stojkovic, J. Klofas, and D. Kalinich (Eds.), *The Administration and Management of Criminal Justice Organizations.* Prospect Heights, IL: Waveland.

Josi, D., and D. Sechrest. 1998. *The Changing Career of the Correctional Officer: Policy Implications for the 21st Century.* Boston: Butterworth-Heinemann.

Kauffman, K. 1988. *Prison Officers and Their World.* Cambridge, MA: Harvard University Press.

Philliber, S. 1987. "Thy Brother's Keeper: A Review of the Literature on Correctional Officers." *Justice Quarterly* 4: 9–35.

Stojkovic, S., and M. Farkas. 2003. *Correctional Leadership: A Cultural Perspective.* Belmont, CA: Wadsworth.

REFERENCES

American Correctional Association. 2000. *Vital Statistics in Corrections.* Lanham, MD: American Correctional Association.

American Correctional Association. 1993. *Correctional Standards.* Lanham, MD: American Correctional Association.

Bales, D. 1997. *Correctional Officer Resource Guide,* 3d ed. Lanham, MD: American Correctional Association.

Bass, B. 1990. *Bass and Stogdill's Handbook of Leadership.* New York: Free Press.

Bennis, W. 1989. "Why Leaders Can't Lead." *Training and Development Journal* 43, 4: 35–40.

Black, R. 1982. "Stress and the Correctional Officer." *Police Stress* 5, 1: 10–16.

Breed, A. 1981. "Women in Correctional Employment." In B. H. Olsson (Ed.), *Women in Corrections,* pp. 37–44. Lanham, MD: American Correctional Association.

Brodsky, C. 1982. "Work Stress in Correctional Institutions." *Journal of Prison & Jail Health* 2, 2: 74–102.

Calhoun, A., and H. Coleman. 2002. "Female Inmates' Perspectives on Sexual Abuse by Correctional Personnel: An Exploratory Study." *Women and Criminal Justice* 13: 101–126.

Camp, C., and G. Camp. 1998. *Corrections Yearbook.* Middletown, CT: Criminal Justice Institute.

Camp, S., W. Saylor, and K. Wright. 2001. "Racial Diversity of Correctional Workers and Inmates: Organizational Commitment, Teamwork, and Worker Efficacy in Prisons." *Justice Quarterly* 18, 2: 411–428.

Carlson, N., K. Hess, and C. Orthmann. 1999. *Corrections in the 21st Century: A Practical Approach.* Belmont, CA: West/Wadsworth.

Champion, D. 2005. *Corrections in the United States: A Contemporary Perspective.* Englewood Cliffs, NJ: Prentice Hall.

Cheek, F., and M. Miller. 1983. "The Experience of Stress for Correction Officers: A Double-Bind Theory of Correctional Stress." *Journal of Criminal Justice* 11: 105–120.

Cheeseman, K. 2004. *Training Day: A Typology of Correctional Trainees and the Academy Experience*. School of Criminal Justice. Sam Houston State University.

Cheeseman, K., J. Mullings, and J. Marquart. 2001. "Inmate Perceptions of Staff Across Various Custody Levels of Security." *Corrections Management Quarterly* 5: 41–48.

Cherniss, C. 1980. *Professional Burnout in Human Service Organizations*. New York: Praeger.

Cornelius, G. 1994. *Stressed Out: Strategies for Living and Working with Stress in Corrections*. Laurel, MD: American Correctional Association.

Crouch, B. 1980. "The Book v. the Boot: Two Styles of Guarding in a Southern Prison." In B. Crouch (Ed.), *The Keepers*, pp. 207–224. Springfield, IL: Charles Thomas.

Crouch, B., and J. Marquart. 1990. "On Becoming a Prison Guard." In S. Stojkovic, J. Klofas, and D. Kalinich (Eds.), *The Administration and Management of Criminal Justice Organizations*, pp. 37–45. Prospect Heights, IL: Waveland.

Cullen, F., and B. Link 1985. "The Social Dimensions of Correctional Officer Stress." *Justice Quarterly* 2, 4: 505–533.

DiIulio, J. 1987. *Governing Prisons: A Comparative Study of Correctional Management*. New York: Free Press.

Ethridge, R., C. Hale, and M. Hambrick. 1984. "Female Employees in All-Male Correctional Facilities." *Federal Probation* 48: 54–65.

Finn, P. 1998. "Correctional Officer Stress: A Cause for Concern and Additional Help." *Federal Probation* 62: 6574–6577.

Fok, L., S. Hartman, A. Patti, and J. Razek. 2004. "The Relationship Between Equity Sensitivity, Growth, Need, Strength, Organizational Citizenship Behavior, and Perceived Outcomes in the Quality Environment: A Study of Accounting Professionals." *Journal of Social Behavior and Personality* 15, 1: 99–120.

Gerstein L., C. Topp, and G. Correll. 1987. "The Role of the Environment and Person When Predicting Burnout Among Correctional Personnel." *Criminal Justice and Behavior* 14, 3: 352–369.

Goffman, E. 1959. *The Presentation of Self in Everyday Life*. Woodstock, NY: Overlook.

Grossi, E. and B. Berg. 1991. "Stress and Job Dissatisfaction Among Correctional Officers: An Unexpected Finding." *International Journal of Offender Therapy and Comparative Criminology* 35: 73–81.

Grossi, E., T. Keil, and G. Vito. 1996. "Surviving 'the Joint': Mitigating Factors of Correctional Officer Stress." *Journal of Crime and Justice* 19: 103–120.

Grosswald, B. 2003. "Shiftwork and Negative Work-to-Family Spillover." *Journal of Sociology and Social Welfare* 30, 4: 31–57.

Hepburn, J. 1985. "The Exercise of Power in Coercive Organizations: A Study of Prison Guards." *Criminology* 23, 1: 145–164.

Hunter, S. 1993. "On the Line: Working Hard with Dignity." *Corrections Today* 55: 12–13.

Jacobs, J. 1978. "What Prison Guards Think: A Profile of the Illinois Force." *Crime and Delinquency* 25: 185–196.

Jacobs, J., and M. Grear. 1977. "Dropouts and Rejects: An Analysis of the Prison Guard's Revolving Door." *Criminal Justice Review* 2: 57–70.

Johnson, T. 1995. "Stressing the Value of Targeted Recruitment in Corrections." *Corrections Today* 57: 96–99.

Josi, D., and D. Sechrest. 1998. *The Changing Career of the Correctional Officer: Policy Implications for the 21st Century.* Boston: Butterworth-Heinemann.

Jurik, N., and G. Halemba. 1984. "Gender, Working Conditions, and the Job Satisfaction of Women in Non-traditional Occupations: Female Correctional Officers in Male Prisons." *Sociological Quarterly* 25: 551–556.

Jurik, N., and R. Winn. 1987. "Describing Correctional Security Dropouts and Rejects: An Individual or Organizational Profile?" *Criminal Justice and Behavior* 14, 1: 5–25.

Kauffman, K. 1988. *Prison Officers and Their World.* Cambridge, MA: Harvard University Press.

Kotter, J. 1985. *Power and Influence: Beyond Formal Authority.* New York: Free Press.

Levinson, M. 1982. "Corrections Training: Beyond Bar Tapping One." *Corrections Magazine* 8, 6: 40–47.

Likert, R. 1976. *New Ways of Managing Conflict.* New York: McGraw-Hill.

Lombardo, L. 1989. *Guards Imprisoned.* New York: Elsevier.

Mintzberg, H. 1979. *The Management of Organizations.* Upper Saddle River, NJ: Prentice Hall.

Morton, J. 1981. "Women in Correctional Employment: Where Are They Now and Where Are They Headed?" In B. H. Olsson (Ed.), *Women in Corrections,* pp. 7–16. Lanham, MD: American Correctional Association.

Moshavi, D., F. Brown, and N. Dodd. 2003. "Leader Self-Awareness and Its Relationship to Subordinate Attitudes and Performance." *Leadership and Organization Development Journal* 24: 407–418.

Philliber, S. 1987. "Thy Brother's Keeper: A Review of the Literature on Correctional Officers." *Justice Quarterly* 4: 9–35.

Poole, E., and R. Regoli, 1983. "Professionalism, Role Conflict, Work Alienation, and Anomia: A Look at Prison Management." *Social Science Journal* 20: 63–67.

Potter, J. 1980. "Female Correctional Officers in Male Institutions." *Corrections Magazine* (October): 30–38.

Rothmiller, M., and I. Goldman. 1992. *L.A. Secret Police.* New York: Pocket Books.

Shamir, B., and A. Drory. 1982. "Occupational Tedium Among Prison Officers." *Criminal Justice and Behavior* 9: 79–99.

Stojkovic, S., and M. Farkas. 2003. *Correctional Leadership: A Cultural Perspective.* Belmont, CA: Wadsworth.

Sykes, G. 1958. *The Society of Captives.* Princeton, NJ: Princeton University Press.

Texas Department of Criminal Justice. 2005. Personal communication, February 10, 2005.

Toch, H., and J. Klofas. 1982. "Alienation and Desire for Job Enrichment Among Correctional Officers." *Federal Probation* 46: 35–44.

Triplett, R., J. Mullings, and K. Scarborough. 1996. "Work-Related Stress and Coping Among Correctional Officers: Implications from Organizational Literature." *Journal of Criminal Justice* 24: 291–308.

United States Bureau of Labor. 2005. *Correctional Officer.* Retrieved on August 27, 2005, from www.bls.gov/oco/ocos156.htm earnings.

VanVoorhis, P., F. Cullen, G. Link, and N. Wolfe. 1991. "The Impact of Race and

Gender on Correctional Officers' Orientation to the Integrated Environment." *Journal of Research in Crime and Delinquency* 28: 472–500.
Webb, G., and D. Morris. 2002. "Working As a Prison Guard." In T. Gray (Ed.), *Exploring Corrections: A Book of Readings*. Boston: Allyn and Bacon.
Weber, M. 1958/1964. *Essays in Sociology.* New York: Oxford University Press.
Whitehead, J., and C. Lindquist. 1986. "Correctional Officer Job Burnout: A Path Model." *Journal of Research in Crime and Delinquency* 23: 23–42.
Worley, R., J. Marquart, and J. Mullings. 2003. "Prison Guard Predators: An Analysis of Inmates Who Establish Inappropriate Relationships with Prison Staff, 1995–1998." *Deviant Behavior* 24, 2: 175–198.
Wright, K., W. Saylor, E. Gilman, and S. Camp. 1997. "Job Control and Occupational Outcomes Among Prison Workers." *Justice Quarterly* 14, 3: 525–546.
Zimmer, L. 1986. *Women Guarding Men*. Chicago: University of Chicago Press.

CASES CITED

Everson v. Michigan Department of Corrections, 222 F. Supp. 2d 864 (2002)

8 Prisoners' Rights

John McLaren
Texas State University–San Marcos

Chapter Objectives

- Be aware of the history of prisoners rights litigation.
- Be able to articulate the mechanisms used to bring grievances to court.
- Understand the rights associated with the 1st Amendment.
- Understand the rights associated with the 8th and 14th Amendments.
- Be aware of emerging issues in prisoners rights litigation.

The prisoners' rights era emerged during the civil rights movement of the 1960s. As public support has declined for using the law to advance rights in education, the workplace, and other institutions, the use of the courts as a tool for reform of prison conditions has also declined. The prison reform movement is, however, of great symbolic significance in *understanding* the fundamental rights of all people in our society, and it has served as a vehicle for public debate regarding those elusive standards.

Era of the Hands-Off Doctrine

Prior to the civil rights movement of the 1960s, prisons held little interest for the courts, the press, or the general public. Federal courts had assumed a comfortable and convenient posture with regard to the occasional grievances that prisoners brought to their attention. Although *Ruffin v. Commonwealth* (62 Va. 790 [1871]) was a case decided in state court, its holding that a prisoner had the status of a "slave of the state" with no rights was embraced by both state and federal courts well into the 20th century. *Ruffin* offered the judiciary a convenient doctrine to justify dismissal of any prisoner claim of unlawful treatment.

The *Ruffin* doctrine later gave way to a more liberal rule that recognized that "a prisoner retains all the rights of an ordinary citizen except those expressly, or by necessary implication, taken from him by law," as stated in *Coffin v. Reichard* (143 F. 2d 443 [6th Cir. 1944]). Although this appears to be a departure from the uncompromising perspective of the *Ruffin* case, *Coffin* initially had little impact on the rights of prisoners. The **retention-of-rights theory** was initially applied in such a narrow manner that its advantage to prisoners challenging conditions of confinement was illusory. In almost every case of prisoners who alleged a grievance related to the conditions of confinement, the holding was that "lawful incarceration brings about necessary withdrawal or limitation of many privileges and rights, a retraction justified by the considerations underlying our penal system" (*Price v. Johnston*, 68 S. Ct. 1049 [1948]).

retention-of-rights theory idea that prisoners retained all the rights of free people except for those rights that were inconsistent with their status as prisoner.

The **hands-off doctrine** refers to the practice of the state and federal courts to reject prisoner appeals. It was supported by several different rationales:

hands-off doctrine represented the court's unwillingness to become involved in prisoner rights issues.

1. **Deferral to the expertise of administrators:** This refers to the view that correctional administration is a technical matter best left to the discretion of experts rather than courts.
2. **Definition of rights:** This refers to the definition of grievances as relating to privileges rather than rights. Prisoners were deemed to have forfeited privileges by virtue of conviction.
3. **Federalism:** The conditions of confinement experienced by prisoners were viewed as matters properly left to the legislative and executive branches of government, which are equipped with superior resources for fact-finding, budgeting, and policy. Federal courts did not want to get into the business of telling states how to run their prisons.
4. **"Opening the door":** It was believed that even a slight erosion of the hands-off doctrine would invite a tidal wave of litigation that could overwhelm the resources of the judicial branch and drain the energy of states and governmental agencies brought into court to defend their policies and practices (National Advisory Commission 1973, 18).

The hands-off doctrine appeared to erode rapidly in the face of an avalanche of legal arguments for its abandonment (Branham and Krantz 1997; Smith 1999). Federal courts, increasingly activist, became a welcome harbor for the politically disenfranchised. Not surprisingly, prisoners (a bitterly alienated and ever-volatile category of citizen) joined other dissident groups in using federal legal remedies, specifically the civil rights statutes, resulting in an "inmate litigation explosion" (Roleff 1996). Federal prisoners filed about 1,500 petitions in 1962. That number grew in 25 years to about 63,000 petitions. In 1971, state prisoners filed slightly more than 12,000 petitions. By 1997, the number of state-prisoner filings exceeded 48,000. (Maguire and Pastore 1999, 442). **Figure 8-1** shows the drop in prisoner petitions after the Prison Litigation Reform Act, which we will discuss later.

Prisoner success in federal litigation to reform prisons peaked and began to decline before the Reagan presidency of 1980. The years after activism faded have been

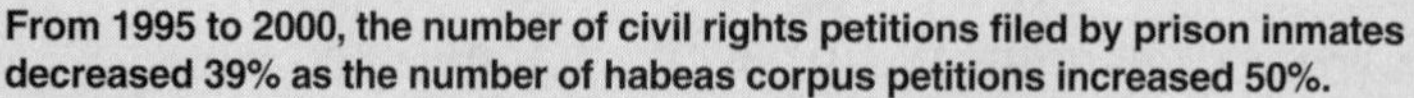

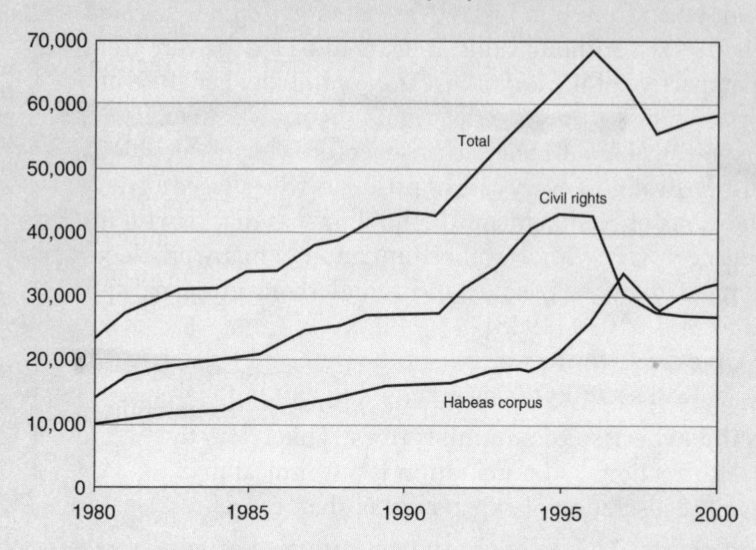

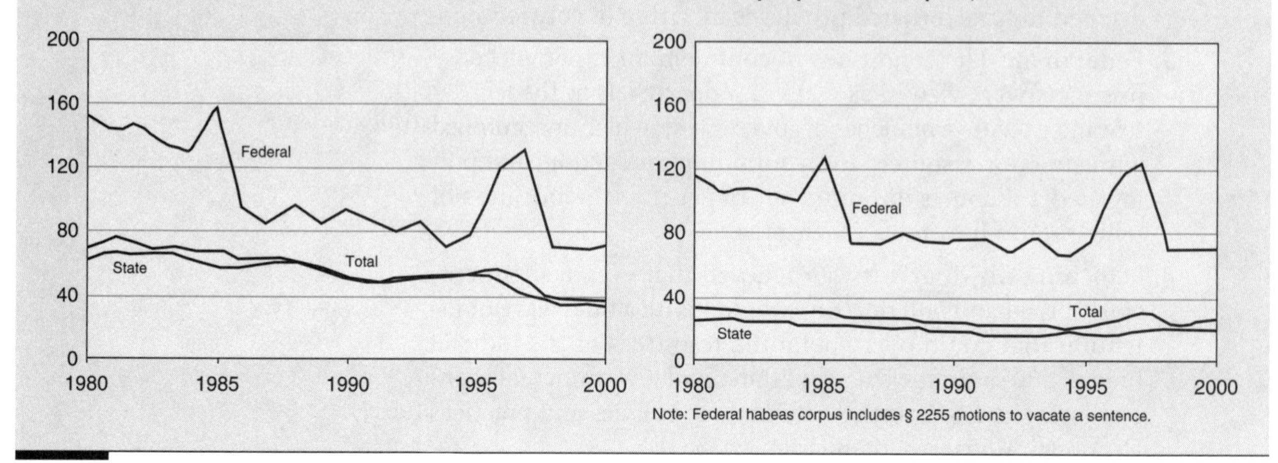

Figure 8-1 Trends in Filing of Prisoner Lawsuits, 1980–2000. ***Source:*** *Scalia 2002.*

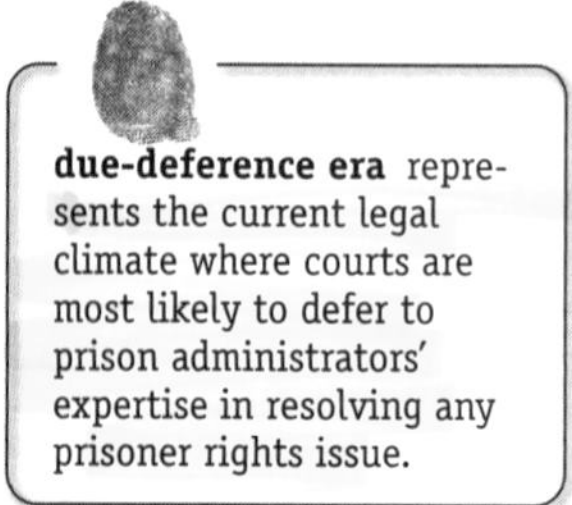

due-deference era represents the current legal climate where courts are most likely to defer to prison administrators' expertise in resolving any prisoner rights issue.

called the **due-deference era**. Courts did not completely abdicate their involvement in prisoner-grievance litigation, but there was a renewed emphasis on a presumption of good faith on the part of prison administrators, and courts came to intervene with greater caution, usually deferring to the expertise of the administrators.

Since the 1979 Supreme Court decision in *Bell v. Wolfish* (99 S. Ct. 1861 [1979]), the federal courts, guided by the Supreme Court, have consistently restricted prisoners' opportunities to file civil suits and judges' abilities to remedy violations. The paramount need for maintenance of order and security usually prevails over any constitutional challenges made by the prisoner to policies and procedures of the prison.

Mechanics of Litigation

Most litigation addressing conditions of confinement or treatment is brought under the federal Civil Rights Act (42 U.S.C.A. § 1983) in federal court. **Section 1983 actions** allow individuals to sue public officials for alleged violations of civil rights. This mechanism is attractive to prisoner litigants for a number of reasons: the Act embraces a wide variety of official misconduct regarding prison conditions and procedures; it offers comprehensive remedies, including money damages, injunctive relief, and attorney's fees; and it is a relatively easy cause of action to file. **Damages** are monies that a court orders paid to a person who has suffered a loss or injury at the hands of the person whose fault caused the injury or loss. Damages may be actual or punitive.

Section 1983 actions allows individuals to sue public officials for alleged violations of civil rights.

damages monies that a court orders paid to a person who has suffered a loss or injury by a defendant.

Prisoners' rights claims, while perhaps not receiving a friendly reception in any judicial forum, at least found a more receptive atmosphere in the federal courts. In theory, a state inmate could file a civil rights lawsuit in either state or federal court. The lack of filings in state courts has been well documented, and is attributable to several factors (Jacobs 1983). State-court judges do not enjoy the insulation from politically explosive rulings that federal judges do, and state-court juries, especially those located in counties in which there are significant inmate populations, are believed to be less favorably inclined to the claims of the prison population. The lack of state-court claims is further attributable to the fact that prisoners have not generally been required to exhaust available state administrative or judicial remedies before filing a civil rights cause of action (*Morgan v. LaVallee*, 526 F. 2d 221 [2d Cir. 1975]). The lack of litigation in state courts may change in the future as some progressive jurisdictions adopt statutes that approach the functional equivalent of federal civil rights laws. The increasingly conservative character of the federal judiciary may also divert some cases into state forums.

Section 1983 Actions

The single most important development in the abandonment of the "hands-off doctrine" was the Supreme Court decision in *Monroe v. Pape* (81 S. Ct. 473 [1961]). That case did not deal with the rights of prisoners, but rather, arose from a claim filed after 13 Chicago police officers allegedly entered the plaintiffs' home and conducted a warrantless search and arrest. The case imposed a revolutionary interpretation of a federal law enacted originally to discourage lawless activities by state officials in the aftermath of the Civil War. The statute states:

> *Every person who, under color of any statute, ordinance, regulation, custom, or usage, of any State or Territory or the District of Columbia, subjects or causes to be subjected, any citizen of the United States or other person within the jurisdiction thereof to the deprivation of any rights, privileges, or immunities secured by the Constitution and laws, shall be liable to the party injured in an action at law, suit in equity, or other proper proceeding for redress (42 U.S.C.A. § 1983).*

The central issue in the case was the requirement that a person, to be liable, act "under color of" state law. The Supreme Court, in an opinion by Justice Douglas, held that for activities to have taken place under color of state law, it was not necessary that state law authorize the activities. The statute was intended to protect against "misuse of power, possessed by virtue of state law and made possible only because the wrongdoer is clothed with the authority of state law" (*Monroe v. Pape*, 81 S. Ct. 473, 482 [1961]). In 1964, the U.S. Supreme Court held that state inmates could file suit against their keepers under Section 1983 (*Cooper v. Pate*, 378 U.S. 546 [1964]).

States themselves, as well as state agencies, remain immune from liability, although since 1978, the states' political subdivisions (particularly municipal corporations) can be sued. In most cases, the named defendants are individual employees who allegedly have harmed the claimed right of a person in custody (*Monell v. Department of Social Services of the City of New York*, 98 S. Ct. 2018 [1978]). To be held liable under Section 1983, a defendant must be acting "under color of state law," as courts have interpreted that phrase. Often private parties who contract with state agencies to provide medical, psychological, or other services will be deemed to be acting under state authority (*West v. Atkins*, 108 S. Ct. 2250 [1988]).

deliberate-indifference test a standard to be met to determine whether a constitutional violation has occurred; when prison officials are deliberately indifferent to the violation or loss of a constitutional right.

The state of mind of the party allegedly inflicting the injury is extremely relevant to the disposition of these lawsuits. Mere negligence (minor deviance from the standard of care a reasonable, prudent person would provide), a state of mind usually adequate to impose liability in traditional personal injury lawsuits, is clearly not an adequate foundation in civil rights cases (*Estelle v. Gamble*, 97 S. Ct. 285 [1976]). Later cases have created the **deliberate indifference test**—a standard of *deliberate indifference to the exercise of a known right*, comparable to gross negligence or recklessness to the inmate's risk of injury. Meeting this test is necessary to establish a valid claim. This is consistent with the theory that the statute was intended to protect people from deprivations associated with an abuse of official power or authority. Thus, its protections are not triggered solely by a failure to give due care that an ordinary, prudent person would provide.

sovereign immunity shield from liability that historically had attached to governmental entities and their agents.

The doctrine of **sovereign immunity** is a significant shield from liability in suits claiming deprivation of civil rights. Judges, legislators, prosecutors, and parole-board members still have almost absolute immunity from liability. Members of the executive branch involved in activities such as planning and budgeting are also cloaked with similar immunity. Other officials of state and local governments enjoy "qualified immunity"—that is, immunity from liability for actions undertaken in "good faith," with no desire to maliciously deprive an individual of constitutional rights (*Scheuer v. Rhodes*, 94 S. Ct. 1683 [1974]).

A prisoner who is successful in establishing a claim under the Civil Rights Act may qualify for a variety of remedies—including monetary damages, injunctive relief, attorney's fees, and relief in the form of a judgment defining the legal rights and responsibilities of the parties to the litigation, known as a declaratory judgment. Three types of monetary damages may be awarded a successful plaintiff: (1) actual damages to compensate for expenses incurred and for mental suffering; (2) nominal damages to vindicate the plaintiff's rights even where no actual damages were sustained; and (3) punitive damages if the wrongful acts were done maliciously, intentionally, or with "evil motive or intent" (*Smith v. Wade*, 103 S. Ct. 1625 [1983]).

Habeas Corpus

Another significant legal tool often employed by prisoners is the writ of **habeas corpus** (28 U.S.C. § 2544). Since 1944, it has been available to contest not only the legality of confinement, but also the conditions of confinement (*Coffin v. Reichard*, 143 F. 2d 443 [6th Cir. 1944]). "The Great Writ," once properly filed, requires any governmental official holding another person in custody to come forward and show why the person should not be released. Unlike its counterpart in Section 1983, there is a general requirement that the prisoner first exhaust available state remedies before being eligible for federal postconviction consideration. States have habeas corpus procedures that parallel the remedies provided in federal courts.

habeas corpus a writ that demands the government show why the person should not be released; used to challenge the legality of confinement.

The ability of state prisoners to successfully utilize habeas corpus procedures in the federal courts were severely restricted by the holding in *Stone v. Powell* (428 U.S. 465 [1976]). The Court declared that such petitions cannot be examined via the federal habeas corpus procedure if they were already "fully considered" by a state appellate court. This initiated a significant decline in utilization of the remedy by state prisoners. The Supreme Court continues to hold that state prisoners must exhaust all available state-court remedies before a federal court may consider granting habeas corpus relief. Petitioners must also place *all* constitutional claims in one petition (*McCleskey v. Zant*, 499 U.S. 467 [1991]), and are barred from filing new petitions even if they identify other ways in which their rights may have been violated in the process leading to conviction (*Coleman v. Thompson*, 501 U.S. 722 [1991]).

In 1996, Congress further limited habeas proceedings in the Antiterrorism and Effective Death Penalty Act. The statute imposed limits on the number of habeas corpus petitions that can be filed, requiring prisoners to include all known claims in the original petition or waive them. Bluntly stated, the act was aimed specifically at expediting the death-penalty cases in which defendants, as a life-preservation strategy, tried to prolong litigation by filing only one claim per petition. Claims based on new facts are eligible for review only if it appears they would inevitably have changed the result of the original trial. Furthermore, before a second petition can be filed, it must be approved by a three-member panel of a U.S. court of appeals.

Criminal Remedies

In addition to these civil, or personal, remedies, there are also criminal prosecution alternatives available in both state and federal jurisdictions. Criminal prosecution to enforce the constitutional rights of prisoners is rarely employed. Prosecutorial discretion, the lack of a financial incentive to complain, the credibility of the complainants, jury skepticism regarding the victims, and the paramount requirement that willfulness or intent be proved to gain a conviction are all factors contributing to this fact. Prisoners and their advocates have not elected to use state-court remedies to any great degree, and the body of law defining the legal status of prisoners has been shaped almost exclusively by the federal civil-rights statutes, especially the ubiquitous Section 1983, and the federal statutory habeas corpus mechanism.

Americans with Disabilities Act

In 1998, in *Pennsylvania Department of Corrections v. Yeskey* (118 S. Ct. 1952 [1998]), the Supreme Court found that the Americans with Disabilities Act (ADA) applied to prisoners with disabilities. The statute, adopted by Congress in 1990, is intended to protect people with disabilities from discrimination or exclusion from government programs and services. Actions filed under the **ADA** usually seek access and accommodation so that disabled people may participate in and take advantage of government services and programs. Although this development seems promising for advocates for prisoners, it is based on a statute, not a constitutional right, and the possibility remains that the ADA could be amended to exclude prisoners. The potential financial burdens for compliance with the ADA are anticipated to be very high as the prison population ages.

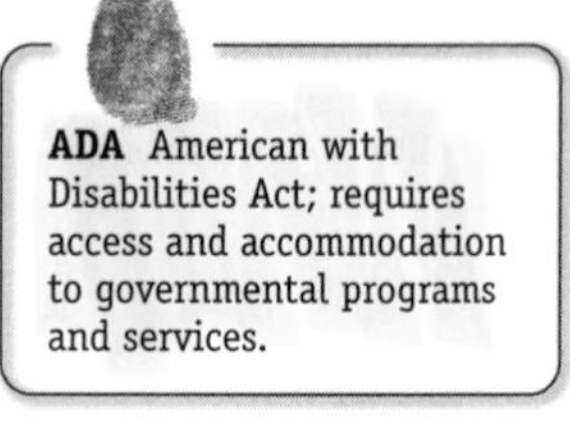

ADA American with Disabilities Act; requires access and accommodation to governmental programs and services.

Prison Litigation Reform Act

For most inmates, successful prisoners' rights litigation is improbable, given the low rate of functional literacy among this population. However, both state- and federal-government officials have voiced their concerns about the burden of litigation and the filing of frivolous, resource-consuming lawsuits. Congress responded by signing into law on April 26, 1996 the *Prison Litigation Reform Act (PLRA)* (42 U.S.C. 3626). Significant features of the Act include:

PLRA Prison Litigation Reform Act; severely restricted prisoners' ability to file writs to challenge their confinement.

1. The establishment of specific requirements for findings that judges must make before relief can be granted or consent decrees imposed
2. Diminished authority of judges to order release of overcrowded prisoners
3. Short time limits on remedial orders
4. A requirement that prisoners use all administrative remedies before filing a lawsuit
5. Granting state officials a role in the appointment of special masters who supervise the implementation of court orders
6. The requirement that the costs of a special master be borne by the court, not the defendant
7. A requirement that prisoners make at least partial payment of court costs

Injunctive relief, also referred to as equitable relief, is expressed in a judicial order that a person refrain from doing a particular act or (rarely) that a person perform a particular act (an **injunction**). Injunctive relief is a common and useful remedy in prison litigation because it protects the complaining prisoner(s) from continued future deprivation of the right(s) specifically addressed in the court order. Provisions of the PLRA diminish the scope of available injunctive relief. Consent decrees are agreements between the parties to settle a lawsuit without admission of liability. A **consent decree** may contain extensive relief, but is a resolution of the issue without a trial. The provisions of the PRLA have also restricted consent decrees.

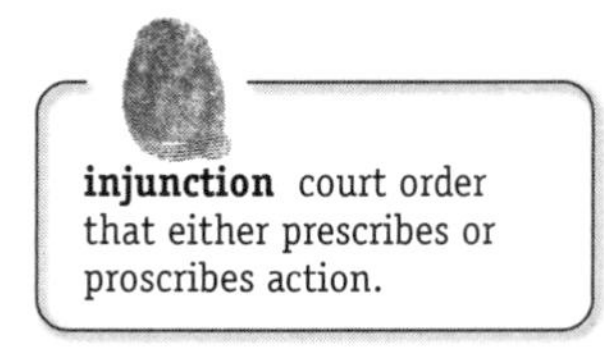

injunction court order that either prescribes or proscribes action.

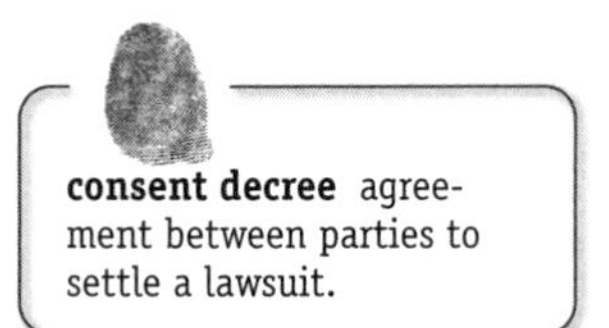

consent decree agreement between parties to settle a lawsuit.

The Act effected changes in the awarding of attorney's fees, specifying that they may not exceed 25 percent of the judgment. Any money ordered to be paid by defendant prison officials must be used to pay any outstanding restitution (42 U.S.C. § 1997e[d][2]).

In general, the PLRA was instrumental in drastically reducing the flow of prisoners' rights cases to the federal appellate courts. Supporters argue that it was necessary; critics argue that the draconian nature of the act has made it difficult, if not impossible, for inmates with legitimate constitutional grievances to have their "day in court" (Palmer 2003).

The Eighth Amendment: Corporal Punishment and the Use of Force

Legitimate uses of force include self-defense, defense of third parties, prevention of escape, and prevention of crime. Illegitimate use of force has come to be understood in the Anglo-American legal system as the implementation of force, beatings, and whippings (**corporal punishment**) as retaliation or punishment for inmate misconduct, or the use of excessive and unnecessary force in the course of legitimate application of force. It is well documented that physical force has been utilized for legitimate and illegitimate purposes in the past (Clemmer 1958).

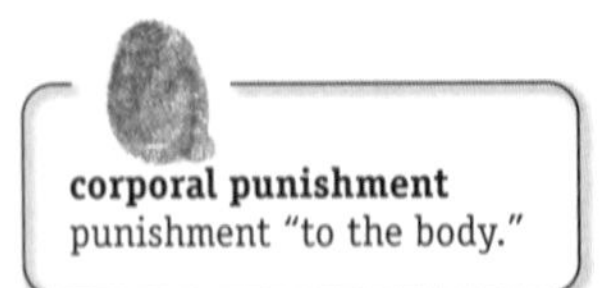

corporal punishment punishment "to the body."

Not until 1968, in *Jackson v. Bishop* (404 F. 2d 571 [8th Cir. 1968]), did a court specifically hold that whipping a prisoner, as a technique of discipline, violated the Eighth Amendment prohibition of cruel and unusual punishment. Only three years earlier, a federal district court in the same judicial circuit had approved the use of that punishment, with the provision that its use be carefully controlled (*Talley v. Stephens*, 247 F. Supp. 683 [E.D. Ark. 1965]). Numerous cases subsequent to *Jackson* have addressed this issue, and there is unanimity in support of it. The American Correctional Association and other professional organizations, such as the American Bar Association, support the prohibition.

Despite the apparent clarity of the prohibition, there are some correctional practices that raise intriguing related issues. These include the use of tranquilizing drugs for punishment purposes, verbal abuse and insults, and reckless failure to protect inmates from assaultive conduct by other inmates. Another type of case is where the **totality of circumstances**, specifically numerous shocking and degrading practices in the prison, result in findings that the conditions of confinement as a whole constitute cruel and unusual punishment.

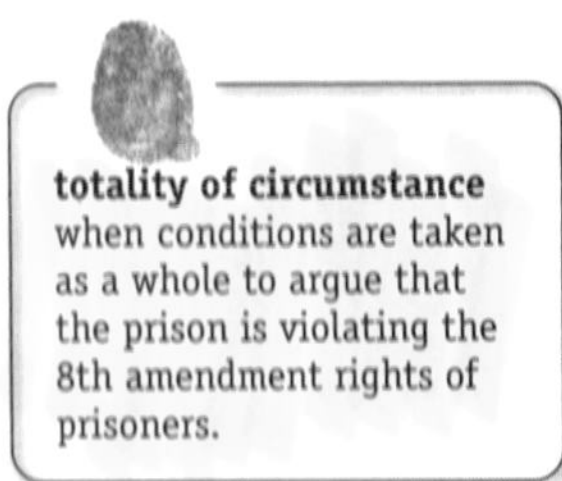

totality of circumstance when conditions are taken as a whole to argue that the prison is violating the 8th amendment rights of prisoners.

The landmark case that determined that a prison system as a whole could be administered in a manner violative of the Eighth Amendment standard was *Holt v. Sarver* (309 F. Supp. 362 [E.D. Ark. 1970], *aff'd* 442 F. 2d 304 [8th Cir. 1971]). Among the shocking list of violations enumerated in *Holt* were the following:

1. *A virtual absence of professional staff to supervise the inmate population*
2. *A prison system administered primarily by inmate trustees*

3. *An atmosphere of hatred and mistrust maintained by brutal use of physical force*
4. *An open-barracks system that invited physical assault*
5. *Unsanitary isolation cells*
6. *A complete absence of rehabilitation or training programs*
7. *Inadequate diet and medical care*
8. *Access to prison records, prescription drugs, contraband alcohol and drugs, weapons, and vehicles by some inmate trustees*

The findings were accompanied by remedial orders grounded in the belief that the legislative and executive branches of the state would rush to remedy the situation. These orders were so ineffective that, in subsequent years, reviewing courts concluded that the conditions described by the trial court had actually deteriorated (*Finney v. Arkansas Board of Correction,* 410 F. Supp. 251 [E.D. Ark. 1976]). Although *Holt* maintains a perhaps unique distinction as the nadir of prison conditions within the United States, Arkansas was not that unique. Cumulative substandard conditions of prisons and jails in many states were determined to be cruel and unusual punishment.

In some cases, the conditions of confinement experienced by prisoners have provided them with a viable defense to the felony of escape from prison. A criminal defendant charged with escape who claims the defense of duress or necessity must offer *bona fide* evidence that conditions within the institution constituted an immediate threat of death or serious bodily injury. Further, he or she must show that the escape was accomplished without threat of force or injury to third parties, and that an effort to surrender to law-enforcement authorities was made as soon as "the claimed duress or necessity had lost its coercive force" (*U.S. v. Bailey,* 100 S. Ct. 624, 638 [1980]). The recognition of this defense in both state and federal jurisdictions is an implicit acknowledgment that intolerable conditions may exist within penal institutions (Gardner and Anderson 1992).

There is consensus that the Eighth Amendment protects inmates from assault not only by correctional officials, but also at the hands of other inmates. Perhaps, in part, because it is a commonplace occurrence within penal institutions, courts have been reluctant to impose liability on prison officials for failure to protect inmates from physical and sexual assaults by other inmates. The protective umbrella of "good-faith" sovereign immunity has been widespread in the cases.

The inmate in these cases must demonstrate state officials' deliberate indifference to a known and substantial risk of violent harm. However, in at least one case, failure of prison officials to establish procedures to determine compatibility of cellmates was found to constitute deliberate indifference when members of competing criminal gangs were housed together and injury resulted (*Walsh v. Mellas,* 837 F. 2d 789 [7th Cir. 1988]).

In *Whitley v. Albers* (106 S. Ct. 1078 [1986]), the prisoner had been severely injured by a close-range shotgun blast during efforts to quell a disturbance. His position was that the gunshot wound, which resulted in permanent disability, constituted cruel and unusual punishment because it was not a justified use of force under the circumstances. The Supreme Court found no Eighth Amendment violation because the Eighth Amendment prohibits only the "unnecessary and wanton infliction of pain" in a prison setting. The application of extreme, even unnecessary, force to restore discipline and order violates the constitutional standard only if it was applied "maliciously" and "sadistically" and, specifically to cause harm.

Thus, it appears that only *deliberate* application of excessive force for the purpose of maliciously or sadistically causing harm will trigger a remedy under the Civil Rights Act in the context of disturbance of prison order and security. The use of physical force as a method of institutional control and discipline has largely been abandoned in the contemporary American penal institution. Remnants of the use of corporal punishment remain, but only as unsanctioned, informal power by some correctional officers.

Other Applications of the Eighth Amendment

Some cases have raised interesting issues about the constitutionality of more-"advanced" and more-contemporary applications of force, clothed as therapy. In *Knecht v. Gillman* (48 F. 2d 1136 [8th Cir. 1973]), the complaint stated that prisoners had been subjected to injections of the drug apomorphine without their consent. The evidence demonstrated that the injections were administered as "aversive stimuli." The injections, which were characterized by the state as "therapy" based on "Pavlovian conditioning" techniques, induced vomiting that lasted from 15 minutes to an hour, as well as changes in blood pressure and heart function. The simple question confronted by the court was whether this process, deliberately administered, constituted an acceptable therapeutic treatment or a cruel and unusual punishment. The practice was held to be a type of cruel and unusual punishment, clearly analogous to the infliction of an uncontrolled physical beating. Only with elaborate protections to ensure the voluntary consent of the participants could such novel therapy be lawfully employed in the future. The use of involuntarily administered drugs to control prisoner conduct is an intriguing issue for future study and litigation.

Unfortunately, the nation's prisons house large numbers of mentally ill inmates. Application of psychotropic or tranquilizing drugs over the objection of a prisoner—not for punishment, but rather, to protect the prisoner or others—was unsuccessfully challenged in a number of cases (*Sconiers v. Jarvis*, 458 F. Supp. 37 [D. Kan. 1978]; *Gilliam v. Martin*, 589 F. Supp. 680 [W.D. Okla. 1984]; and *U.S. v. Bryant*, 670 F. Supp. 840 [D. Minn. 1987]). The Supreme Court addressed procedural protections required before psychotropic (tranquilizing) drugs may be administered involuntarily to mentally ill inmates in *Washington v. Harper* (494 U.S. 210 [1990]). Although the Court found that the administration of the drugs deprived inmates of constitutionally protected freedom interests, it rejected the prisoner's claim that a judicial hearing with appointed counsel should be required. The Court found that an administrative hearing with an impartial trier of fact applying a standard of ". . . clear, cogent, and convincing. . ." evidence would suffice. Thus, it seems clear that the Court is comfortable with a relatively low level of due process before prison officials can administer drugs to a resisting inmate, as long as it is for treatment and not punishment.

Access to the Legal System

due-process clauses no person shall be deprived of life, liberty, or property without due process of law (this clause is in both the 5th and 14th amendments).

The U.S. Constitution does not specifically guarantee the right of access to the courts. The U.S. Supreme Court and the courts below it have nonetheless repeatedly affirmed that it is a fundamental right implied in the due-process clauses of both the 5th and 14th Amendments. Prisoners' rights to access were first acknowledged by the Supreme Court in 1941 in *Ex parte Hull* (61 S. Ct. 640 [1941]). In that case, the Court declared that state-prison regulations that required that all legal documents a prisoner might attempt to file with a court be first submitted to prison officials for examination and censorship were unconstitutional. Despite the *Hull* case, oppression and interference with the fundamental right of judicial access, consistent with the belief that an "iron curtain" should exist between convicted offenders and the free world, persisted for decades after the decision.

Despite prevailing in a series of legal challenges to official obstruction of prisoners' right of access to the courts, the need for legal services for prisoners in America remains largely unmet. Prisoners are often immersed in a variety of legal problems independent of criminal conviction. These include disputes such as child custody, divorce, child support, distribution of governmental benefits, and consumer matters.

Jailhouse Lawyers and Law Libraries

jailhouse lawyers inmates who have some legal knowledge and help other inmates with legal matters.

In 1969, in *Johnson v. Avery* (89 S. Ct. 747 [1969]), the Court declared unconstitutional a Tennessee regulation prohibiting jailhouse lawyers (inmates who use their legal knowledge for other inmates) from assisting other inmates in the preparation of habeas corpus petitions (absent any other effective form of legal assistance). In the 1974 opinion in *Procunier v. Martinez* (94 S. Ct. 1800 [1974]), the Court invalidated a California regulation that prohibited law students from entering prison to assist in case interviews and investigations. The Court noted that most prisons are located away from major population centers, handicapping the ability of lawyers to perform those tasks. Similar rules that inhibit attorney-client relationships by excessively restricting personal visits or telephone contact between prisoners and attorneys or their staff members have met a similar fate. *Bounds v. Smith* (97 S. Ct. 1491 [1977]) affirmed the holding of *Johnson* by establishing that the Constitution requires that prisoners have access either to adequate law libraries or to legal services to aid them in cases involving their convictions, prison conditions, or other prison problems.

In recent decisions, the Supreme Court of the United States has restricted the right of access to the courts. In *Lewis v. Casey* (116 S. Ct. 2174 [1996]), the Court declared that, in order to gain relief, prisoners must show that an "actual injury" resulted from the lack of legal materials. In *Shaw v. Murphy* (121 S. Ct. 1475 [2001]), the Court held that the right of prisoners to provide legal assistance does not receive First Amendment protection beyond that which is normally accorded prisoners' speech. Therefore, inmate-to-inmate correspondence must be judged by the proper constitutional test—that is, whether the restrictions are reasonably related to legitimate penological interests. In this case, a letter offering assistance was intercepted and read by prison officials, who then, on the basis of the letter's content, sanctioned the prisoner offering legal assistance.

Legal Correspondence

Freedom of correspondence between attorney and client was addressed in *Wolff v. McDonnell* (94 S. Ct. 2963 [1974]), which invalidated a Nebraska prison regulation authorizing the inspection of all incoming and outgoing inmate-attorney mail. Finding that inspection for purposes of intercepting contraband is distinguishable from censorship, the Court substituted a practice allowing the opening of letters from attorneys in the presence of the prisoners. Similar results have been reached in cases involving correspondence with governmental officials in the executive and legislative branches. More-recent cases suggest that the right of the inmate to be present may be overcome by a showing of probable cause or reasonable suspicion that incoming attorney mail contains impermissible material (*Proudfoot v. Williams,* 803 F. Supp. 1048 [E.D. Pa. 1992]).

First Amendment Rights

The following issues derive from the First Amendment: (1) access to prisons and prisoners by the press, (2) freedom of speech and communication, (3) freedom of association and visitation, and (4) freedom of religion.

Press and Media Access

During the Vietnam War, federal courts confronted a series of challenges to prison rules that barred or curtailed access to the press. Some inmates, such as draft resisters, were characterized as political martyrs rather than as conventional offenders. Other charismatic inmates—such as George Jackson, a black militant, and Charles Manson, a notorious murderer—openly courted and encouraged a high degree of media attention and celebrity.

The Federal Bureau of Prison's regulations, which completely denied press interviews with individual inmates, withstood a First Amendment violation challenge in a federal appellate court in *Seattle-Tacoma Newspaper Guild v. Parker* (480 F. 2d 1062 [9th Cir. 1973]). The Ninth Circuit decided that access to the press was not necessary to publicize legitimate inmate grievances because there were viable alternative means of communication. That court noted that the regulation in question would not interfere with the inmate's right to visit with relatives and friends, counsel with clergy, confer with legal counsel, enjoy access to the courts, or engage in correspondence. The court also noted that free access by the press to individual inmates could contribute to the evolution of inmate celebrity status, undermining constructive rehabilitation. Given the existence of viable alternatives to personal interviews, the court concluded that any burden on the media and the First Amendment in the challenged policy was justified.

The Supreme Court used the same analysis to decide the companion cases *Pell v. Procunier* (94 S. Ct. 2800 [1974]) and *Saxbe v. Washington Post* (94 S. Ct. 2811 [1974]). The Court majority held that the press enjoyed no *greater* right of access than did the general public. The principle of these cases was extended to embrace the electronic media and detainees awaiting trial in *Houchins v. KQED* (98 S. Ct. 2588 [1978]). Thus, even though prison officials may allow television interviews with convict celebrities such as Charles Manson, they do not have to.

Correspondence and Censorship

least-restrictive-means test test used under early 1st amendment caselaw; administrators had to use the means necessary to accomplish their purpose that was least intrusive toward Constitutional rights.

In *Palmigiano v. Travisono* (317 F. Supp. 776 [D. R.I. 1970]), the appellate court held that prison officials must employ the **least-restrictive-means test** with outgoing mail. Specifically, outgoing mail should not be read or otherwise interfered with absent a search warrant. Incoming mail, unless it was from a public official or attorney, was subject to inspection for contraband. Further, incoming mail that came from sources other than an approved addressee list could be inspected, not only for contraband, but also could be reviewed and censored for pornography or highly inflammatory writings.

The Supreme Court considered other institutional restrictions on correspondence in *Procunier v. Martinez* (94 S. Ct. 1800 [1974]). The Court sidestepped a determination of the First Amendment rights of prisoners by focusing its analysis on the rights of citizens with whom a prisoner might correspond. The California regulations under scrutiny in the case prohibited inmate correspondence that "unduly complained of" or "magnified" grievances, two fatally vague standards. The expression of "inflammatory political, racial, religious, or other views or beliefs" in writing was likewise prohibited. Third, inmates could not mail letters that pertained to criminal activity, contained foreign matter, or were lewd, obscene, defamatory, or otherwise inappropriate. The sweeping and vague nature of the prohibitions doomed these regulations.

Procunier declared that restrictions on prisoners' right of correspondence must demonstrably further a substantial governmental interest: specifically, security, order, or rehabilitation. Any limitation on freedom of expression had to be carefully drawn so that it would be no greater than necessary for protection of the governmental interest. Thus, even if a regulation furthers an important, legitimate governmental interest, it is invalid if its coverage is unnecessarily broad.

This approach was substantially relaxed in the subsequent case of *Pell v. Procunier* (94 S. Ct. 2800 [1974]) in which the Court pronounced that First Amendment restrictions need only be demonstrated to be "reasonably related to a legitimate security concern," a standard that indicated substantial deference to the expertise of prison administrators. *Martinez* and *Pell* were hollow victories for prisoners. They established that prisoners and those with whom they correspond retain rights to communicate, but the decisions imposed a relatively light burden to justify mail censorship.

Later, the Supreme Court accepted Missouri regulations that gave prison administrators the power to completely prohibit correspondence between inmates in different prisons as reasonably related to security interests (*Turner v. Safley*, 107 S. Ct. 2254 [1987]). Four dissenters, led by Justice Stevens, were not persuaded and expressed the belief that the regulations were an "exaggerated response" to the problem.

Distinct from the issue of interpersonal communications is censorship of publications. It is common for prison authorities to suppress reading materials unless they come directly from the publisher (the "publisher-only rule"). Officials also censor publications deemed inflammatory, obscene, racially divisive, or likely to incite criminal or other inappropriate conduct. The publisher-only rule, grounded in a concern about contraband and weapons concealed in books and magazines, was affirmed in *Bell v. Wolfish* (99 S. Ct. 1861 [1979]), a case involving pretrial detainees in jail. It may be assumed that any restrictions on the rights of pretrial detainees apply to the convicted, since pretrial detainees should enjoy more-expansive rights.

Some clarification of the standards in this delicate area was produced in *Thornburgh v. Abbot* (109 S. Ct. 1874 [1989]). At issue were regulations promulgated by the Federal Bureau of Prisons. Federal administrators attempted to give fair notice to the inmate population of the circumstances under which a publication might be intercepted and withheld. Publications determined to be "detrimental to the security, good order, or discipline of the institution" or that might "facilitate criminal activity" were targeted. The Court relied upon the four-pronged *Turner v. Safley* test:

1. Is the regulation rationally related to a legitimate governmental objective, and is that objective a legitimate and neutral one, without regard to content?
2. Are there open to prison inmates alternative means of exercising the right?
3. What adverse impact will the asserted right have on guards and other inmates?
4. Is the regulation an exaggerated response to the problem—that is, are less-restrictive alternatives available?

This four-pronged approach is the current standard for analysis, not only of First Amendment claims of prisoners, but of other claims of rights as well. Opponents of the majority approach often cite it as evidence of a gradual rediscovery of the hands-off doctrine. At the least, it represents an increased deference to administrative expertise.

Interference with, and restraints on, the ability of inmates to prepare and publish articles and books while incarcerated is another area of First Amendment dispute. It is unclear what governmental interest is served by preventing inmate writings from access by the general public. Such censorship would have to be defended on the tenuous ground that the act of creation is "antirehabilitative." Certainly there may be varieties of fiction or quasi-fiction highly reminiscent of the offender's criminal history that would cause concern, but proving that they undermine rehabilitation is a legal challenge that raises concerns about the limits of state authority to control thoughts and beliefs. Legislatures in recent years have attempted to remove the financial incentive to inmate literary enterprise by creating statutes that require the proceeds from a contract with a prisoner who had admitted guilt of or been convicted of a specific crime to be placed in escrow with the state and then distributed as restitution if the depiction was of that specific offense. Any depiction of the person's crime in book, magazine, film, or other medium of communication would be subject to the statutory provisions. Such legislation is known popularly as a "Son of Sam" law, after David Berkowitz, a serial murderer convicted in New York.

In *Simon & Schuster v. Members of New York State Crime Victims Board* (112 S. Ct. 501 [1991]), a publisher who had contracted with an organized-crime figure for a book about his life filed suit, challenging the "Son of Sam" law on First Amendment grounds. The Court recognized that the statute in question served two compelling state policies: (1) that criminals not profit from their crimes, and (2) that victims receive compensation. Balancing the gravity of a substantial interference with freedom of expression and the public's interest in having access to information against those two legitimate interests resulted in a finding that declared the statute unconstitutional. The statute was determined to be defective because of its sweeping scope, and the fact that it would encompass accounts of crime that had occurred many years earlier and could no longer be prosecuted because the statute of limitations had run out.

Freedom of expression within prisons is viewed differently because the rights of free-world citizens to communicate are not involved. Rigorous restrictions on prisoners' freedom of expression within the institution have generally received favorable treatment by courts. Issues include sanctions against the use of unsavory language, the right to solicit organization membership, the right to solicit funds in support of political agendas, and the right to petition grievances to the administration or legislature. Discipline imposed against prisoners for angry racial comments, peaceful work stoppage, and petitions protesting prison conditions are usually sustained. Strong restraints on speech and heavy penalties for violation of those restraints may be necessary in penitentiaries. Maintenance of order and security are always foremost.

Although prisoners have the right to communicate their grievances to prison officials, spontaneous colorful speech by inmates remains a different matter, as evidenced by *Ustrak v. Fairman* (781 F. 2d 573 [7th Cir. 1986]). In this case, the appellate court sustained prison regulations prohibiting speech that was disrespectful to prison employees or was considered ". . . vulgar, abusive, insolent, threatening or improper language toward any . . . resident or employee."

Attempts by prisoners to communicate with each other and form organizations not approved by prison administrators were addressed definitively in *Jones v. North Carolina Prisoners' Labor Union* (97 S. Ct. 2532 [1977]). Although prisons actively encourage many organizations within their walls (such as Rotary International, the Junior Chamber of Commerce, Boy Scouts, and Alcoholics Anonymous), each has a relationship to a valid penological purpose—rehabilitation. The ability of prisoners to form and maintain organizations that openly challenge conditions of confinement is another matter. In *Jones,* North Carolina had adopted regulations that prohibited inmates from soliciting other inmates to join the union, barred all union meetings, and denied use of the mail to deliver applications and information within the prison. The Supreme Court sustained the state regulations, finding that First Amendment rights must give way when an association possesses a likelihood of disruption of order or stability or otherwise interferes with legitimate penological objectives. Despite vigorous dissent by Justice Marshall, the majority's concerns for inmate exploitation, power struggles within the prison, and creation of unrealistic expectations prevailed. *Jones* and *Bell v. Wolfish* are widely regarded as determinative cases in stifling the blossoming prisoners' rights movement. The Court rejected the traditional "clear and present danger" test for First Amendment issues in prison matters in favor of a much more elastic "reasonable likelihood of danger" approach. Later, the *Turner v. Safley* test all but assured prison officials' victory in any challenge to prison regulations unless the regulation was blatantly abusive.

Visitation

It is widely believed that visitation is essential for the offender who will reenter society. Maintenance of social bonds between family, friends, and community are essential elements of successful reintegration into society. There is virtual unanimity on the need for visitation (National Advisory Commission 1973; American Bar Association 1981; United Nations 1956). Recognition of a right to, and encouragement of, visitation is widely reflected in professional commentary.

Despite arguments favoring visitation, numerous cases have categorized visitation as a *privilege* that may be curtailed or denied for just cause, not a *right*. This categorization is applied to furloughs as well. Justifying this are concerns for institutional security (weapons and contraband infiltration), for associations that undermine punishment or rehabilitation, and the logistical burdens of administering visitation. Administrative controls over visitation are accepted unless a clear abuse of discretion is shown. The courts have held that a protected liberty interest in visitation occurs only when state regulations or statutes create it with "explicitly mandatory language." Prison administrators may tightly regulate visitation privileges by prohibiting physical contact, even between parents and minor children. They may also require intrusive physical searches of inmates before and after a contact visit without probable cause or reasonable suspicion. Since visitation is rarely defined as a right, the remote location of prisons also acts as a significant barrier to visitation. Likewise, the discretionary transfer of inmates to remote institutions and its adverse impact on visitation was determined not to raise a constitutional issue in *Olim v. Wakinekona* (103 S. Ct. 1741 [1983]).

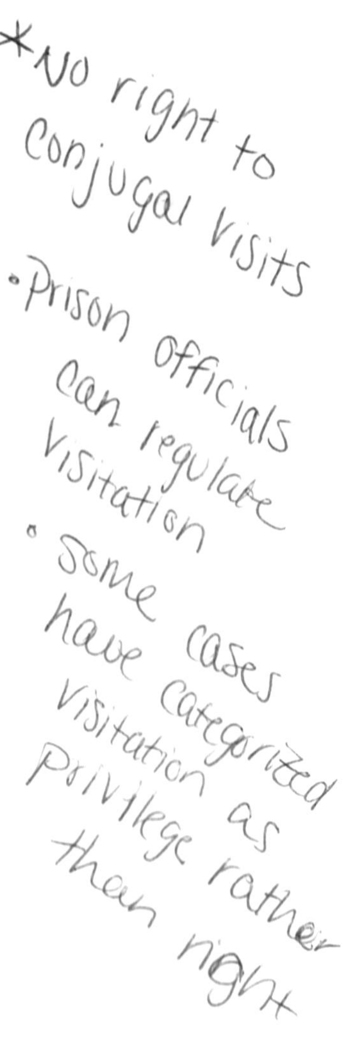

The most significant decisions regarding visitation rights have considered the status of pretrial detainees rather than convicted offenders. In *Block v. Rutherford* (104 S. Ct. 3227 [1984]), the Supreme Court held that jail inmates could be denied contact visits with spouses, children, relatives, and friends on security grounds. *Bell v. Wolfish* (99 S. Ct. 1861 [1979]), defined jail inmate search standards before and after receiving a visit. Conjugal visitation is not a popular practice within American correctional facilities, although it is fairly commonplace in Europe and Latin America. In *Polakoff v. Henderson* (370 F. Supp. 690 [D. Ga. 1973]), the court held that no constitutional right to such visitation exists.

Women who bear children while incarcerated or are mothers of infants or small children pose different (and difficult) problems for analysis of visitation rights. Few courts to date have recognized any special or unique rights at stake between mother and child. A humane system of corrections, if for no other reason than interest in preventing future criminal conduct, should make efforts to minimize damage to the parental bond. Minimal attention has been given this issue in the courts, and it has been left to the discretion of administrators working within the confinement of their budgets. In *Women Prisoners of the District of Columbia Department of Corrections v. District of Columbia* (877 F. Supp. 634 [D.D.C. 1994]), the trial court found an Eighth Amendment (cruel and unusual punishment) violation in the lack of opportunity and appropriate facility for child visitation. It further found that lack of adequate child-placement counseling for female prisoners immediately after giving birth resulted in an unacceptable risk of psychological trauma.

Despite these lower-court holdings, the Supreme Court's view of rights to visitation is predictably narrow, and reflects due deference to administrative concerns and expertise. *Overton v. Bazzetta* (123 S. Ct. 2162 [2003]) is a recent case that concerns rigorous Michigan restrictions on visitation. The Supreme Court applied the *Turner v. Safley* test and predictably upheld the prison officials' restrictions.

Religious Practices

Freedom of religion is a primary constitutional right guaranteed by the First Amendment. It provides not only a barrier to state regulation of religious beliefs and practices, but also to state promotion of religious doctrine. Judicial interpretation of the intent and scope of the First Amendment has resulted in a policy that the guarantee of free religious *belief* is an absolute, while freedom to *act* in the exercise of religious belief is subject to regulation.

A primary issue in the consideration of prisoner religious practices is determination of whether the practice in dispute is the product of a sincerely held religious belief or an attempt to gain privilege and advantage through subterfuge veiled in the First Amendment. There has been substantial litigation on the issue of freedom of religion within the confines of penal institutions, a significant portion of it revolving around the exercise of religious practices by confined Black Muslims.

The Black Muslim movement within prisons was linked directly to social strife and change external to prison. Demands for separate religious services, special diets, alternative times of worship, specialized religious tracts and books, and alteration of personal appearance were at first vigorously resisted by prison administrators. Restrictions on religious practices were upheld at first by judicial deference to administrative concern for, and expertise in, security and discipline (*Ex parte Ferguson*, 361 P. 2d 417 [Cal. 1961]).

In later challenges, however, prisoners prevailed, and religious diversity has been accommodated, within boundaries, in American penal institutions. Once again, however, the Supreme Court has closed the door on further religious challenges by its devotion to the **rational relationship test**. The current legal standard, or test, to be applied in religious-freedom cases was announced in *O'Lone v. Estate of Shabazz* (107 S. Ct. 2400 [1987]). That case, in which Muslim inmates challenged work assignments that made it impossible for them to attend Friday-evening religious services, resulted in a victory for state interest when the Court applied the rational-relationship test from *Turner v. Safley*. The majority, in a narrow 5-4 decision, determined that the restrictions were valid because they were "reasonably related to legitimate penological interests."

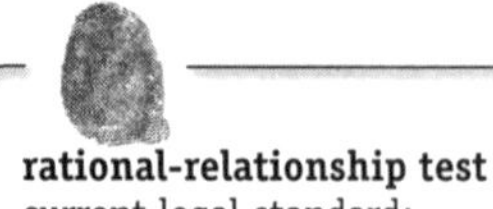

rational-relationship test current legal standard; states have to show a legitimate state interest and prove that the rule, policy, or practice in question is reasonably related to that state interest.

Despite *O'Lone*, major advances in the free exercise of religion have occurred (Smith 1999; Palmer 2003). A summary list of such advances includes the following:

1. A Buddhist inmate must have a reasonable opportunity to pursue his faith if other inmates are allowed to participate in conventional religious practices
2. An Orthodox Jewish inmate has the right to a diet that would sustain her in good health and not be offensive to her religious beliefs
3. A Black Muslim cannot be punished with administrative segregation for refusal to handle pork
4. A Cherokee Indian inmate may have the right to have long hair
5. The government is required to pay a Muslim minister at the same hourly rate paid chaplains

Two recent statutes are pertinent to prisoner's religious freedom. The Religious Freedom Restoration Act of 1993 (**RFRA**) (42 U.S.C.S. § 2000bb et seq.) sought to more fully detail the power of the government to pass laws that affected religious freedoms. In *City of Boerne v. Flores* (521 U.S. 507 [1997]), the Supreme Court concluded that Congress lacked the authority to enact the legislation. It was unclear whether or not Congress meant for the act to apply to prisoners, but then Congress responded with another statute, the Religious Land Use and Institutionalized Persons Act (RLUIPA) (42 U.S.C. §§ 2000cc to 2000cc[5]). Like its predecessor, this act prohibits state and local governments from imposing any "substantial burden" on anyone's—including an inmate's—free exercise of religion unless the restriction furthers a compelling governmental interest and is also the least restrictive means to achieve that objective. RLUIPA is confined to programs receiving federal funding in which the restriction on religious freedom affects interstate/international commerce or commerce with Native American tribes.

RFRA Religious Freedom Restoration Act; Congress passed the Act to protect religious rights, but the Supreme Court struck it down as unconstitutional.

In the recent case of *Cutter v. Wilkinson* (349 F. 3d 257 [6th Cir. 2003]), the RLUIPA was challenged. Ohio prisoners filed suit against the prison system, arguing that various rules and restrictions violated their right to practice religion as protected by RLUIPA. The state challenged the constitutionality of RLUIPA, alleging that it exceeds the powers of Congress, violates states' rights under the 10th amendment, and improperly advances religion in violation of the establishment clause. The Court of Appeals for the Sixth Circuit agreed with the state's argument that if the act was applied, inmates would be able to have certain freedoms because of their religion while others would not have such freedoms, and that would be a violation of the establishment clause. The Supreme Court was expected to decide this case during its 2005 term. The future of prisoner freedom-of-religion cases is uncertain at present.

Personal Appearance

An intriguing issue litigated under color of the First Amendment is personal appearance. There is disagreement among penologists about the effect of prison regulations that restrict inmate appearance. Prison life is dehumanizing. Prisoners are denied control of symbols of identity such as clothing, hairstyle, facial hair, and decorative accessories. This stripping of personality is reinforced by the sterility and monotonous decor of prison life. Tolerance of diversity in clothing and appearance is largely a matter of administrative discretion and correctional philosophy because courts have generally refused to disturb regulations governing appearance. Although some lower courts have dissolved such regulations when appearance is linked to religious beliefs, the general rule is to the contrary (*Fromer v. Scully,* 817 F. 2d 227 [2d Cir. 1987]). Restriction on appearance is an example of the many areas in which courts do not generally interfere with administration of the institutions. The theory that appearance is a form of speech entitled to constitutional protection within the confines of penitentiaries has minimum viability.

Summary of First Amendment Protections

In summary, litigation of issues related to the First Amendment has resulted in marginal, but important, gains for inmates in free expression and communication. Prisoners have been successful in forging rights related to religious practices so long as they are not overly burdensome to accommodate and are based in sincerely held, recognized religious beliefs and practices. The courts have been astute in rejecting prisoner requests where they detect that the prisoner is manipulating First Amendment principles to gain special privilege, mock institutional authority, or engage in subterfuge (*Theriault v. Carlson,* 495 F. 2d 390 [5th Cir. 1974]).

In general, all First Amendment claims are now determined by applying the *Turner v. Safley,* or "rational-relationship," test. If prison officials can identify a legitimate state interest and show that the rule or regulation in question is rationally related to such interest, they are likely to win against any prisoner challenge.

The Right to Privacy

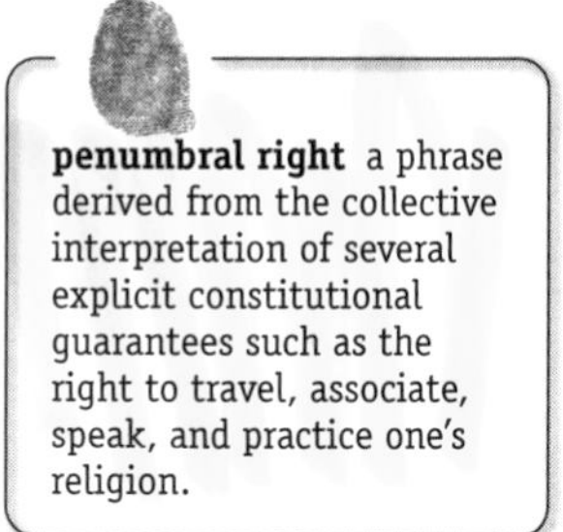

penumbral right a phrase derived from the collective interpretation of several explicit constitutional guarantees such as the right to travel, associate, speak, and practice one's religion.

The right of privacy, a phrase found nowhere in the U.S. Constitution, is a curious alchemy sometimes referred to as a **penumbral right**. It is a phrase derived from the collective interpretation of several explicit constitutional guarantees such as the rights to travel and associate freely, freedom of speech and religion, and freedom from unreasonable search and seizure. It symbolizes a respect for individual integrity and minimal state regulation of behaviors, including procreation, reasonably deemed to be "private." The concept that prisoners might enjoy some vestigial expectation of privacy, a theory associated with the rehabilitative ideal, has been dashed on the rocks of administrative convenience and necessity.

The death knell for even a minimal expectation of privacy in prison cells was sounded in *Bell v. Wolfish* (99 S. Ct. 1861 [1979]), a case in which pretrial detainees (awaiting trial and presumed innocent) had won a lower-court victory establishing that they had a right to be present at and observe cell searches and be free from body-cavity searches conducted after visitation, absent probable cause to believe the inmate was concealing contraband. Those holdings were reversed by the Supreme Court, which held that institutional-security concerns were paramount. The only concession to the right of privacy and freedom from unreasonable search recognized in the majority opinion was that prison administrators could not use searches abusively for purposes of harassment and punishment.

Many lower courts resisted the *Bell v. Wolfish* holding, and the Supreme Court further clarified its position in *Hudson v. Palmer* (104 S. Ct. 3194 [1984]), maintaining that a prisoner has no "reasonable expectation of privacy" in a prison cell. Searches and seizures—of inmates themselves or of their living quarters—that inflict injury, are more intense than necessary, or are conducted exclusively for purpose of harassment remain viable civil-rights violations, but the presumption is that discretionary searches are valid.

The Supreme Court decision in *Turner v. Safley* inferred that prisoners retain some constitutionally protected privacy interests when it ruled that a regulation restricting inmate marriage violated a fundamental right. The right of marriage and selection of a marriage partner have been recognized as within the boundaries of the right to privacy. Courts may infer that inmates have residual privacy rights, as the Supreme Court did, without specifically identifying the source of the rights.

Several cases have addressed sexually discriminatory search policies in regard to arrestees or those temporarily detained in jails. Prominent among these is *Mary Beth G. v. City of Chicago* (723 F. 2d 1263 [7th Cir. 1983]). Although a number of states have placed statutory limits on strip searches of people arrested for minor offenses, the court's invalidation in *Mary Beth G.* of strip searches and visual body-cavity inspections of women arrested for minor infractions rested, in part, on the fact that such inspections were not routinely performed on men arrested on minor charges. There is a presumption against the validity of degrading and humiliating searches of arrestees, unless based at least upon reasonable suspicion that reinforces the 14th Amendment equal-protection policy against sexually discriminatory treatment of arrestees. Any situation in which female prisoners experience greater intrusion of personal privacy than that experienced by their male counterparts would probably be violative of equal protection.

Substantial litigation has resulted from the collision of rights guaranteed by the Equal Employment Opportunity Act and the privacy interests of prisoners. These challenges have presented difficult problems. For instance, in an early case, the court accepted a rationale that justified the exclusion of female guards from contact positions in Alabama's maximum-security male institutions. In *Dothard v. Rawlinson* (97 S. Ct. 2720 [1970]), the Court decided that exposing female guards to the male maximum-security prisoner population in Alabama, *under the conditions then existing*, posed an unacceptably high security risk. The Court characterized Alabama's prisons as having intolerable levels of rampant violence and disorganization. Therefore, they concluded that male gender was a "*bona fide* occupational qualification" despite the willingness of some female guards to expose themselves to risk.

As to the inmates' rights to privacy, courts have generally concluded, as in *Brasch v. Gunter* (unpublished opinion, D.S.D.C. Nebraska 1985), that inmates are protected from unrestricted viewing of genitals or bodily functions by members of the opposite sex in the workforce of the prison. This restriction on the performance of routine duties of guards may impede the career advancement of female guards in male penitentiaries and vice versa. Accidental or infrequent observations of private functions and nudity by a member of the opposite sex of the prisoner have been classified as a minimal intrusion on privacy interests, inadequate to rise to the level of a constitutional violation. Inmates' constitutionally protected privacy interests are not violated when members of the opposite sex are present in inmate living quarters for a brief and predictable amount of time (*Avery v. Perrin*, 473 F. Supp. 90 [D.N.H. 1979]).

In addition to privacy issues raised by visual observation of prisoners, gender issues have been raised about the performance of more-intrusive searches of inmates. Pat-down or "frisk" searches of inmates by members of the opposite sex are generally condoned if they are conducted in a manner that does not include touching of genital areas. In *Smith v. Fairman* (678 F. 2d 52 [7th Cir. 1982]), the frisk by female correctional officers of male inmates' necks, backs, chests, stomachs, waists, buttocks, and outer legs was approved. In *bona fide* emergency situations, not only can correctional officers of the opposite sex pat down the genital areas of inmates, but they can also participate in more-intrusive strip or body-cavity searches. In *Lee v. Downs* (641 F. 2d 1117 [4th Cir. 1981]), the court found no invasion of the female inmate's privacy protections when two male correctional officers restrained the arms and legs of the woman while a female nurse conducted a body-cavity search for matches after the inmate had set her clothing afire.

The courts have had to balance the employment interests of opposite-sex correctional officers with inmates' privacy interests. The most thorough analysis of the issue is found in *Forts v. Ward* (621 F. 2d 1210 [2d Cir. 1980]). The court confronted the clash between principles of equal-employment opportunity and female-inmate privacy interests, and fashioned a remedy partially accommodating the competing interests. In *Galvan v. Carothers* (855 F. Supp. 285, 291 [D. Alaska 1994], *cert. denied* 515 U.S. 1149 [1995]), the court held that ". . . minimal standards of privacy and decency include the right not to be subject to sexual advances, to use the toilet without being observed, and to shower without being viewed by members of the opposite sex." Also, in *Everson v. Michigan Department of Corrections* (222 F. Supp. 2d 864 [2002]), the court agreed that women in prison, because of their history of sexual abuse, should be granted greater privacy rights against male officers supervising them in showers or private functions. The many cases addressing gender issues illustrate the difficulty of detailed supervision of correctional practices by courts (Wooster 2002).

Medical Treatment in Correctional Institutions

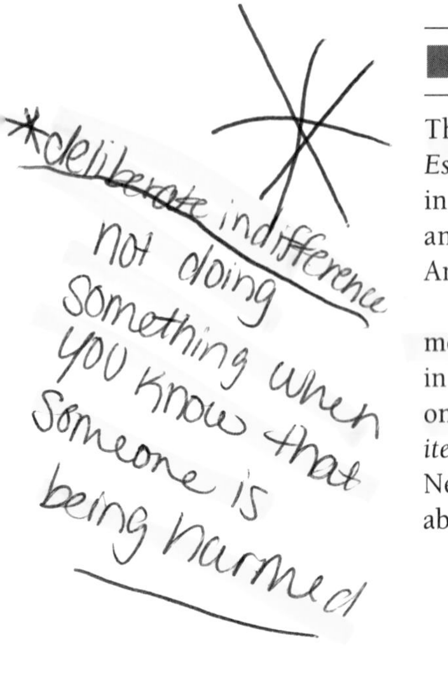

The landmark case defining the appropriate level of medical care for prisoners is *Estelle v. Gamble* (97 S. Ct. 285 [1976]). It was preceded by cases, such as those in Arkansas and Alabama, in which grossly inadequate medical care was cited as an element of penitentiary conditions that, as a whole, violated the Eighth Amendment.

In *Gamble*, the Supreme Court clearly established that there is a right to medical care and a corresponding duty of the state to provide it. It was established in *Gamble* that ". . . deliberate indifference to the serious medical needs of prisoners constitutes unnecessary and wanton infliction of pain proscribed *(prohibited)* by the Eighth Amendment" (*Estelle v. Gamble*, 97 S. Ct. 285, 291 [1976]). Negligent or careless failure to provide adequate medical care, while far from laudable, would not constitute a viable cause of action under the civil-rights statute.

Accidents, negligence, or professional disagreement on diagnosis or treatment of illness or injury are not within federal civil-rights jurisdiction. Nonetheless, agreed settlements of prison-reform lawsuits routinely addressed and remedied deficiencies in staff and medical facilities, and there has been widespread and systematic improvement in the health care extended to prisoners.

The special issues of women's health services, the needs of disabled and geriatric prisoners, and the provision of care to the terminally ill must be addressed as these issues proliferate when mandatory-sentencing schemes are installed. Legal issues of prisoners with AIDS are rapidly emerging. They include the right to treatment, quality of care to be administered, protection of inmates and staff, the right of HIV-positive individuals and AIDS patients to remain in the general population, mandatory HIV testing, the relevance of medical condition to parole and furlough decisions, and the issue of notification of spouses and sexual partners on release (Branham 1990; Hammett and Moiri 1990).

In *Nolley v. County of Erie* (776 F. Supp. 715 [W.D.N.Y. 1991]), automatic segregation of a female prisoner known to be HIV positive in a wing otherwise reserved for mentally ill inmates was held to violate the constitutional right to privacy. However, in *Harris v. Thigpen* (941 F. 2d 1495 [11th Cir. 1991]), the court found that segregation of all HIV-positive inmates was constitutionally permissible.

A particularly thorough case that paints a grim picture of health care in the California state-penitentiary system is *Madrid v. Gomez* (889 F. Supp. 1146 [1995]). The court found systemic inadequate health-care deficiencies in staffing levels, training and supervision of medical employees, medical-records maintenance, initial screening for serious contagious diseases such as tuberculosis and AIDS, access to medical care, and quality control of medical care when it did occur. The trial court found the medical-care system at Pelican Bay grossly inadequate, and the administration deliberately indifferent to the deficiencies. The language of the court in this case is reminiscent of the shock experienced by judges in the earliest prison-reform cases.

Recent cases such as *Madrid v. Gomez* remind us that the issue of adequate care of prisoners is a continuing dilemma, particularly in health care, with its attendant expenses. The case also revealed a pattern of application of excessive physical force to inmates. Many troublesome issues that transcend provision of health care await resolution. For example, what individuals, within and without the institutional setting, are entitled to notification of the HIV-positive status of a prisoner? The prevalent theory is that other prisoners and prison staff in non-health-related capacities have no right to notice. According to the Health Insurance Portability and Accountability Act of 1996, family and associates of the inmate have no right to receive notice of his or her infectious status.

As previously noted, the policies of retribution, isolation, and deterrence are increasingly the foundation of sanctions, while treatment and rehabilitation rationales are diminishing in significance. Nonetheless, courts continue to recognize the dependency imposed by incarceration. Closely related to the issue of adequate medical treatment of physical ailments is that of treatment for emotionally unhealthy inmates. The serious medical ailments implicated in *Estelle v. Gamble* are not restricted to physical ailments. Unattended mental illness may cause pain and suffering that equals or exceeds that attributable to physical injury.

A prisoner, by virtue of confinement, will be depressed during some or all of a sentence. Because depression is so common but not easily measured, issues arise as to when a mental or emotional problem imposes a constitutional duty requiring the custodian of the inmate to provide care. In *Parte v. Lane* (528 F. Supp. 1254 [N.D. Ill. 1981]), the court found that the inmate's complaint that he felt depressed and was not receiving adequate attention to his condition did not give rise to a claim of sufficient magnitude. According to the *Parte* case, the constitutional entitlement to psychological or psychiatric care is triggered when a physician or other health-care provider concludes, with reasonable medical certainty, that: (1) the prisoner's symptoms evidence a serious disease or injury, (2) such disease or injury is curable or may be substantially alleviated, and (3) the potential for harm to the prisoner by reason of delay or denial of care would be substantial.

The right to treatment under these circumstances is limited to that which may reasonably be provided under constraints of time and cost (*Bowring v. Godwin*, 551 F. 2d 44 [4th Cir. 1977]). The Supreme Court extended the deliberate-indifference standard derived from *Estelle v. Gamble* to medical, dental, and psychiatric care and defined it narrowly. In *Farmer v. Brennan* (114 S. Ct. 1970 [1994]), it was held that a prison official can be liable under the Eighth Amendment for denying an inmate humane conditions of confinement only if the official both knows of and disregards an excessive risk to inmate health and safety. The current posture of the courts appears to be that prisoners are entitled to adequate, rather than exemplary or even good, health care.

Women's needs for health care may exceed the customary needs of male prisoners. In everyday life, women are less reluctant than men to seek medical services; the same is probably true in institutional settings. Women's needs for medical care exceed those of men because of the more complex biology of their reproductive systems (Pollock 2002). If greater need is acknowledged and inferior delivery of health care exists, a violation of the principle of equal protection exists.

Recent literature indicates that substance abuse is more common among women than men; that female arrestees are more likely to test positive for drugs, especially cocaine and heroin; that a greater percentage of women are incarcerated for drug-related offenses; and that women are more likely than men to have been using drugs daily at the time of their arrest (Fletcher, Shaver, Dixon, and Moon 1993). This implies not only an enhanced need for medical services related to substance abuse, but also an increased risk factor for HIV and other diseases associated with drug use (Greenspan 1990).

A class-action suit brought on behalf of all female prisoners in the District of Columbia found Eighth Amendment violations in sexual harassment, inferior living conditions, and lack of medical care (*Women Prisoners of the District of Columbia Department of Corrections v. District of Columbia*, 877 F. Supp. 634 [D.C. 1994]). The court, applying the deliberate-indifference standard, found liability for inadequate health care, especially gynecologic services, health education, prenatal care, ineffective prenatal education, and overall inadequate prenatal protocol. In a scathing portion of the opinion, the judge found that shackling of female prisoners in the third trimester of pregnancy and immediately after childbirth violated contemporary standards of decency.

Equal Protection

In *Glover v. Johnson* (478 F. Supp. 1075 [E.D. Mich. 1979]), the court held that female inmates were provided educational and vocational programs dramatically inferior to those offered male counterparts. The court, finding an equal-protection violation, prohibited such differential treatment as discriminatory. Equal access to educational and vocational programs must be provided unless the state could demonstrate an important, legitimate governmental objective in the provision of inferior treatment to female offenders. This case has continued coming back for rehearings for more than 20 years, and only recently has the state of Michigan been released from court monitors (*Glover v. Johnson*, 199 F. 3d 310 [6th Cir. 1999]). Since *Glover* was first decided, the analysis of equal-protection cases concerning gender has changed, as have the courts' treatment of prisoners' rights cases. Some would argue that today the case might have been decided differently because of relaxed tests to determine gender discrimination ("suspicious" scrutiny rather than "strict scrutiny") and the extremely generous test for prisoner challenges ("rational relationship").

The American Bar Association Standards Relating to the Legal Status of Prisoners (1981) states in Standard 26-6.14 that prisoners should be free from discriminatory treatment based solely on race, sex, religion, or national origin. Furthermore, the standards recommend that prisoners of either sex may be assigned to the same facility or assigned to separate facilities so long as there is essential equality in living conditions and access to community and institutional programs, including education, employment, and vocational training.

Since maintenance of order and safety is the highest priority of the prison administrator, and since racial violence seems to be endemic in prisons, officials have segregated inmates of different races. Such segregation may appear as a legislative mandate, administrative regulation, de facto phenomenon, or as a temporary response to a disturbance. It may also assume subtle camouflage in any area of prisoner treatment, such as classification, work assignments, and access to therapeutic programs.

Statutorily imposed racial discrimination in prison facilities was first addressed in *Washington v. Lee* (263 F. Supp. 327 [M.D. Ala. 1966]). Both white and black prisoners filed a complaint alleging violation of the equal-protection clause of the 14th Amendment. The district court brushed aside the state defense of institutional security and held the practice unconstitutional. The district court was affirmed per curiam (which means a hearing before the full court) by the U.S. Supreme Court in *Lee v. Washington* (88 S. Ct. 994 [1968]). Three justices wrote a concurring opinion in which they observed that an extremely narrow exception to the prohibition on racial segregation might exist where there is a clear and immediate threat to institutional security in jails or prisons, a result only implied in the main opinion. Even a specific and factually grounded concern for institutional security is an impermissible reason for racial segregation until other alternatives—such as reduction of the prison population, increased supervision, or disciplinary action against sources of conflict—have been fully explored (*Blevins v. Brew*, 593 F. Supp. 245 [W.D. Wis. 1984]). Claims that inmates prefer racial segregation and that it is thus a voluntary election by the inmates were char-

per curiam hearing before the whole court.

acterized as a "gauze for discrimination" in the Fifth Circuit case *Jones v. Diamond* (636 F. 2d 1363 [5th Cir. 1981]).

Regarding First Amendment issues, Mexican-American inmates have been generally successful in gaining access to publications catering to ethnic issues unless they are extremely inflammatory and conducive to violence. Courts have not been sympathetic toward demands for accommodation of cultural tastes in food or in popular entertainment such as movies (*U.S. v. State of Michigan*, 680 F. Supp. 270 [W.D. Mich. 1988]). While racial tension and friction continue to be society's overriding social problem within and outside the walls, when called on to address the legal issue of the rights of prisoners, the courts have spoken in an unmuffled voice in favor of the principles of desegregation and equal protection of the law.

Procedural Due Process

The due-process clauses of the U.S. Constitution prohibit both state and federal authority from taking life, liberty, or property without due process of law. The meaning and scope of procedural due process is constantly evolving. Its most familiar application is in the criminal trial, where defendants are guaranteed an impartial trier of fact, a jury, counsel, right to remain silent, the right to confront and cross-examine accusers, and so on. In the past three decades, governmental agencies have experienced a dramatic "due-process revolution," imposed by courts, to curtail the abuses of governmental discretion. The case that initiated this trend was *Goldberg v. Kelly* (90 S. Ct. 1011 [1970]), in which the Court prohibited a public-assistance agency from terminating benefits without a pretermination evidentiary hearing. Since that time, any allegedly "grievous loss" imposed on a person by state action has been subject to judicial challenge as to the need for and adequacy of procedural due-process protection.

Prisoners challenged the adequacy of due-process protections in disciplinary hearings, with modest success. In *Wolff v. McDonnell* (94 S. Ct. 2963 [1974]), the Supreme Court imposed a minimal due-process model on prison disciplinary proceedings involving punitive isolation or diminished eligibility for early release. Three safeguards to check abuse of discretion were imposed:

1. *Advance written statement of the claimed violation (24 hours minimum)*
2. *A written statement by an impartial fact finder as to evidence relied on and reasons for the discipline*
3. *The right to testify and the right to call witnesses and present documentary evidence, unless the fact-finder concludes that these rights would undermine institutional security or valid correctional goals*

The Court majority specifically denied appointment of counsel and confrontation or cross-examination, classifying such procedures as hazards to valid correctional goals. Many advocates of correctional reform were dismayed by the decision, but the Supreme Court has subsequently reduced even the minimal procedural due-process requirements of *Wolff*.

In 1985, the Court stated in *Superintendent, Massachusetts Correctional Institution, Walpole v. Hill* (105 S. Ct. 2768 [1985]) that disciplinary revocation of a prisoner's accrued good time (which adversely affects date of release) could be based on "some evidence" contained within the administrative record. This low threshold of evidence to sustain disciplinary sanctions has greatly reduced the hearings requested by prisoners. In *Ponte v. Real* (105 S. Ct. 2192 [1985]), the Court continued its reduction of due-process burdens on prison officials, determining that it was unnecessary for disciplinary officials to prepare written reasons for refusing to allow the prisoner to call a witness on his behalf unless and until the matter is appealed to the courts. Some states, through statute or regulation, establish more-generous procedural safeguards for disciplinary hearings than the minimal standards found in Supreme Court cases.

Challenges to reclassification or physical transfer of prisoners to other penal institutions have likewise been rejected. In *Meachum v. Fano* (96 S. Ct. 2532 [1976]), the Court concluded that prisoners had no right to be in any particular prison, and, therefore, there are no due process protections before transfer from one prison to another, even if the transfer involved substantially inferior living conditions. This has been an important case, frequently cited for the Court's analysis regarding the source of rights. Some justices argue for a narrow model of rights. If a right is not specifically defined in the Constitution or by state statute, it does not exist. *Meachum* is also a bellwether case because it unequivocally states that the Court will defer to the states in matters of prisoners' rights classified as insubstantial. Not just any deprivation of liberty, but only those deemed significant and in violation of a specific constitutional guarantee, will invoke federal-court jurisdiction in the foreseeable future.

The availability of a due-process hearing for prison discipline was further reduced in *Sandin v. Conner* (115 S. Ct. 2293, 2303 [1995]), in which the Court observed that a hearing is appropriate only when the discipline "imposes atypical and significant conditions on inmates in relation to ordinary incidents of prison life." In that case, Conner had been placed in segregated confinement for 30 days without benefit of a hearing. The Court majority reasoned that because the confinement did not inevitably affect the duration of the sentence, no due-process rights attached to the prisoner's grievance. This case, and numerous others, clearly indicates a return to wide discretion for administrators, as well as reluctance to make federal courts available to review prison disciplinary proceedings.

Conclusions

The fundamental understanding that has evolved from judicial review of the conditions of penal confinement in the United States is that prisoners retain all the rights of free people except those that are inconsistent with institutional needs of order, security, and rehabilitation, and must, of necessity, be withdrawn or diminished during incarceration. Evolving standards of decency, inside and outside correctional institutions, require constant reevaluation in a dynamic and pluralistic society that claims to value individual integrity and potential. Judicial evaluation of the rights of prisoners is the established forum for continuing dialogue, and there is little reason to believe that, despite numerous defeats, prison litigation will disappear.

The concerns of convicted individuals struggling to retain or shape individual identity and integrity are not that much different from those of the unconvicted. Each issue raised in prison litigation has precedent in some other area of civil rights law. The treatments afforded those who have offended society's laws are a fair measure of that society's ethics. Although conviction and incarceration necessarily involve fundamental change in constitutional status, Dostoyevsky's observation that a society can be judged by the state of its prisons retains validity now just as it has throughout history.

KEY TERMS

ADA—American with Disabilities Act; requires access and accommodation to governmental programs and services.

consent decree—agreement between parties to settle a lawsuit.

corporal punishment—punishment "to the body."

damages—monies that a court orders paid to a person who has suffered a loss or injury by a defendant.

deliberate-indifference test—a standard to be met to determine whether a constitutional violation has occurred; when prison officials are deliberately indifferent to the violation or loss of a constitutional right.

due-deference era—represents the current legal climate where courts are most likely to defer to prison administrators' expertise in resolving any prisoner rights issue.

due-process clauses—no person shall be deprived of life, liberty, or property without due process of law (this clause is in both the 5th and 14th amendments).

habeas corpus—a writ that demands the government show why the person should not be released; used to challenge the legality of confinement.

hands-off doctrine—represented the court's unwillingness to become involved in prisoner rights issues.

injunction—court order that either prescribes or proscribes action.

jailhouse lawyers—inmates who have some legal knowledge and help other inmates with legal matters.

least-restrictive-means test—test used under early 1st amendment caselaw; administrators had to use the means necessary to accomplish their purpose that was least intrusive toward Constitutional rights.

penumbral right—a phrase derived from the collective interpretation of several explicit constitutional guarantees such as the right to travel, associate, speak, and practice one's religion.

per curiam—hearing before the whole court.

PLRA—Prison Litigation Reform Act; severely restricted prisoners' ability to file writs to challenge their confinement.

rational-relationship test—current legal standard; states have to show a legitimate state interest and prove that the rule, policy, or practice in question is reasonably related to that state interest.

retention-of-rights theory—idea that prisoners retained all the rights of free people except for those rights that were inconsistent with their status as prisoner.

RFRA—Religious Freedom Restoration Act; Congress passed the Act to protect religious rights, but the Supreme Court struck it down as unconstitutional.

Section 1983 actions—allows individuals to sue public officials for alleged violations of civil rights.

sovereign immunity—shield from liability that historically had attached to governmental entities and their agents.

totality of circumstance—when conditions are taken as a whole to argue that the prison is violating the 8th amendment rights of prisoners.

REVIEW QUESTIONS

1. Until the 1960s, the courts generally followed a hands-off philosophy toward prison matters. What factors may have influenced the courts to abandon this policy and intervene in prison administration?
2. Why do prisoners almost always file suit in federal court rather than state court? What are the mechanisms for filing grievances in court?
3. What are some of the rights at issue under the First Amendment? What tests are used to determine whether or not the alleged restrictions are constitutional?
4. Where do privacy rights come from? Should prisons employ male guards in institutions for women? Should prisons employ female guards in institutions for men? How do the courts balance these rights against the inmates' rights of privacy?
5. What is the standard applied by the Supreme Court in medical cases? Are mental-health issues treated differently from physical-health issues?
6. How important is freedom of religion in penitentiaries, and what accommodations should be made to protect it?
7. If a state decides to permit individuals of the same gender to marry, may prisoners demand the right to marry and be cell mates while incarcerated?
8. Where does the right come from for a prisoner to receive rehabilitative treatment, if such a right exists?
9. Explain the current test used in almost all prisoners' rights cases. What case does it come from?
10. Do prisoners have too many "rights?" Defend your answer using case decisions discussed in this chapter.

FURTHER READING

Branham, L., and S. Krantz, S. 1997. *The Law of Sentencing, Corrections, and Prisoners' Rights*, 5th ed. Minneapolis: West.

Palmer, J. 2003. *Constitutional Rights of Prisoners*, 7th ed. Cincinnati, OH: Anderson.

Smith, C. 1999. *Law and Contemporary Corrections*. Belmont CA: West /Wadsworth.

REFERENCES

Allen, H., and C. Simonson. 1982. *Corrections in America*, 5th ed. New York: Macmillan.

American Bar Association. 1981. *American Bar Association Standards Relating to the Legal Status of Prisoners.* Chicago, IL: ABA.

Branham, L. 1990. "Out of Sight, Out of Danger? Procedural Due Process and the Segregation of HIV-Positive Inmates." *Hastings Constitutional Law Quarterly* 17: 193–351.

Branham, L., and S. Krantz, S. 1997. *The Law of Sentencing, Corrections, and Prisoners' Rights,* 5th ed. Minneapolis: West.

Clemmer, D. 1958. *The Prison Community*. New York: Holt, Rinehart, and Winston.

Fletcher, B. R., L. Shaver, L. Dixon, and D. Moon. 1993. *Women Prisoners: A Forgotten Population.* Westport, CT: Praeger.

Gardner, T., and T. Anderson. 1992. *Criminal Law,* 5th ed. St. Paul, MN: West.

Greenspan, J. 1990. "States Move Toward Mainstreaming of HIV-Infected Prisoners." *National Prison Project Journal* 22: 18–52.

Hammett, T., and S. Moiri. 1990. "Update on AIDS in Prisons and Jails." U.S. Department of Justice. Washington, DC: U.S. Government Printing Office.

Jacobs, J. 1983. *Perspectives on Prisons and Imprisonment.* Ithaca, NY: Cornell University Press.

Maguire, K., and A. Pastore. 1999. *Bureau of Justice Statistics Sourcebook Criminal Justice Statistics 1997.* Albany: Hindelang Criminal Justice Research Center, State University of New York at Albany.

McCarthy, C. M. 1989. "Experimentation on Prisoners: The Inadequacy of Voluntary Consent." *New England Journal on Civil and Criminal Confinement* 15: 55–80.

National Advisory Commission on Criminal Justice Standards and Goals. 1973. "Report on Corrections." Washington, DC: U.S. Government Printing Office.

Palmer, J. 2003. *Constitutional Rights of Prisoners,* 7th ed. Cincinnati,OH: Anderson.

Pollock, J. 2002. *Women, Prison, and Crime,* 2d ed. Belmont, CA: Wadsworth.

Robertson, J. E. 1993. "Fatal Custody: A Reassessment of Section 1983 Liability for Custodial Suicide." *University of Toledo Law Review* 24: 807–830.

Roleff, T. (Ed.). 1996. *The Legal System: Opposing Viewpoints.* San Diego: Greenhaven Press.

Scalia, J. 2002. *Prisoner Petitions Filed in U.S. District Courts, 2000, with Trends 1980–2000.* Washington, DC: Bureau of Justice Statistics, U.S. Department of Justice.

Smith, C. 1999. *Law and Contemporary Corrections.* Belmont, CA: West /Wadsworth.

United Nations. 1956. *Standard Minimum Rules for the Treatment of Prisoners.* First United Nations Congress on the Prevention of Crime and Treatment of Offenders. Annex I (A); U.N. Doc.A/Conf. 6/1.

Wooster, A. K. 2002. "Title VII Sex Discrimination in Employment—Supreme Court Cases." *A.L.R. Fed.* 182: 61–125.

CASES CITED

Avery v. Perrin, 473 F. Supp. 90 (D.N.H. 1979)
Bell v. Wolfish, 441 U.S. 520, 99 S. Ct. 1861 (1979)
Blevins v. Brew, 593 F. Supp. 245 (W.D. Wis. 1984)
Block v. Rutherford, 468 U.S. 576, 104 S. Ct. 3227 (1984)

Bounds v. Smith, 430 U.S. 817, 97 S. Ct. 1491 (1977)
Bowring v. Godwin, 551 F. 2d 44 (4th Cir. 1977)
Brasch v. Gunter (unpublished opinion, D.S.D.C. Nebraska 1985)
City of Boerne v. Flores, 521 U.S. 507, 117 S. Ct. 2157 (1997)
Coffin v. Reichard, 143 F. 2d 443 (6th Cir. 1944)
Coleman v. Thompson, 501 U.S. 1277, 112 S. Ct. 27 (1991)
Cooper v. Pate, 378 U.S. 546, 84 S. Ct. 1733 (1964)
Cutter v. Wilkinson, 349 F. 3d 257 (6th Cir. 2003)
Dothard v. Rawlinson, 433 U.S. 321, 97 S. Ct. 2720 (1970)
Estelle v. Gamble, 429 U.S. 97, 97 S. Ct. 285 (1976)
Everson v. Michigan Department of Corrections, 222 F. Supp. 2d 864 (2002)
Farmer v. Brennan, 511 U.S. 825, 114 S. Ct. 1970 (1994)
Ferguson, Ex parte, 361 P. 2d 417 (Cal. 1961)
Finney v. Arkansas Board of Correction, 410 F. Supp. 251 (E.D. Ark. 1976)
Forts v. Ward, 621 F. 2d 1210 (2d Cir. 1980)
Fromer v. Scully, 817 F. 2d 227 (2d Cir. 1987)
Galvan v. Carothers, 855 F. Supp. 285 (D. Alaska 1994), *cert. denied* 515 U.S. 1149 (1995)
Gilliam v. Martin, 589 F. Supp. 680 (W.D. Okla. 1984)
Glover v. Johnson, 478 F. Supp. 1075 (E.D. Mich. 1979)
Glover v. Johnson, 199 F. 3d 310 (6th Cir. 1999)
Goldberg v. Kelly, 397 U.S. 254, 90 S. Ct. 1011 (1970)
Harris v. Thigpen, 941 F. 2d 1495 (11th Cir. 1991)
Holt v. Sarver, 309 F. Supp. 362 (E.D. Ark. 1970), *aff'd* 442 F. 2d 304 (8th Cir. 1971)
Houchins v. KQED, 438 U.S. 1, 98 S. Ct. 2588 (1978)
Hudson v. Palmer, 468 U.S. 517, 104 S. Ct. 3194 (1984)
Hull, Ex parte, 312 U.S. 546, 61 S. Ct. 640 (1941)
Jackson v. Bishop, 404 F. 2d 571 (8th Cir. 1968)
Johnson v. Avery, 393 U.S. 483, 89 S. Ct. 747 (1969)
Jones v. Diamond, 636 F. 2d 1363 (5th Cir. 1981)
Jones v. North Carolina Prisoners' Labor Union, 433 U.S. 119, 97 S. Ct. 2532 (1977)
Knecht v. Gillman, 48 F. 2d 1136 (8th Cir. 1973)
Lee v. Downs, 641 F. 2d 1117 (4th Cir. 1981)
Lee v. Washington, 390 U.S. 333, 88 S. Ct. 994 (1968)
Lewis v. Casey, 518 U.S. 343, 116 S. Ct. 2174 (1996)
Madrid v. Gomez, 889 F. Supp. 1146 (1995)
Mary Beth G. v. City of Chicago, 723 F. 2d 1263 (7th Cir. 1983)
McCleskey v. Zant, 499 U.S. 467, 111 S. Ct. 1454 (1991)
Meachum v. Fano, 427 U.S. 215, 96 S. Ct. 2532 (1976)
Monell v. Department of Social Services of the City of New York, 436 U.S. 658, 98 S. Ct. 2018 (1978)
Monroe v. Pape, 365 U.S. 167, 81 S. Ct. 473 (1961)
Morgan v. LaVallee, 526 F. 2d 221 (2d Cir. 1975)
Nolley v. County of Erie, 776 F. Supp. 715 (W.D.N.Y 1991)
O'Lone v. Estate of Shabazz, 482 U.S. 342, 107 S. Ct. 2400 (1987)
Olim v. Wakinekona, 461 U.S. 238, 103 S. Ct. 1741 (1983)
Overton v. Bazzetta, 539 U.S. 126, 123 S. Ct. 2162 (2003)

Palmigiano v. Travisono, 317 F. Supp. 776 (D.R.I. 1970)
Parte v. Lane, 528 F. Supp. 1254 (N.D. Ill. 1981)
Pell v. Procunier, 417 U.S. 817, 94 S. Ct. 2800 (1974)
Pennsylvania Department of Corrections v. Yeskey, 524 U.S. 206, 118 S. Ct. 1952 (1998)
Polakoff v. Henderson, 370 F. Supp. 690 (D. Ga. 1973)
Ponte v. Real, 471 U.S. 491, 105 S. Ct. 2192 (1985)
Price v. Johnston, 334 U.S. 266, 68 S. Ct. 1049 (1948)
Procunier v. Martinez, 416 U.S. 396, 94 S. Ct. 1800 (1974)
Proudfoot v. Williams, 803 F. Supp. 1048 (E.D. Pa. 1992)
Ruffin v. Commonwealth, 62 Va. 790 (1871)
Sandin v. Conner, 515 U.S. 472, 115 S. Ct. 2293, 2303 (1995)
Saxbe v. Washington Post, 417 U.S. 843, 94 S. Ct. 2811 (1974)
Scheuer v. Rhodes, 416 U.S. 232, 94 S. Ct. 1683 (1974)
Sconiers v. Jarvis, 458 F. Supp. 37 (D. Kan. 1978)
Seattle-Tacoma Newspaper Guild v. Parker, 480 F. 2d 1062 (9th Cir. 1973)
Shaw v. Murphy, 532 U.S. 223, 121 S. Ct. 1475 (2001)
Simon & Schuster v. Members of New York State Crime Victims Board, 502 U.S. 105, 112 S. Ct. 501 (1991)
Smith v. Fairman, 678 F. 2d 52 (7th Cir. 1982)
Smith v. Wade, 461 U.S. 30, 103 S. Ct. 1625 (1983)
Stone v. Powell, 428 U.S. 465, 96 S. Ct. 3037 (1976)
Superintendent, Massachusetts Correctional Institution, Walpole v. Hill, 472 U.S. 445, 105 S. Ct. 2768 (1985)
Talley v. Stephens, 247 F. Supp. 683 (E.D. Ark. 1965)
Theriault v. Carlson, 495 F. 2d 390 (5th Cir. 1974)
Thornburgh v. Abbot, 490 U.S. 401, 109 S. Ct. 1874 (1989)
Turner v. Safley, 482 U.S. 78, 107 S. Ct. 2254 (1987)
U.S. v. Bailey, 444 U.S. 394, 100 S. Ct. 624 (1980)
U.S. v. Bryant, 670 F. Supp. 840 (D. Minn. 1987)
U.S. v. State of Michigan, 680 F. Supp. 270 (W.D. Mich. 1988)
Ustrak v. Fairman, 781 F. 2d 573 (7th Cir. 1986)
Walsh v. Mellas, 837 F. 2d 789 (7th Cir. 1988)
Washington v. Harper, 494 U.S. 210 (1990)
Washington v. Lee, 263 F. Supp. 327 (M.D. Ala. 1966)
West v. Atkins, 487 U.S. 42, 108 S. Ct. 2250 (1988)
Whitley v. Albers, 475 U.S. 312, 106 S. Ct. 1078 (1986)
Wolff v. McDonnell, 418 U.S. 539, 94 S. Ct. 2963 (1974)
Women Prisoners of the District of Columbia Department of Corrections v. District of Columbia, 877 F. Supp. 634 (D.D.C. 1994)

IV

Convicts and the Community

Risen from Prison

Risen from prison
Back from the dead
Released convicts
Rejoin the living.

Alleluia! Alleluia!
They have returned!
It is a miracle!

Every day
A Miracle

Our prodigal sons
And daughters
Return
Every day,

From graveyards
We call prisons,

Each release
A resurrection
A quest for
Grace,

For life
To begin
Anew

Amen.

Robert Johnson

Jails

9

Dennis Giever
Indiana University of Pennsylvania

Chapter Objectives

- Be able to describe how jails are differentiated from prisons and lockups.
- Describe the history of jails.
- Describe the population one can find in jails.
- Discuss the three phases of jail architecture.
- Discuss the special challenges of jail management.

Jails remain one of the most understudied and widely misunderstood agencies within the criminal-justice system. Adding to this plight is the fact that jails are almost totally dependent on other agencies within the system and have little control over their own destiny (Thompson and Mays 1991b, 1). The population that enters and is housed in our jails is diverse, including pretrial detainees (those who have not been convicted of any crime) and those convicted of misdemeanors who are serving a short sentence (usually under one year). Irwin (1985, 2) described jail inmates as society's **rabble**. According to Irwin (1985, 2), the rabble are those individuals who are perceived by society as irksome, offensive, threatening, capable of arousal, even protorevolutionary. Jails house inmates with a wide array of problems, including mental illness, alcoholism, and homelessness (Klofas 1990, 69). Since jails often house individuals that no other agency within the criminal-justice system will take, they are often referred to as the "dumping ground" for society's problems (Moynahan and Stewart 1980, 104).

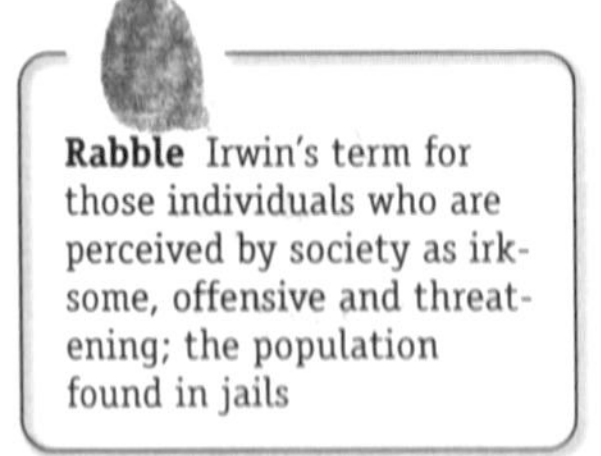
Rabble Irwin's term for those individuals who are perceived by society as irksome, offensive and threatening; the population found in jails

Jails suffer from an identity crisis as well. Often the public and the media use the words "jail" and "prison" interchangeably. When the media do pay homage to a local jail, it is usually after some major event—such as a riot, a fire, or a death—has occurred within the facility's walls. This chapter will examine the unique role that jails play within the criminal-justice system. We will begin by addressing a number of unique features of jails that distinguish them from other places of confinement.

Jails must be distinguished not only from prisons, but also from lockups. **Lockups**, also known as *police lockups*, are facilities authorized to hold individuals awaiting court appearances. The holding periods usually do not exceed 48 hours (Clear and Cole 1994, 144). Such facilities include drunk tanks and holding tanks, which are usually administered by local police agencies. Lockups are found in most police stations, and serve the purpose of holding individuals during questioning and intake until they can be transported to the local jail. These facilities may be little more than a steel cage set up inside the police station and offering no toilet facilities or beds.

Jails are distinguished from prisons by a number of factors. First, the vast majority of prisons in the United States are designed as single-gender institutions that house only adults convicted in a court of law of having committed a crime, except that they may also now house juveniles waived to the adult system. Jails have a very heterogeneous population made up of both men and women, both adults and juveniles, who are often housed within the same facility. Women accounted for about 11.9 percent of the jail population in the United States in 2003, up from about 7.1 percent in 1983 (Bureau of Justice Statistics 2004, 1995). While there is a mandate in the federal Juvenile Justice and Delinquency Prevention Act calling for the removal of all juveniles from adult facilities, thousands are nonetheless incarcerated in such adult facilities yearly. On June 30, 2003, there were a total of 6,869 individuals under the age of 18 being detained in adult jails (Bureau of Justice Statistics 2004). Of these, 5,484 were held as adults, and 1,385 were held as juveniles. As pointed out above, jails house both convicted criminals as well as those who have not been convicted of a crime, thus adding to the heterogeneity of the population.

Although jails house a very diverse group of individuals, the jail population is characterized by a number of commonalities. With only a few exceptions, the individuals housed in jails are poor, undereducated, unemployed, and disproportionately members of minority groups (Irwin 1985, 2). Later in this chapter, we will look at the jail population in more detail.

A second distinguishing factor between jails and prisons is how the facilities are administered. Elected county sheriffs usually administer jails, while prisons are state or federal institutions (Zupan 1991, 47–48). Often jails play a secondary role to the law-enforcement responsibilities of the local sheriff. In six states—Alaska, Connecticut, Delaware, Hawaii, Rhode Island, and Vermont—jails are administered by state officials rather than by county sheriffs (Mays and Thompson 1991, 12; Zupan 1991, 47). In only a small number of other jurisdictions are jails administered by a specific local department devoted to corrections. The fact that county sheriffs administer most jails in the United States is an important one. Most sheriffs have law-enforcement backgrounds, with little or no correctional-management experience. There are a number of problems related to this factor, all of which will be covered in detail later in this chapter.

Within the United States, there are 11 federal detention facilities that function as jails (Bureau of Justice Statistics 2001, 7). According to the Bureau of

Justice Statistics (2001, 7), these facilities housed 11,209 individuals who were either awaiting adjudication or serving a sentence of less than one year. These federal jails had a rated capacity of 8,040 on June 30, 1999, resulting in an occupancy rate that was 39 percent over capacity (Bureau of Justice Statistics 2001, 7). The federal government relies on local jail facilities for the vast majority of its short-term needs, holding about 29,000 prisoners in 1999. That year, about 74 percent of these federal inmates were unconvicted, up from about 53 percent in 1993 (Bureau of Justice Statistics 2001, 7).

The third factor that distinguishes jails from prisons is the term of confinement. Jails are institutions for short-term confinement, usually less sthan one year. Prisons are places of long-term confinement, ranging from one year to life, and house those awaiting a death sentence as well. Jails have a large turnover in population. While the number of individuals being held in jail at any one time is smaller than the prison population in this country, about 17 million inmates pass through our nation's jails each year. Prisons, by contrast, have a rather stable population. While jails, by definition, do not house those sentenced to longer terms of confinement, many jails nonetheless find themselves holding convicted felons for increasingly longer periods of time due to overcrowding within state- and federal-prison facilities.

Another important distinction between jails and prisons is the type and number of programs that are available to the inmates. While many prisons have only limited programs, they are far more prevalent than those offered in jails. Much of this is related to a number of the factors mentioned above. Due to the large turnover, it would be difficult to offer classes or technical training to many prisoners in jail. However, as jails house more and more inmates waiting transfer to prison facilities, the need to keep such individuals busy has become a concern to administrators.

Finally, jails are most often located in or near the central business districts of most cities. Often these jails are located in the same building as the county courthouse. Such an arrangement facilitates the quick transfer of the pretrial detainees to court hearings. Prisons are, in most cases, located in remote locations.

Just as there are a number of unique features that distinguish jails from prisons, the histories of those facilities are in stark contrast as well. In the section that follows, we will take a brief look at the history of jails, both in Europe and in the United States.

History of Jails

The development of jails in America can be traced to England. In fact, the American jail has been classified as a "curious hybrid between the tenth century **gaol** with its principal function being to detain arrested offenders until they were tried, and the fifteenth and sixteenth century **houses of correction** with their special function being punishment of minor offenders, debtors, vagrants, and beggars" (Flynn 1973, 49). The term *jail* comes from the English term *gaol* (pronounced "jail"), which can be traced back to the year 1166, when King Henry II ordered the reeve (the official law-enforcement officer for the crown) of each shire (county)

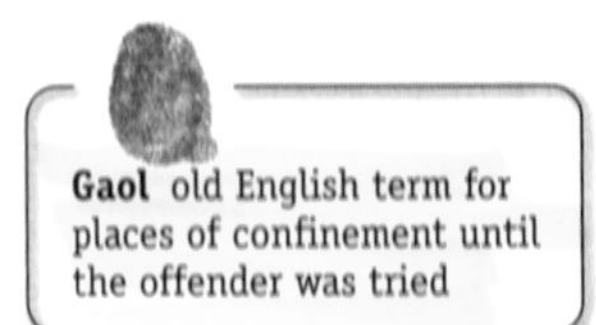

Gaol old English term for places of confinement until the offender was tried

Houses of correction in 15th and 16th century these facilities held minor debtors, vagrants and beggars

to establish a place to secure offenders until the next appearance of the king's court. The shire reeve (sheriff) had a variety of law-enforcement responsibilities, only one of which was to maintain a jail. These jails fulfilled a single function of detaining the accused until such time as a trial could be held. While the original purpose of the jail was for this single function, there is evidence that, in some instances, these jails were used to house those convicted of crimes as well (Moynahan and Stewart 1980, 13). Prisoners had to wait for long periods of time, often years, before a trial was held. The conditions of these early jails were dismal at best, and frequently the sheriff would utilize existing structures such as dungeons, cellars, or towers to serve the county's needs (Moynahan and Stewart 1980, 15). In other cases, jails were constructed of wood, and amounted to little more than a shed set under the city wall (Zupan 1991, 11).

At this time in the history of jails, no real attempt was made to separate prisoners by gender, age, or seriousness of offense. The young were housed with the hardened criminals, and women were often housed in the same cells as men. No provisions were made for either the construction or the operation of the facilities. As such, the sheriff relied on the **fee system** for income. Inmates were often charged a fee upon entering the jail. If they were unable to pay, "other prisoners would literally strip the clothes from the back of the new inmate" (Zupan 1991, 12). Prisoners or their families had to pay for the privilege of being housed in such a facility, and if prisoners did not have the resources, or at least friends or relatives who would provide the moneys for such necessities as food, they would often perish long before a court appearance was made.

Fee system early jailers charged inmates for their room and board

Real change in the conditions of these early jails did not occur until the beginning of the 16th century, and even then such change was due more to economic conditions than to any real attempt to rectify the many problems. During that time period, a large number of people were moving to the cities and looking for work. This influx of vagrants and beggars caused an economic strain on these cities. Largely as a result of this influx, and the failure of more-serious measures such as branding and mutilation in controlling the masses, houses of corrections or **bridewells** were established. The first, St. Brigit's Well (Bridewell), was established in London in 1553 in a mansion that was originally built to house royal visitors. As discussed in Chapter 2, these houses of corrections, or workhouses, were for sentenced criminals and provided places for inmate labor.

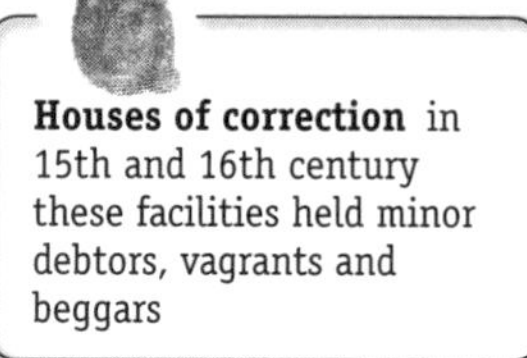

Houses of correction in 15th and 16th century these facilities held minor debtors, vagrants and beggars

Bridewells another name for House of correction, named for the location of the first established facility at St. Brigit's Well

Further attempts to improve the conditions of these jails did not occur until the 18th century. Although attempts were made as early as 1702, such efforts were never published and had little, if any, impact (Moynahan and Stewart 1980, 21; Zupan 1991, 14). It was not until the publication of John Howard's *The State of the Prisons in England and Wales with Preliminary Observations and an Account of Some Foreign Prisons* (1777) that true reform began. Howard, who was at that time the sheriff of Bedford County, wrote about the deplorable conditions of jails in Europe in the latter half of the 18th century. In 1779, Howard, along with Sir William Blackstone and William Eden, drafted the Penitentiary Act, which was then passed by Parliament. This act enumerated four principles of reform: secure and sanitary structures, systematic inspections, abolition of fees, and a reformatory regime. Even after stepping down as sheriff, Howard continued touring confinement facilities throughout Europe. It is ironic that Howard, who worked so tirelessly to improve the sanitary conditions in jails, died of a disease (jail fever or typhus) that claimed the lives of many who either worked or were housed in these facilities (Moynahan and Stewart 1980, 22). Although many of the reforms Howard recommended did not take shape until long after his death, his contribution had a vast impact on the condition of facilities in Europe as well as those in the American colonies.

The History of Jails in America

Jails in colonial America were basically an extension of those found in England. As had been the case in England, jails remained the responsibility of the local government, most notably the sheriff. The fee system was retained as well, and the deplorable conditions found in England were also common in the colonies. The first such jail was thought to be in Jamestown, Virginia, and was established in the early part of the 17th century. As new communities began to spring up, each began establishing punishments for those who committed criminal acts. Some of the more common methods of punishment were stocks, pillories, dunking stools, and whipping posts. Stocks were little more than a bench or stump to which the offender was shackled; they usually were located in a public area where the townspeople could jeer and throw garbage.

Moynahan and Stewart (1980, 27) claim that two important practices emerged from colonial times. First, the colonies used the most direct and least expensive form of punishment possible. Second, the punishment imposed in the colonies was more humane than that found in Europe. The latter fact was due to economic conditions. Unemployment was high in Europe, and often death sentences were imposed for rather minor offenses. In contrast, the colonies were still growing and were in need of labor; the citizens were reluctant to execute anyone except those who committed the most serious or heinous crimes.

The jails in colonial America, as in Europe, were used largely to house those awaiting trial and corporal punishment. The first real reform in America occurred in Pennsylvania under the direction of William Penn, a leader of the Quakers. The cornerstone of the Quaker movement in America was penal reform. Penn and his followers believed that hard labor in a house of corrections, rather than corporal or capital punishment, was more effective in handling crime. In 1682, the Quaker code, or Great Law, was enacted in Pennsylvania. It emphasized fines and hard labor in a house of corrections as punishment for most crimes. At the same time, the first jail designed exclusively to house convicted offenders serving a sentence was opened in Philadelphia. The reforms established by Penn and his followers were idealistic at best, for soon after the High Street Jail was built in Philadelphia, it, too, became overcrowded and deteriorated to deplorable conditions. The reforms did not last, for in 1718, when William Penn died, so did the Great Law. Pennsylvania reverted back to using whipping, mutilation, branding, and other forms of corporal punishment for criminal offenses. The conditions and role of the jails in America remained largely unchanged until about the time of the American Revolution.

Soon after the end of the Revolutionary War, a number of reformers, Benjamin Franklin among them, led a movement to change the English Criminal Code of 1718, which had been in effect since the death of William Penn. The new law enacted on September 15, 1786, allowed prisoners to be put to work out in the streets cleaning and repairing roads. Prisoners had their heads shaved to distinguish them from others and were encumbered with iron collars and chains (Takagi 1975, 20). Shortly after this law was enacted, the Quakers formed the Philadelphia Society for Alleviating the Miseries of Public Prisons. The society's main goal, aside from introducing religious services in the Walnut Street Jail, was to amend the new law (Takagi 1975, 20). The society felt that a more private or solitary labor would be more successful.

Many of the efforts of the Quakers were centered on the Walnut Street Jail. The society was able to get women segregated from the male population, and was able to abolish the fee system at the Walnut Street facility. They provided food and clothing for all inmates, and medical care was offered weekly (Zupan 1991, 18). As was previously the case, while the reforms were widely heralded, the conditions soon began to deteriorate. By 1816, the conditions in the Walnut Street Jail had deteriorated back to those found before the reforms. The only exceptions were that inmates were still segregated by sex and offense, and liquor was still prohibited (Zupan 1991, 19).

In the late 1700s and early 1800s, jails began to be used not only to detain those awaiting trial, but also to hold those who had been convicted (Moynahan and Stewart 1980, 41). If available, those sentenced for more-serious crimes were sent to the newly developed state prisons. States that did not have prisons either executed those convicted of serious crimes or confined them to time in the county jail (Moynahan and Stewart 1980, 41). The whipping post, stocks, and pillory had begun disappearing from the American scene. These changes did not occur overnight, and did not take place in all jails at the same time. Jails then, as they are now, were locally operated, and the speed at which changes occurred was due largely to the local climate. Some jurisdictions brought about change rather quickly, and others took decades (Moynahan and Stewart 1980, 41).

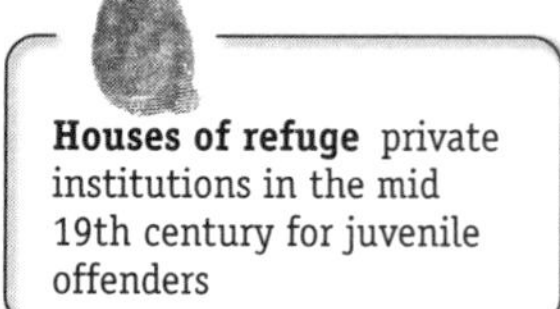

Houses of refuge private institutions in the mid 19th century for juvenile offenders

During the mid-19th century, the jailing of individuals for debt was generally abolished (Moynahan and Stewart 1980, 43). At about this time in U.S. history, new facilities dedicated to housing juveniles were opening around the country. A number of private **houses of refuge** had opened in Boston in 1826, and in Philadelphia in 1828, but not until 1847 did the first state—Massachusetts—establish a reformatory. (For a discussion on the plight of juveniles in such early institutions, see Platt 1969.) There was no typical jail in the United States during the 19th century, just as there is no typical jail today. Their sizes and purposes varied greatly (Moynahan and Stewart 1980, 46).

The 20th-Century Jail

The growth of jails in America through the 20th century is sketchy at best. In 1880, the Census Bureau began compiling information on individuals housed in jails, workhouses, and prisons. During this time, and to a certain degree today, the definition of a jail was ambiguous. While the data are open to ques-tion, one can look at the trend in confinement by studying the population trends.

Champion (1990) compiled data collected from a number of sources and developed a table of jail populations. The figures he developed are given in **Table 9-1**.

As Table 9-1 shows, the jail population in the United States almost doubled from 1880 through 1890. The population is still exploding, with the latest survey estimating 691,301 incarcerated in jails (Bureau of Justice Statistics 2004, 1).

Local county governments today operate about 85 percent of the jails in America. In most cases, large municipalities operate jails that are not operated by county government. The federal government, through the Federal Bureau of Prisons, operates a number of pretrial detention facilities that house those awaiting trial on federal charges.

While the number of inmates housed in jails is on the rise, the number of jails has actually remained rather constant. In 1983, there were 3,338 local-jail facilities, but by 1988, that number dropped to 3,316, and further declined in 1993 to 3,304. By 1999, the number of jails had risen again to 3,365 (Bureau of Justice Statistics 2001, 2). This fact can be attributed to the concept of regionalization, whereby two or more governments join forces to build one large regional jail serving multiple jurisdictions (Mays and Thompson 1991, 13; Cox and Osterhoff 1991, 237–238). Many jurisdictions, faced with the need to build a new facility, to either alleviate overcrowding or to replace a physical plant that is old and deteriorating, find it cost-effective to pool resources with a neighboring county to build one facility. In some areas, as a way to encourage regionalization, the state will reimburse a large portion of the construction cost of the new jail when three or more jurisdictions are involved in such a joint venture (Leibowitz 1991, 42–43).

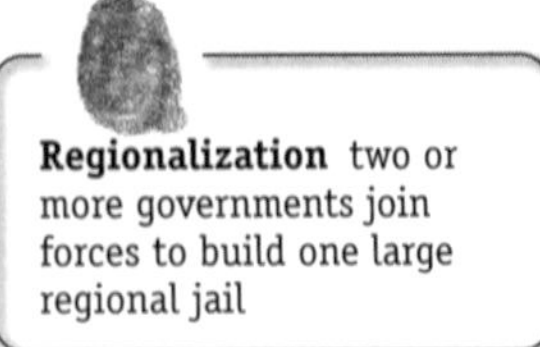

Regionalization two or more governments join forces to build one large regional jail

Table 9-1 Jail Populations

Year	Number of Jail Inmates
1880	18,666
1890	33,093
1940	99,249
1950	86,492
1960	119,671
1970	129,189
1980	163,994
1983	223,551
1986	274,444
1988	343,569
1993	459,804
1999	605,943

Sources: Cahalan, M. 1986. *Historical Corrections Statistics in the United States, 1850–1984*. Washington, DC: U.S. Department of Justice. Champion, D. 1990. *Corrections in the United States: A Contemporary Perspective*, pp.164–165. Englewood Cliffs, NJ: Prentice Hall.

Bureau of Justice Statistics. 2001. *Census of Jails, 1999*, p. iii. Washington, DC: U.S. Department of Justice.

There are a number of problems associated with these arrangements—for example, transportation of inmates (most notably pretrial detainees) and multijurisdictional problems associated with both the funding of the facility and the determination of a location and management of the facility (McGee 1975, 11). Given the number of benefits of multijurisdictional jails, many communities have been able to work through these problems and to develop solutions allowing them to take advantage of these benefits.

Functions of Jails

As was discussed earlier, jails were originally conceived as places to hold pretrial detainees until their appearance in court. This single purpose has evolved into a multitude of differing roles, which often cause a strain on the system. Jails are often thought of as the criminal-justice agency of last resort—and, by default, deal with a large number of individuals that no other agency can handle. Jails house a wide variety of sentenced and unsentenced individuals. According to the Bureau of Justice Statistics, jails perform the following functions:

1. *Receive individuals pending arraignment and hold them awaiting trial, conviction, and sentencing*
2. *Readmit probation, parole, and bail-bond violators and absconders*
3. *Temporarily detain juveniles pending transfer to juvenile authorities*
4. *Hold mentally ill individuals pending their movement to appropriate health facilities*
5. *Hold individuals for the military, for protective custody, for contempt, and for the courts as witnesses*
6. *Release convicted inmates to the community upon completion of sentence*
7. *Transfer prisoners to state, federal, and other local authorities*
8. *House inmates for federal, state, or other authorities because of crowding of their facilities*
9. *Relinquish custody of temporary detainees to juvenile and medical authorities*
10. *Operate community-based programs with day reporting, home detention, electronic monitoring, or other types of supervision*
11. *Hold inmates sentenced to short terms (generally less than one year) (Bureau of Justice Statistics 2001, 2)*

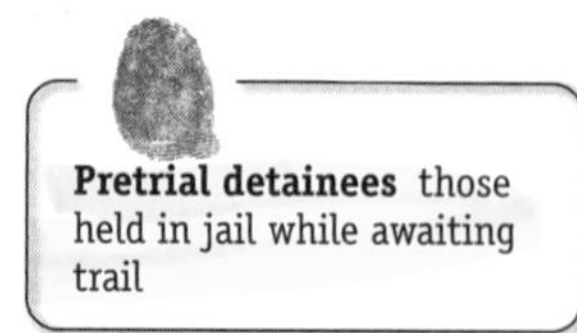

Pretrial detainees those held in jail while awaiting trail

Pretrial Detainees

About 54 percent of all inmates housed in jails are **pretrial detainees** (Bureau of Justice Statistics 2001, 4). These individuals are arrested for a wide assortment of offenses and, for one reason or another, are either unable to afford or are denied bail. As a result, they are housed in jails until their trial. As the jail population increases and administrators are faced with lawsuits, they are constantly searching for ways to relieve the crowding. With little control over the number of convicted misdemeanants sent to their facilities, jail administrators are faced with the challenge of decreasing their population by reducing the number of pretrial detainees held or by reducing the time the detainees spend in jail.

Attempts have been made to limit the amount of time pretrial detainees have to spend in jail. In 1974, Congress passed the Speedy Trial Act, which mandated that federal charges must be filed against a defendant within 30 days of the arrest. A preliminary hearing must be held within 10 days of that date, and the trial must begin within 60 days of the arraignment. While the states are not specifically bound to the Speedy Trial Act, the Sixth Amendment of the Constitution guarantees that "the accused shall enjoy the right to a speedy and public trial."

Of obvious concern is who is held as a pretrial detainee. The Eighth Amendment to the Constitution states that "excessive bail shall not be required. . . ." The purpose of setting bail is to ensure that the defendant will appear for trial. The obvious consequence of this is that those who can afford bail will pay it, while those unable to pay will have to wait in jail. In reality, while the purpose of bail is to ensure appearance at trial, often those who are forced to stay in jail until trial are the individuals who pose the least risk of flight.

The Manhattan Bail Project was conducted by the Vera Institute of Justice in New York City from 1961 through 1964. Law students would interview defendants before their arraignment, ask questions about the defendants' ties to the community, and then assign points based on the defendants' responses. If a defendant obtained a sufficient number of points, the Vera Institute would recommend to the court that he be released on his own recognizance (Goldfarb 1975, 40). It was found that those released on the recommendation of the Vera Institute returned to trial in greater numbers than those who were released on bond through a bail bondsman. This project evolved into the practice of release on recognizance (**ROR**).

ROR Release on Recognizance

The Bail Reform Act of 1984 has also had an impact on the number of pretrial detainees. One of the provisions in the Bail Reform Act permitted judges to deny bail to those charged with a violent crime, to those charged with crimes that carried a possible life sentence, to those charged with crimes that carried a death sentence, to those charged with some major drug offenses, and in those cases in which the defendant is charged with a felony and has a serious past criminal record (Bureau of Justice Statistics 1994).

Misdemeanants

A second function of jails is to house offenders serving short-term sentences. Those convicted of misdemeanors and sentenced by the local courts to incarceration for a period of less than one year typically serve their sentence in jail. Almost half of all inmates in jail are serving sentences, the number fluctuating very little in the 1990s—from a low of about 46 percent in 1999 to a high of about 49 percent in 1994 (Bureau of Justice Statistics 2001, 4).

The average maximum sentence of misdemeanants who are serving their sentence in jail was estimated to be 24 months in 2002, up from 23 months in 1996. While the sentence length increased during that six-year period, the actual time served by such inmates actually decreased—from 10 months in 1996 to 9 months in 2002 (Bureau of Justice Statistics 2002, 5). These drops in time served were across the board—including violent offenses (16 months in 1996 to 14 months in 2002), property (11 months to 9 months), and drug offenses (13 months to 11 months) (Bureau of Justice Statistics 2002, 5). In 2002, half of the inmates had received sentences of 8 months and spent 5 months in jail serving time (Bureau of Justice Statistics 2002, 5).

Felons

Jails must also house convicted felons who are awaiting transfer to a state or federal prison. Some of these inmates are waiting for the court to sentence them; in some cases, a separate sentencing hearing is held once a presentence investigation (PSI) is undertaken. In such cases, either the state or federal government usually reimburses the jail for housing the individual.

As previously discussed in Chapter 3, jails in a number of jurisdictions find themselves housing convicted felons due to overcrowding in either state or federal prisons. If sufficient bed space is not available to accommodate an inmate, the state or federal government enters into a leasing agreement with the local jail to house the inmate. According to the Bureau of Justice Statistics (2003c, 6), in 2002 approximately 71,256 jail inmates were being housed in local facilities due to crowding in state or federal prisons. This number represented about 8.6 percent of all jail prisoners. Once again, when the local jurisdictions house such convicts, the state or federal government reimburses them locality.

The courts are increasingly using jails as their sentencing option for convicted felons. Between 1986 and 1992, the number of convicted felons who were sentenced to confinement in local jails almost doubled (Bureau of Justice Statistics 1995, 13). Those who were sentenced to jail amounted to 28 percent of all convicted felons in this country in 2000, up from 21 percent in 1986 (Bureau of Justice Statistics 2003b, 1).

Others

According to Irwin (1985, 18), "the vast majority of the persons who are arrested, booked, and held in jail are not charged with serious crimes." Sometimes the mentally ill or homeless are held in jails for their own protection and welfare. Such inmates are the "rabble" of society, meaning the "disorganized" and "disorderly," the "lowest class of people" (Irwin 1985, 2). There are a growing number of mentally ill inmates who are incarcerated in jail (Gibbs 1986). Jails are ill-equipped to handle the unique needs of such prisoners. Many smaller jails do not provide full-time medical staff to deal with ongoing problems, but have to make do with doctors on call.

Jails hold individuals wanted for crimes in other states. If a warrant is issued in another jurisdiction and the person is detected or stopped for a traffic violation, he or she is arrested and held until extradition proceedings can take place. Jails hold alleged probation and parole violators. Such individuals are entitled to a hearing to determine whether their probation or parole should be revoked. If suspected of committing a serious crime, however, they often are held in jail. Such violators share many of the characteristics of the pretrial detainee.

Finally, jails hold a number of inmates who do not fit neatly into any of the above categories. They may be material witnesses who are reluctant to testify, or possibly a witness in need of protection. As Goldfarb stated: "American jails operate primarily as catchall asylums for poor people" (1975, 27). It is only rarely that they hold a person who fits the popular image of a criminal, one who is a serious threat to society (Irwin 1985, 1).

Juveniles

Jails may house juvenile offenders when no juvenile-detention facility is available in a jurisdiction. These underage individuals are often housed in adult facilities despite a mandate in the federal Juvenile Justice and Delinquency Prevention Act that banned the jailing of juveniles (Schwartz 1991, 216-217). Many such youths are incarcerated for relatively minor offenses.

The law requires that any juvenile being housed in a jail receive sight and sound separation from the adult population. The requirement stems from the Juvenile Justice and Delinquency Prevention Act of 1974, which states that juveniles "shall not be detained or confined in any institution in which they have regular contact with adult persons incarcerated because they have been convicted of a crime or are awaiting trial on criminal charges" (U.S. Department of Justice 1980, 400). This provision was interpreted to mean that juveniles could be held in adult jails as long as they had sight and sound separation from the adults, which meant that youths would often find themselves housed in solitary confinement away from everyone else, including jail personnel. Such an arrangement can have an impact on the number of suicides in jails, as we will see below.

There have been lawsuits filed to control the practice of housing juveniles with adults (Soler 1988). On June 30, 2003, there were a total of 6,869 inmates under the age of 18 housed in adult jails (Bureau of Justice Statistics 2004, 8). In 1991, there were more than 60,000 juvenile admissions and more than 56,000 juvenile releases from jails. Boys made up 53,257 of the admissions, while girls accounted for 6,954. As for releases, boys made up 49,571, while girls accounted for 6,728 (Bureau of Justice Statistics 1999, 2). These data are more than a decade old, and a major study in 2000 was critical of the fact that "there are no current estimates of the number of youths admitted to jails each year" (Bureau of Justice Assistance 2000, x).

The average number of juveniles (those under the age of 18) held in local jails on any given day in 2003 was 6,869 (Bureau of Justice Statistics 2004, 8). Of these youths, 79.8 percent were being held as adults, and approximately 20.2 percent were held as juveniles. These numbers represent a one-day total, and tell nothing of the thousands of underage individuals who pass through adult jails each year. A statistic that has become available in the past couple of years is the number of youths held in jails *as adults*. Such juveniles have been tried, or await trial, as an adult. According to the annual survey of jails in 1994, there were 5,139 juveniles who were being held in adult jails as adult inmates. In 1993, there were 3,300 such youths. The latest figures show that about 5,484 such juveniles are being held as adults in jail (Bureau of Justice Statistics 2004, 8). Thus, in 2003, there were 1,385 juveniles held in adult jails, with an additional 5,484 defined as "adults," for a total of 6,869 underage detainees in jail.

Schwartz, Harris, and Levi (1988, 146–148) advocate passing legislation that would make the jailing of juveniles a crime. The same group of researchers also recommended the development of alternatives to adult jails for female juveniles. In many areas, girls account for as many as one quarter of the youths admitted to adult facilities, but the vast majority of them are of no risk to the community.

Population Characteristics

Of the 596,500 adult inmates housed in American jails in 1999, 321,000, or 53.81 percent, were unconvicted or pretrial detainees (Bureau or Justice Statistics 2001, 4). As **Table 9-2** shows, 67,487, or 11.1 percent, of the inmates held during that year were women. The female population in American jails is on the rise. Women accounted for 6.9 percent of the adult population in jails in 1983; by 1999, that number had risen to 11.1 percent, with almost 52,000 additional female inmates housed in jails.

Women in jail present a number of problems. According to the Bureau of Justice Statistics (1999, 7), about 70 percent of women in jails have children under the age of 18. Two thirds of these women were living with their children before entering jail. In addition to facing the shock of incarceration, another problem that women face in jail is the loss of contact with their children. While the vast majority of these children are placed with either their father or a close relative, almost 10 percent are placed in foster care or some other institutional setting (Bureau of Justice Statistics 1999, 9). Many women worry about the arrangements available for their children, and their ability to cope in a jail environment may depend on whether they know their children are being cared for. Another concern for many mothers is the ability to keep in contact with their children, either through the mail or by periodic visits. The children, too, must deal with the loss of their mother. Whether the children are placed with relatives or in foster care, sanctions imposed on their mothers may act as a punishment for them as well.

The relatively small number of women housed in jails presents a second set of more-ominous problems. Jails rarely have programs and services for women to the extent offered to male offenders. Women spend an average of 17 hours

Table 9-2 Average Daily Population in Local Jails

	June 30 1983	June 30 1988	June 30 1983	June 30 1999
Average daily population	227,541	336,017	466,155	607,978
Number of confined inmates	223,551	343,569	459,804	605,943
Adults	221,815	341,893	455,500	596,485
Male	206,163	311,594	411,500	528,998
Female	15,652	30,299	44,100	67,487
Juveniles	—	—	4,300	9,458
Held as adults	—	—	3,300	8,598
Held as juveniles	1,736	1,676	1,000	860

Source: Bureau of Justice Statistics. 2001. *Census of Jails, 1999,* p. 1. Washington, DC: U.S. Department of Justice.

per day in their cells, and are less likely than their male counterparts to have work assignments (Bureau of Justice Statistics 1999, 8). Because women represent only a small number of the total inmates, their medical needs often are not adequately addressed (American Correctional Association 1985, 24).

Most women serving time in jail serve their sentences in facilities that also house male inmates. In a study done to examine jails that are exclusively women's facilities, Gray, Mays, and Stohr (1995) collected data (surveys and interviews) from 5 of 18 facilities that house only female inmates. They explored the extent to which these jails meet the particular needs of female inmates. With the exception of treatment programs for drug and alcohol abuse, they found that the programming in these women's jails was "woefully inadequate" (Gray et al. 1995, 199).

Jails house a disproportionate number of black inmates. In 2003, more than 39 percent of the inmates housed in jails were black, with Hispanics accounting for an additional 15.4 percent (Bureau of Justice Statistics 2004, 8). In the past 10 years, the gap has widened, with the incarceration rate for blacks climbing from about 400 black inmates per 100,000 black residents in 1987 to almost 750 black inmates per 100,000 black residents in 2003.

This explosion in the jail population is apparent in the number of prisoners and in building trends. The percentage of convicts grew on average by 5.4 percent per year during the period 1993 to 1999; during that same period, jail occupancy grew to 93 percent from 85 percent. The percentage growth in the inmate population has far outpaced the growth in the general population, with the number of jail inmates per 100,000 residents almost doubling from 96 in 1983 to 222 in 1999 (Bureau of Justice Statistics 2001, 1).

Jail-population numbers represent a one-day census undertaken on June 30 of each year. Such numbers do not represent the many individuals who pass in and out of jail each year, some staying for only a few hours. Much of the turnover is concentrated during peak periods, such as Friday and Saturday evenings. In 1993, there were more than 13 million people booked into jails. This number represents an increase of 63.8 percent from 1983. The vast turnover of inmates each year represents an administrative burden on the staff of jail facilities (BJS 2001).

Architecture

There are two distinct designs present in American jails today: the traditional, or linear, design, and what are now being characterized as new-generation, or podular, jails. Each of these designs is based on a number of underlying philosophies, and presents a number of important distinctions. Note, however, that most U.S. jails are small. More than 46 percent of American jails house fewer than 50 inmates (Bureau of Justice Statistics 2001). About 10 percent of the nation's jails house 10 or fewer inmates (Mays and Thompson 1988, 437). While we address the differences in architectural styles of jails, it is important to remember that alternative approaches are limited in many jurisdictions due to these size restrictions.

Traditional Jails

Most jails in America are traditional in design. This design, also referred to as the **linear design**, has a long history dating back to the time of the Walnut Street

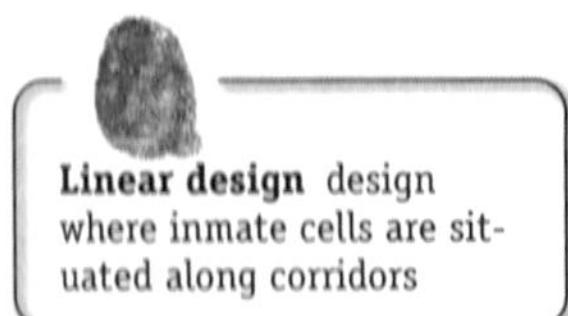

Linear design design where inmate cells are situated along corridors

Jail in Philadelphia. In the linear design, inmate cells are situated along corridors. The staff monitors the cells of these traditional jails by walking the corridors. The staff cannot monitor the entire inmate-housing unit at one time; they can provide only intermittent surveillance (Nelson 1988, 2). During the period when no staff are in the corridor, the inmates enjoy freedom from scrutiny in the cells. Most behavior problems occur between the intermittent patrols. The familiar image of inmates holding mirrors from their cells to observe the corridor is a striking example of this reality. Often inmates watch out for the jailer and warn others when he is coming.

Traditional jails are built to last. Prisoners, even with time on their hands and very little supervision, can cause very little damage to the physical structure. The toilet facilities and beds are designed to hold up to the most intense punishment, and as such are costly to purchase. They are not designed for aesthetics, but rather, for function. Therefore, traditional jails are noisy, with concrete floors and walls that do not absorb sound. The metal doors clang shut, causing the sound to reverberate throughout the facility. Lighting fixtures are either encased in metal or outside the actual cell to keep inmates from having access to them.

The lack of supervision combined with the noise level adds to the stress that convicts face. Inmates are already experiencing stress, and when they find themselves locked up in a noisy and often hostile environment, they frequently exhibit signs of psychopathology soon after their arrival in jail (Gibbs 1987, 308).

Second-Generation Jails

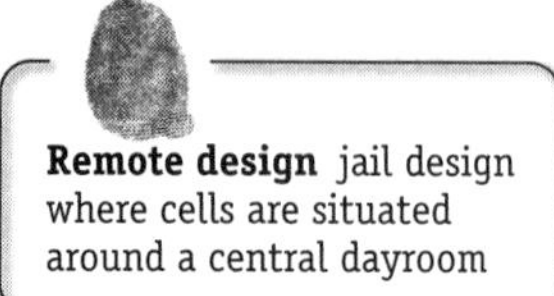
Remote design jail design where cells are situated around a central dayroom

A second generation of jails has been designed and put into use in many areas. In this design, jail personnel have what is termed remote, or indirect, surveillance of inmates. Under the **remote design**, cells are situated around a central dayroom, and jail personnel occupy a secure control room that overlooks the dayroom and also the individual cells. While such a design actually increases the visual surveillance by correctional staff, it limits verbal interactions between inmates and jail personnel, who often communicate through an intercom. While the prisoners are more closely monitored in second-generation jails, most such facilities still employ high-security fixtures, furnishings, and finishes (Nelson 1988, 2).

New-Generation Jails

New-generation jails stand in stark contrast to the more traditional linear jails and their new cousin, the remote-supervision jail, both in design and in their philosophy of operation (**Figure 9-1**). A number of important distinctions must be made between the new-generation jail and its traditional counterpart. First, traditional jails, by design, require the staff to constantly patrol inmates' cells, usually offering only intermittent supervision at best. Under the new-generation design, the jail staff are in direct contact with the inmates. Jail personnel actually occupy a space within the dayroom or housing pod (Nelson 1988, 170). The fundamental goal of the new-generation jail is to provide a safe, violence-free environment for both inmates and staff, an environment that treats inmates in a humane fashion (Zupan 1991, 73).

This style of architecture originated with an effort by the U.S. Bureau of Prisons to open three federal detention facilities in Chicago, San Diego, and New York

City in the mid-1960s. The government commissioned three architectural firms to design these new facilities, and gave only three stipulations: single cells for inmates, direct supervision by staff, and functional living units (Zupan and Menke 1991, 185). The concept of **functional living units** places all sleeping, food, hygiene, and recreational facilities in a self-contained unit. A number of advantages to these new designs have added to their appeal. First, the concept of direct supervision, where staff and inmates share a common area, has a very practical significance. No longer are inmates locked in cells with other inmates out of sight from jail personnel. In the new-generation jail, inmates and staff share the facility, with staff members in constant contact with the inmates. It forces the staff to run the facility, rather than allowing the inmates to do so.

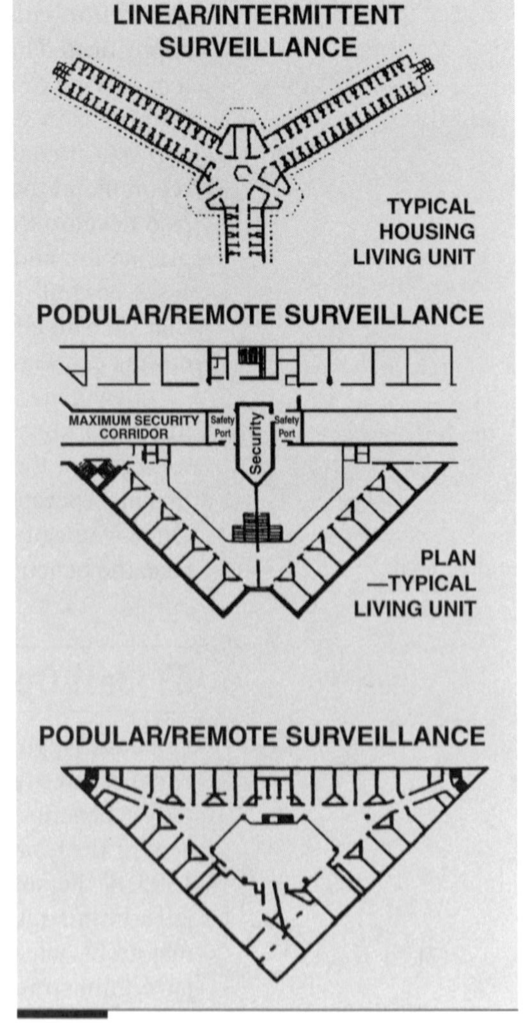

Figure 9-1 Jail Architecture. ***Source:*** *U.S. Government Printing Office 1993.*

Another advantage of the new-generation design is that the furnishings, fixtures, and finishes of these jails are of normal commercial grade, as opposed to high-priced security fixtures. This can result in a substantial savings in the cost of building such a facility. For example, Nelson (1988) determined that the savings in building costs for a single unit housing 48 inmates would be more than $200,000 compared to a traditional jail. There are savings in the initial construction of the facility, and substantial savings in operating expenses and upkeep. With inmates in constant contact with jail personnel, incidences of vandalism and graffiti are greatly decreased. The new-generation design also incorporates carpeting and acoustic tile to reduce the noise level and to add to the aesthetics of the environment. Instead of gates and steel bars, solid walls and doors with impenetrable glass are used. Once again, this adds to the aesthetics and reduces the noise of metal clanging against metal often found in traditional jails.

Another important issue with new-generation facilities is staffing. Do new-generation jails require a larger staff than their traditional counterparts? There is no definite answer to this important question, since often it is dependent on the level of staffing present in the other facilities. In some states with jail standards, the inmate-to-staff ratio is set, and little savings will be found. Much depends on the efficiency and size of the traditional jail. The working conditions are another important factor. The new-generation concept is, without question, a quantum leap, with jail personnel now occupying living space with the inmates.

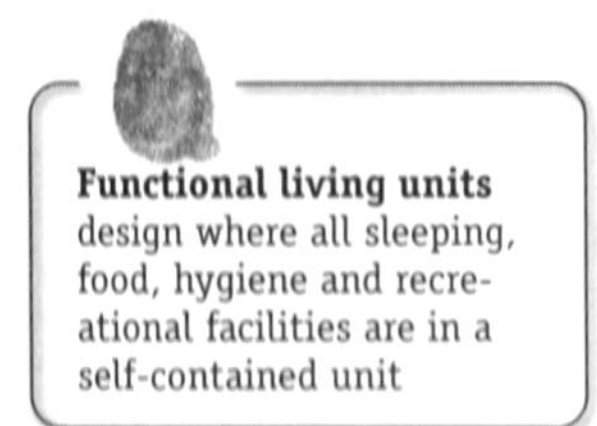

Functional living units design where all sleeping, food, hygiene and recreational facilities are in a self-contained unit

How do correctional staff feel about such an arrangement? One way of measuring this is to look at the number of sick days taken by the staff. In one study reported by Nelson (1988, 4), the National Institute of Corrections found that sick leave in the new-generation jails was significantly less than the average in the four other houses of detention. The savings amounted to 1,810 staff days, or the equivalent of eight full-time positions. This number would have amounted to $250,000 in overtime expenditures if overtime had been used to fill the vacancies.

Another concern with staffing is that, to effectively manage inmates in a new-generation jail, the correctional officers must be trained to use a number of sophisticated human-relations skills, ranging from conflict management to problem solving to interpersonal community (Zupan and Menke 1991, 193). Such training is in stark contrast to the traditional custodial skills now taught, such as physical-control techniques and firearms usage. Zupan and Menke (1991, 193) recommend that, for the new-generation philosophy to really succeed, we must also develop the correctional officer's career orientation. The costs of such an orientation and of training officers in conflict management have not been fully assessed, but the literature available on new-generation jails seems to indicate that such an effort will have a substantial savings over the long run, both in operating costs and in the reduction in the number of lawsuits.

New-generation jails offer an opportunity for cost savings, an improvement in the working environment for jail personnel, and an improvement in the environment for the inmates. As of 1991, there were about 80 direct-supervision jails that incorporated podular designs (Stohr, Lovrich, Menke, and Zupan 1994, 473). Once we begin looking at a number of concerns that jail administrators must face, the benefits of the new-generation philosophy will become more apparent.

Jail Operations and Administration

Personnel represent the single largest expenditure in local jails. Total operating expenditures—which include salaries, wages, employer contributions to employee benefits, food, supplies, and contractual services—accounted for 71.3 percent of the total expenditures for local jails in 1993 (Bureau of Justice Statistics 1995, 8). Personnel expenditures represent only half of the picture. When asked, jail administrators rank personnel concerns as second only to crowding as the major obstacle facing them (Guynes 1988). Staff shortages are one problem that jail administrators face. Shortages often result from the "poor image of jail work and inadequate career incentives" (Poole and Pogrebin 1991, 163).

As pointed out above, in many jurisdictions, the county operates the local jail. The person most often responsible for the jail's operation is the sheriff. In most jurisdictions, the sheriff is an elected official who may or may not have law-enforcement experience. The local sheriff has a large number of responsibilities, only one of which is the administration of the local jail. For the most part, the public has one of two attitudes toward jails: at best, indifference; at worst, a strongly negative view (Mays and Thompson 1991a, 10). As such, jails often find themselves at the bottom of the newly elected county sheriff's list of priorities. As might be expected, the local sheriff, as an elected official, wants to appease the residents of his or her jurisdiction, and the most visible way to do this is by putting more deputies out on the street, patrolling the neighborhoods, making arrests, and answering citizens' calls for service. As long as there are no major problems in the jail, such as a riot or escape, the public is often unwilling to commit money for facility improvements.

The local sheriff will appoint a jail administrator, or deputy sheriff, to oversee the jail's day-to-day operation and to supervise the jail staff. Often the staff consists of individuals who, looking to move into the more lucrative and much-higher-status deputy position, are seeing the job in the jail as only a temporary position. In some cases, the jail is staffed with deputies who are being punished for violating a departmental rule (Champion 1990, 173). Sometimes the correctional officer on the night shift also has to handle incoming calls and dispatch officers.

In some jurisdictions, a second career track has been initiated exclusively for correctional officers. There are a number of problems with this approach as well. Many times, these officers are paid less than their counterparts in the sheriff's office, and as such are often considered second-class citizens. Adding to this lack of stature is the fact that jail personnel often receive less training, and the environment in which they work is often less than desirable. The jail employee suffers from a lack of potential advancement as well. Correctional officers make up more than 72 percent of all jail personnel. Promotions to supervisory positions are highly competitive and often nonexistent (Bureau of Justice Statistics 2001, 9).

Salaries in law enforcement and corrections vary greatly from jurisdiction to jurisdiction, but data available on a national level from the U.S. Department of Labor found that the median salaries of state corrections officers and jailers were about $32,670 a year in 2002 (U.S. Department of Labor 2004). The lowest 10 percent earned less than $22,010, while the top 10 percent earned more than $52,370. The same report indicated that sheriff's deputies had a median annual salary of about $42,270 in 2002 (U.S. Department of Labor 2004).

From 1993 to 1999, female staff at local jails more than doubled (from 28,500 to 70,700), but this increase represents only about a 10 percentage point increase in the proportion of female staff (to 34 percent of all officers from 24.2 percent). The race and ethnic origin of jail staff has changed very little in the same time period (to 24 percent from 23.2 percent for blacks, and to 8 percent from 6.7 percent for Hispanics). The number of inmates per jail employee has fluctuated in the past 15 years. In 1983, the inmate-to-correctional-officer ratio was 5.0 to 1. In 1993, that ratio had dropped to 3.9 to 1, but it rose to 4.3 to 1 in 1999 (Bureau of Justice Statistics 2001, 9-10).

The problems with jail staff cannot be understated. Poor working conditions affect employee morale, and consequently affect the treatment of those under their care. Efforts must be made to bring about a number of changes for jail personnel. If changes are made in the training and enrichment of the correctional officer's job, this will also enhance the environment in which we house inmates.

Another important factor in improving the working conditions of jails is the philosophy of the new-generation jail. As pointed out above, some of the end results of the direct-supervision models are improved staff morale, decreased staff tension, reduced sick leave, improved treatment of inmates by staff, decreased staff-inmate conflicts, and reduced employee misconduct (Nelson 1988, 4). New-generation jails offer correctional officers greater control of inmate behavior, and as such, much more interaction and control over what happens. Officers must actively supervise the inmates, resulting in more responsibility and job satisfaction.

Legal Issues

While jail administrators are wrestling with personnel problems, an even bigger concern looms on the horizon—that of inmate lawsuits. Such suits have the potential of costing the local government millions of dollars. Prisoners are filing lawsuits in record numbers for such things as overcrowded conditions, damage to personal property during shakedowns, and injuries received from other inmates and staff. Local jurisdictions do not have the same immunity from civil liability and the collection of damages that states usually enjoy (McCoy 1982).

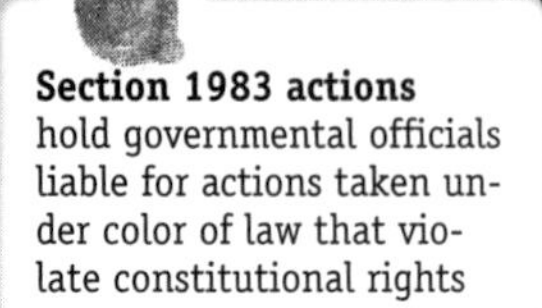

Section 1983 actions hold governmental officials liable for actions taken under color of law that violate constitutional rights

Champion (1991, 205) lists three major avenues that jail inmates follow for filing lawsuits: civil-rights violations, habeas corpus petitions, and mandamus actions. As was discussed in Chapter 8, **Section 1983 actions** hold governmental officials liable for actions taken "under color of law" that violate constitutional rights. Not only can an inmate file suit against an individual employee for a violation of civil rights, but, in many cases, a suit is filed against the jail administrator (often the sheriff), the county, and, quite possibly, the county commissioners court.

The rationale for such suits is that administrators at all levels are responsible for the actions of subordinates. If it can be shown that either through neglect in hiring, training, or retention, the administrator or county officials should have been aware of inappropriate actions by their subordinates, then they, too, can be held liable. The jail administrator hires and is responsible for the training and supervision of all subordinates. If something happens, both the jail employee and the supervisor are responsible. Section 1983 actions are the most prevalent litigation actions brought by jail inmates (Champion 1991, 206).

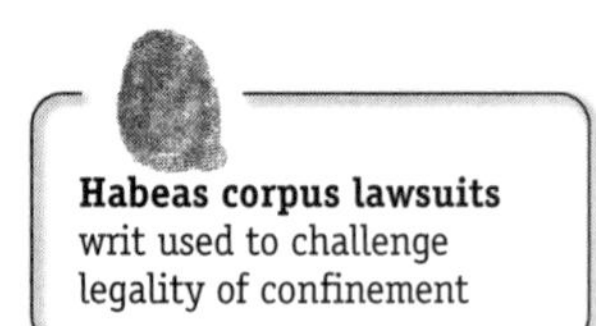

Habeas corpus lawsuits writ used to challenge legality of confinement

Mandamus actions legal action used to compel government officials to perform their duties

Habeas corpus lawsuits are the second most prevalent form of inmate suits filed. Under habeas corpus lawsuits, prisoners challenge either their confinement or the conditions of confinement (Champion 1991, 206). A person who is confined can use this writ to obtain a hearing as to the legality of his or her detention. **Mandamus actions** are often filed to compel jail personnel to perform their administrative duties. If the court orders a jail to provide a service or reduce overcrowding, and the jail staff fails to carry out that order, a mandamus action can be sought. The problem with such actions in a jail is that when the case finally reaches the court, the inmate filing the action may already have been released from confinement, or may no longer be in the facility. As such, lawyers will often file class-action suits rather than individual suits.

The best approach to the litigation crisis is to avoid the conditions in which lawsuits are brought. Jail administrators should design and run their facility in a "preventative" fashion. Sechrest and Collins (1989) offer jail administrators and jail personnel an alternative to simply dealing with lawsuits as they occur. They recommend that steps should be taken to avoid the possibility of lawsuits. They offer recommendations to the administrator, as well as security checklists that should be used to head off problems.

As discussed in Chapter 8, the *Bell v. Wolfish* (441 U.S. 520 [1979]) decision covered a number of issues dealing with crowding, health care, and the rights of pretrial detainees. As to the latter, the court ruled that a pretrial detainee may not be punished prior to adjudication of guilt. However, the decision held that double-bunking and other practices that were necessary to manage the institution and maintain security were not punishment. If jail officials show that conditions are designed to maintain security or for other management concerns, they are not considered to be punishing the inmate.

One positive effect of inmate litigation has been that local jurisdictions have had to take a serious look at the facilities and training afforded their staff. While large settlements are not the norm in these cases, such threats have forced local jurisdictions to look very seriously at both their physical plants and their training procedures. In fact, one of the best mechanisms to head off litigation is in the development of jail standards, which includes a procedure for monitoring compliance (Thompson and Mays 1988).

Jail Standards

The implementation of jail standards has been hailed as potentially the best way to correct many of the deficiencies found in American jails (Mays and Thompson 1991a, 15). Doing so is also seen as one of the best defenses against inmate litigation. A number of agencies have developed and disseminated model standards for jails (American Correctional Association 1981; National Advisory Commission on Criminal Justice Standards and Goals 1973). Although these standards are readily available, their implementation has been slow at best. Mays and Thompson (1991a) have pointed to three possible reasons why these standards have not made a significant impact on the conditions of American jails. First, they point out that adopting these standards may require major capital expenditures (Mays and Thompson 1991b, 15). In many jurisdictions, such large capital outlays are just not possible. The local jurisdictions must also deal with public pressure to put tax money into more-popular public programs such as education and recreation.

A second concern is that smaller jails cannot provide the level of service advocated by the jail standards. Many treatment programs are just not feasible in such small facilities. Small jails also suffer from the problems of economies of scale (Mays and Thompson 1988). The rationale behind economies of scale is that the cost of services and products can be greatly reduced if larger quantities are purchased at one time. The cost to provide medical care in a small jail must be divided by only a small number of inmates, while in large metropolitan jails—such as Illinois's Cook County facility, which can house more than 9,000 inmates—the relative cost per prisoner is small. The same rationale holds for other services such as food, supplies, and, to a certain extent, labor costs. The obvious solution to this problem is to eliminate smaller, economically inefficient jails and consolidate them into regional detention facilities (Mays and Thompson 1991, 16). This trend is already taking place, with the number of small jails (fewer than 50 inmates) declining, and the number of large jails (more than 1,000 inmates) increasing. In 1999, there were 1,573 small jails, which compares to 2,844 in 1978; in 1999, there were 298 large jails, compared to only 10 in 1978 (Bureau of Justice Statistics 2001).

The third and final reason that progress has been slow is because even if the standards are put into place, rarely do mechanisms exist to enforce those standards (Mays and Thompson 1991, 16). Without an effective enforcement body with authority to force compliance, true change is unlikely to take place.

The obvious trend is for jail administrators to move toward the implementations of jail standards on a local level, or face the possibility that the courts would force compliance. An even better approach would be to develop standards at the state level. In 1999, out of 2,838 local-jail jurisdictions, only about 15 percent were under a court order or consent decree to reduce crowding or address other confinement conditions (Bureau of Justice Statistics 2001, 5).

Overcrowding

The rated capacity of local jails in this country has risen from 261,432 in 1984 to 736,461 in 2003, an increase of almost 182 percent (Bureau of Justice Statistics 2003c, 9). Despite this stepped-up attempt to build our way out of a problem, our nation's jails are still overcrowded. In fact, "overcrowded penal facilities represent the single largest dilemma facing modern American penology." About 11 percent of the jurisdictions had at least one jail under court order to limit their population (Bureau of Justice Statistics 2001, 5).

Even though 23,839 new beds were added each year from 1995 to 2003, jail capacity continues to hover very near the 95 percent mark. The Bureau of Justice Statistics (1995, 6) defines rated capacity as "the number of beds or inmates assigned by a rating official to facilities within each jurisdiction." The obvious implication here is that we find bed space for all those held in jail, and the rated capacity tells us little about crowding. In fact, it may just be an indication of the number of beds in a jail rather than an indication of crowding. If this is the case—and there is every indication that it is—then a number of states are beyond the crisis point. In seven states and the District of Columbia, the occupied capacity is above 100 percent. Figures for 1999 indicate that Virginia jails were at 118 percent of their rated capacity, and that jails in Washington, D.C., were at 120 percent of their rated capacity (Bureau of Justice Statistics 2001, 5).

A number of states are well below their rated capacity. Excluding Alaska, six states had occupancies below 80 percent of their rated capacity. The state with the lowest occupancy rate was North Dakota, at 64 percent (Bureau of Justice Statistics 2001, 5). All of these figures may be somewhat deceiving because occupancy fluctuates throughout a week or month in most jurisdictions. These numbers represent averages, and the actual numbers may increase drastically on a Friday or Saturday evening and drop off during the week. Sometimes people serving sentences for minor offenses are given their sentence during the weekend so they can continue to hold down a job during the week.

A second major concern is that the above numbers may be misleading. Due to classification problems, many jails (even if they are below their rated capacity) may be very crowded due to the housing of women and juveniles at the same facility. Federal law mandates that there be sight and sound separation of the sexes and between juveniles and adults. In many small jails, if space is needed to house a female inmate or juvenile, a whole wing of the facility must be emptied to accommodate this person. Thus, you may have one or two individuals staying in an area that is designed to hold 50 inmates, while the other displaced inmates occupy a much smaller area.

In an attempt to deal with the crowding problem, policy makers have decided to house many jail inmates in what have been called makeshift jails (Welch 1991). These facilities may range from converted gas stations or motels to barges or ferryboats converted to floating detention facilities. New York City has taken the lead in the use of decommissioned ferryboats and military troop transports as makeshift jails. The first two such floating detention centers were moored at Rikers Island to add much-needed jail capacity (Welch 1991, 151). One such barge has 800 beds and is moored off the South Bronx opposite Rikers Island (New York City 2004).

Although it seems that the only approach being undertaken to control the crowding problem is to build or fabricate additional beds, attempts are being made on other fronts to reduce the actual number of individuals housed in jails—for example, by home confinement and monitoring. A number of factors must be considered when addressing the problem of crowding. Removing juveniles from adult facilities would reduce crowding, and it would also bring many jails into compliance with the amended Juvenile Justice and Delinquency Prevention Act of 1980. A second widely mentioned solution is to reduce the number of pretrial detainees in jail. Since they account for about 50 percent of the jail population, reducing their number would substantially reduce crowding. However, any attempt to do so should be done with caution. Even with the new provisions of the Bail Reform Act, about 16 percent of those on pretrial release are rearrested, three fifths of whom were rearrested for a felony (Bureau of Justice Statistics 2003a, 22).

The courts could also rely more heavily on alternatives to incarceration for many individuals found guilty of nonviolent offenses. The use of fines, community service, and such could be expanded. Where the public demands jail time, such as in the case of driving while intoxicated or while under the influence, alternative forms of incarceration could be employed. Many such offenders are in little danger of escape, and as such, can often be punished by using a less restrictive and costly approach. The maximum-security jails that we spend millions of dollars to build should be reserved for those who possess the greatest risk to society. Whenever possible, alternatives should be used for all other offenders.

Special-Needs Inmates

Many jails are just not equipped to meet the special needs of some prisoners. Inmates with special medical needs or those suffering from mental illness create additional sets of burdens on an already overtaxed system. In many cases, the physical plant is not equipped to handle those who are visually or hearing-impaired. Older inmates who have trouble walking may find the maze of corridors and stairs insurmountable. In the sections that follow, we will look at a number of these special-needs inmates.

The Mentally Ill

The treatment of mental illness has become an increasingly complex problem for jails in the United States (Gibbs 1986). Much of the problem today is related to the fact that the courts have ruled that mental hospitals must either treat their mental patients or release them (*Wyatt v. Stickney*, 325 F. Supp. 781 [M.D. Ala. 1971]). The proverbial floodgates were opened as previously institutionalized patients were released onto the street. Many such individuals become public nuisances, and the police have few alternatives to deal with such individuals. With a lack of other public facilities to house them, these people subsequently end up in jail. Even if a public or private facility is available, often the mentally ill are housed in jail for short periods of time until they can be committed to such institutions. Studies have found that as many as 30 percent of all jail inmates show some signs of mental disturbance (Guy, Platt, Zwelling, and Bullock 1985, 29).

The question then becomes how to deal with mentally ill individuals in jail. Among the number of studies that addressed this issue, Gibbs (1987) recommended a careful look at both the mentally ill inmates and the jail environment. The actual environment in which we lock up individuals can have a profound effect on the symptoms of psychopathology. When inmates were given a battery of tests to measure symptoms of psychopathology, both at the time of arrest and then again after five days of confinement, it was repeatedly found that the symptoms of psychopathology had increased (Gibbs 1987).

In dealing with the ever-increasing demand, jail administrators must make major operational changes in the delivery of mental-health services (Kalinich, Embert, and Senese 1991, 86; Senese, Kalinich, and Embert 1989). Screening mechanisms must be put into place, with highly trained personnel conducting the intake process. When a need is identified, appropriate mental-health care must be provided. If such care is not available in the community, then the jail must take the lead in providing the service within the facility. Jail personnel must also be trained to recognize the symptoms of mental illness, and to follow the course recommended by Gibbs (1987) in making every effort to reduce the environmental effects that can cause or add to the problem.

A direct-supervision design may offer a partial solution. New-generation jails that employ direct supervision offer prisoners a less stressful environment because inmates feel less threatened from predatory attacks and because the noise level in these facilities is greatly reduced. A study by Zupan (1991, 161–162) that compared the stress levels between a traditional jail and a direct-supervision facility found that inmates in the direct-supervision facility reported fewer symptoms of psychological and physical stress than those prisoners housed in the traditional jail.

Alcoholics

One of the most pressing problems of jail management is what to do with individuals taken into custody who are under the influence of alcohol. The number of people arrested who are under the influence of alcohol has grown rapidly in the past decade, due largely to the efforts of public-interest groups such as Mothers Against Drunk Driving (MADD). Public awareness of the risks presented by drivers who are under the influence of alcohol has forced courts to place more emphasis on this behavior. In some cases, mandatory jail time has been legislated for those convicted of driving under the influence (National Institute of Justice 1984). While the increasing number of individuals housed in jails for alcohol-related charges has added to the crowding problem, an even greater concern is the number of problems that can occur during the first few hours of confinement. Sickness, vomiting, and the possibility of slipping and falling are all very real concerns to jail personnel. Such individuals may become despondent and try to commit suicide, or may have compounding health problems that simply go unrecognized when they enter the jail. It is not uncommon for detainees who seem drunk and are placed in a cell to later fall into a coma and, in some cases, die. As the number of individuals who are housed in jails increases and, more importantly, as the number who flow through jails increases—usually during peak periods such as Friday and Saturday evenings—such problems only compound themselves.

Drug Addicts

Admittees who use and abuse drugs present a special problem to jail personnel. Those who have a long history of drug use will often appear normal when they arrive at jail. The actual physical and medical problems do not begin to materialize until they begin coming off the drugs. The withdrawal symptoms may be very severe and may require the assistance of medical personnel. A recent study undertaken by the U.S. Department of Justice (2003), titled "Arrestee Drug Abuse Monitoring (ADAM)," found that 64 percent or more of adult male arrestees had recently used one of five drugs at the time just prior to their arrest. This is not only a big-city problem; drug use is a problem in all communities. In some cases, the users may be so strung out on drugs that they have to be restrained to ensure that they do not hurt their captors or themselves.

Jail personnel must be trained to look for signs of drug use, and must be prepared to assist the inmates during withdrawals. Correctional staff should be familiar with the side effects of different types of drugs so that appropriate action can be taken. When medical treatment is needed, it is often up to the jailer to administer medication prescribed by doctors.

Another pressing problem in jails is control of the distribution of illicit drugs within the institution. Inmates will often go to extraordinary means to obtain drugs within a jail. Facility and support personnel may be bribed or coerced into supplying drugs, and whenever contact visits are allowed, the potential for drug smuggling exists.

Sex Offenders

Another type of inmate that can present problems in jail is the sex offender. It is impossible to lump all sex offenders into one category, since a wide range of offenses fall under this classification. The jail is often faced with holding a person accused of molesting a child or indecent exposure. Such a prisoner is often quite passive in a correctional environment, and may become a victim of predatory crime within the jail. In fact, other prisoners look down on many sex offenders (especially child molesters), and often will go to great lengths to injure or even kill them. Such offenders may need to be separated from the general population for their own protection. Such arrangements often have an adverse effect because many such offenders become depressed and suicidal, and segregation affords the opportunity to carry out these suicides.

Suicide

The high rate of suicides in jail can be attributed largely to the type of offenders that are housed there. Studies have found that the suicide rate is five to six times higher among jail inmates than among comparable free-world individuals (Winfree 1988). In 1999, 35.3 percent of deaths in local jails were attributed to suicides (Bureau of Justice Statistics 2001, 8), while only 5.1 percent of the deaths in state correctional institutions were attributed to suicides (Bureau of Justice Statistics 2000, 91). During the year ending June 30, 1999, there were 919 inmate deaths that occurred in local jails. More than 35 percent, or 324, of those deaths were attributed to suicides. Suicide deaths per 100,000 jail inmates dropped from 55.6 percent in 1983 to 35.1 percent in 1999, due largely to the increased awareness of the problem and a concern for the possibility of lawsuits (Bureau of Justice Statistics 2001, 8).

Much of this can be attributed to the fact that jails service an at-risk group of individuals—for example, young men abusing drugs or alcohol (Winfree and Wooldredge 1991, 77). In looking at data collected on 344 of the 419 suicides that occurred in jails in 1979, Hayes (1980) identified a number of important victim characteristics. Hayes found that almost 97 percent of suicide victims were men, and about 68 percent were white. About 92 percent were pretrial detainees, and more than half committed suicide during the first 24 hours of incarceration. Twenty-seven percent had taken their life within three hours of being locked up. Sixty percent of those who took their life were being held in isolation (Hayes 1980). Hayes (1983, 467–468) offers a profile of the typical jail suicide:

> *An inmate committing suicide in jail was most likely to be a 22-year-old, white, single male. He would have been arrested for public intoxication, the only offense leading to his arrest, and would presumably be under the influence of alcohol and/or drugs upon incarceration. Further, the victim would not have a significant history of prior arrests. He would have been taken to an urban county jail and immediately placed in isolation for his own protection and/or surveillance. However, less than three hours after incarceration, the victim would be dead. He would have hanged himself with material from his bed (such as a sheet or pillowcase).*

Some words of caution should be issued. First, a reliance on such a profile might cause jail personnel to overlook other inmates who may take their lives (Kennedy and Homant 1988, 452). Second, the use of profiles may allow lawyers to attach tort liability to jail personnel when a death occurs if the individual who committed suicide fits the profile. Due to these concerns, Kennedy and Homant (1988, 453) have suggested to profile alternatives that should be employed by jails. They suggest that we should first develop jails that are suicide-resistant. By developing new designs that allow more-careful monitoring and the use of hardware that does not facilitate hanging, we could reduce the opportunity of suicide. There are a number of potential problems with such an approach, such as the dehumanization of the jail environment, which might just shift the suicides to prisons. The best approach is the direct-supervision design, where inmates are under continuous supervision. Not only are they afforded little or no opportunity to commit suicide, but also jail personnel would be more likely to notice warning signs and could offer assistance.

A second alternative is to become aware of inmates' needs during the first critical 24 hours that they are in custody (Kennedy and Homant 1988, 453). Whether prisoners fit the profile or not, realizing that those first few hours are the most critical, and providing additional supervision during that crucial time period, can reduce a substantial number of suicides. Kennedy and Homant (1988, 454) also noted that often jail suicides are nothing more than suicides that occurred in the facility—meaning that presence in jail was not directly tied to the individual's decision to commit suicide.

Juveniles Suicides

In 1988, a total of 33 youths died while being held in public-custody facilities: 17 were by suicide, 6 were homicides, and the remainder can be attributed to illnesses and other causes (Allen-Hagen 1991, 3). Suicides are just one of many problems jails face with underage inmates. A study undertaken by the Department of Justice on youth suicide in adult jails, lockups, and juvenile-detention centers found that the incidence of youth suicide was 7.7 times greater in adult jails than in juvenile-detention facilities, and 1.4 times greater in adult lockups than in juvenile-detention facilities. The incidence of youth suicides for adult jails and lockups was 4.48 times greater than the risk among juveniles in the general population (Flaherty 1980, 10).

An even more pressing problem is the fact that half of the underage inmates who committed suicide while in adult jails or lockups in 1978 had not committed a felony and appeared to present no real danger to society (Flaherty 1980, 12). These youthful offenders may become victims of abuse or sexual assault by older, more-hardened criminals housed in jails. Such factors can only add to the shock of incarceration. Girls who have a history of sexual or physical abuse are especially vulnerable to suicide and depression (Chesney-Lind 1988, 161–162). Chesney-Lind warns that juvenile girls are vulnerable to sexual assault by jail personnel (1988, 1997). These girls suffer from a "doubly disadvantaged status," being both female and underage (Chesney-Lind 1988, 164).

Jail Socialization and Subcultures

Little research has been undertaken addressing the role of subcultures within local jails. In fact, there is some question as to whether or not subcultures exist within the jail environment. Stojkovic (1986, 32) argues that inmates move into and out of jails with such "rapidity that it is difficult for an identifiable subculture to develop." On the chance that a subculture is found, it is probably "imported" into the facility by those who have experienced prison incarceration. Garafalo and Clark (1985) draw similar conclusions, claiming that if a subculture does exist in jails, it can be attributed to more-experienced inmates who are able to readapt when they find themselves confined in jail again. As early as 1986, Stojkovic brought to light the dearth of research and literature on jail subcultures, and recommended a thorough investigation of all elements of subcultures—such as violence and stress adaptation, sexual relations, race relations, contraband markets, and power relations within the jail setting. To date, little effort has been made toward this effort. Nor has there been much research on the manner of socialization to the jail culture.

This is not to say that socialization does not take place, and we must look at an important factor that is involved here. Jails act as the entry point for virtually all prisoners in the correctional system. Almost everyone who is going to serve time in a state or federal prison will begin the process in a jail. As such, jails act as one's first taste of confinement, and it can be argued that jails begin the actual inmate-socialization process. This process of socialization within a jail is evident in the story told by a 20-year-old who, after being arrested for the first time on a commercial-burglary charge, began his journey into the criminal-justice system within a county jail.

> *My first few months within the county jail can be compared to boot camp in the army. I basically followed the county jail staff's orders and learned to keep my eyes to myself and my mouth shut. I survived through this wicked transition from free-world society into this jungle-like surrounding by adjusting to not only the whims of the jail administrators, but also to the jail inmate code of conduct. I was exposed to many different walks of life: from gang bangers to sex offenders, wealthy inmates to winos, and questionable heterosexuals to full-blown transsexuals. I remember getting initiated into the system my first night by a group of inmates who beat me down to the ground for the sole purpose of stealing my sneakers (Wooden and Ballan 1996, 8).*

According to Wooden and Ballan (1996), socialization begins with jail confinement, where prisoners move from the relative freedom of life on the streets to their first encounter with the correctional system. This important first step in the socialization process was described as a school, with inmates and correctional officers serving as teachers:

> *Day and night I observed the transactions and trends within my world behind bars. I often hid behind a mask to camouflage my true feelings, acting like a chameleon in order to blend into the background. I constantly found myself adapting to fit the rules of each situation. I remember I'd be resting on my bunk late at night and watching inmates quietly move around the bunk area like shadows. A few times I actually caught a glimpse of these shadows stealing, or raping and/or beating other inmates for their own personal gain. I had been a victim to these predators, also, so I used this daily knowledge to protect myself. Almost instantaneously, I learned to sleep with both eyes open. (Wooden and Ballan 1996, 9–10).*

This first stage of inmate socialization—learning the system—leads prisoners into the second stage, in which dysfunctional roles are assumed. This stage takes place when inmates enter prison. Inmates at this stage take on what Goffman has called "impression management," where they attempt to control the impressions that others have of them (Goffman 1959, 208–237). This coping mechanism is part of the adaptation process to the new surroundings.

It seems that as we rely more and more on jails to handle the overflow from state and federal prisons and to house convicted individuals ordered to serve their sentences in jail, subcultures similar to those found in prisons will begin to form. Recent data show that in 1999, more than 29,000 of all inmates housed in jails were being held for federal authorities. In addition, it may be that prisoners are serving longer sentences in jails. Between 1983 and 1989, the average sentence served by convicted inmates in jails increased from 14 months to 17 months (Bureau of Justice Statistics 1995, 13–14). By 2002, the average sentence length had increased to 24 months (Bureau of Justice Statistics 2002, 5). With longer sentences, jail inmates will inevitably be more subject to the jail version of prisonization.

A Day In the Life

A common phrase mentioned by scholars is that "jail time is dead time." Many jails just do not have the resources to provide prisoners with meaningful activities. Part of this problem is due to the the high turnover in population that jails experience. The Bureau of Justice Statistics (1990, 5) found that approximately two fifths of jail inmates are released after spending only one day or less in jail, and three fifths spend only four days or less. The median time that prisoners spend in jail was found to be only three days (Bureau of Justice Statistics 1990, 5). But "without appropriate programs that focus on changing the criminal behavior of inmates, the jail becomes a 'revolving door,' releasing individuals into the community simply to re-admit them in a few months, weeks, days, or even hours, when they are arrested for another crime" (Lightfoot, Zupan, and Stohr 1991, 50).

There is no typical day in jail, but a common theme seems to be inmate idleness. On a typical day, prisoners might be awakened at 6:00 A.M., when they are to clean up their cells and make their bunks. They have breakfast and can then take showers and, when necessary, get ready for court appearances. After 8:00 A.M., the inmates might take part in programs or attend a Bible-study or job-skills class, provided that their facility offers these activities and that space is available. If they choose not to attend, or if no such program is available, prisoners will often spend the morning in the dayroom of their cell block. At about 11:00 A.M., the jail might be closed down so that an inmate count can be taken. When the jail is closed down, visitation is halted, including visits by attorneys, and inmates are required to return to their cells or bunks for a head count. Once the count is complete, lunch is then served. At about 1:00 P.M., the facility is once again open for visitation. If available, inmates can take part in jail programs in the afternoon.

While each jail is different, most prisoners are allowed a limited number of personal visits per week, although unlimited visits are usually allowed with their attorneys. Generally, the only exception is during times when counts are taken or meals are served. During shift change in mid-afternoon, the jail is once again shut down and another count is made. Once this is accomplished, the facility is reopened. At about 5:00 P.M., the jail is shut down for another count and for dinner, then reopens again at 6:00 P.M., The final shutdown may occur as late as 10:00 P.M.

During the week, prisoners are also given some form of recreation time. The American Correctional Association recommends at least four hours of recreation time per week for jail inmates, although the actual time fluctuates from institution to institution. Even with programs and recreation time, inmates have an abundance of idle time on their hands. The challenge for jail administrators and personnel, then, becomes what to do to reduce this prisoner idleness.

One possible solution was addressed in a report developed through the Jail Industries Initiative (Miller, Sexton, and Jacobsen 1991). The purpose of the report was to look at ways to make jails more productive. Four objectives were mentioned:

1. *The development of inmate work habits and skills*
2. *The generation of revenues or reduction of costs for the county*
3. *The reduction of inmate idleness*
4. *The satisfaction of community needs (Miller, Sexton, and Jacobsen 1991, 2)*

While such programs might seem prohibitive to many smaller facilities, the report mentions a continuum of tasks that might fit within any jail. At one end of the continuum, the prisoner might simply cut the grass in front of the jail, earning little more than the privilege to watch television for an extra hour. At the other end of the continuum, the jail inmate might work for private-sector industry to earn real dollars that could be used to offset the cost of confinement (Miller, Sexton, and Jacobsen 1991, 2). In both cases, "the elements of labor, service provision, value, and compensation are all present" (Miller, Sexton, and Jacobsen 1991, 2). The ultimate goal for any such program is to reduce idleness among jail prisoners. Beyond this, such programs can both improve work habits as well as defray the cost of housing inmates.

Conclusions

The importance of jails within the criminal-justice system is growing. Jails act as the entry point of the system. All those convicted of a crime and sentenced to time in a prison pass through the local jail's doors. With prison overcrowding, jails have been saddled with the additional burden of housing convicted felons for longer periods of time until prison space opens up. Increasing conservatism has added jail time to many offenses that just a few years ago would have been diverted with a fine or community-treatment program. The future of jails is largely dependent on how we address the concerns mentioned above.

In looking at the jail crisis, Thompson and Mays (1991a, 244) place much of the blame on the states. They make four specific recommendations. The first recommendation deals with having states provide the necessary resources for construction of new jails or renovation of existing facilities (Thompson and Mays 1991a, 244). States, not local governments, have the ability to finance large capital outlays. Many communities that need to build jails are the very ones that do not have the resources to do so. The states can take the lead, using their superior resources to finance the construction or renovation of jails in small communities.

The second and third recommendations address the problem of mandatory jail standards and their enforcement. According to Thompson and Mays, states should develop mandatory jail standards and the mechanisms to enforce them (1991a, 244). This should be done at the state level, removed from local politics. The fourth recommendation is for states to adopt legislation that enables local jurisdictions to engage in cooperative agreements to build regional jails (Thompson and Mays 1991a, 245). As mentioned above, in many jurisdictions, such arrangements are cost-effective because duplication of services is not a problem.

Local governments must also take the lead in a number of areas. Thompson and Mays (1991a, 245–246) list four recommendations that local officials might employ to handle the jail crisis. The first deals with a community-awareness program, with the goal of informing the community of the jail's functions and of the conditions that exist within the jail. In those communities where the jail is run by the sheriff's department, increased emphasis should be placed on that role during campaigns and elections. If the sheriff has a dual role, then that individual should provide the expertise in both roles, not just in law enforcement.

Once greater public awareness has been established, the local community must develop long-term financial plans for jail construction or renovation, staffing, and operation (Thompson and Mays 1991a, 245). The management-by-crisis philosophy of the past must be replaced by a careful examination of long-term needs and fiscal responsibility. Local officials must realize that there is a cost benefit in building a new facility. The alternative is to continue housing inmates in facilities that may violate constitutional rights, and therefore face losing rather substantial lawsuits.

The third recommendation at the local level deals with requiring local officials to have written policies and procedures (Thompson and Mays 1991a, 245). Having written policies and procedures and following them is at the heart of mandatory state jail standards. Local governments must have explicit guidelines for the jail to follow, and those guidelines should follow the standards adopted by the state. If such policies and procedures are implemented and followed, they can be used to head off lawsuits.

The final recommendation offers probably the greatest hope for the condition of jails today. Communities should look at alternatives to incarceration (Thompson and Mays 1991a, 246). To paraphrase a National Institute of Justice report on alternative sentencing, we cannot build our way out of the current jail crisis; we must instead develop sensible sentencing policies that offer a wide range of sanctions, while at the same time implementing an aggressive public-education program (Castle 1991, 5).

As early as 1974, Hans Mattick (1974, 825) was recommending alternatives to jail incarceration, including suspended sentence, summary probation, and probation without verdict. Other sentencing alternatives might include the use of fines (citation and release), community service, electronic monitoring, day-reporting centers, specialized-treatment facilities, and the use of work-release and weekender sentences (Mattick 1974, 827–828). Local communities might also use early-release or furlough programs that allow the jail to discharge those inmates who show the most promise and who are unlikely to commit further crimes. As noted previously, jails suffer from the fact that they do not have control over their populations: outside forces control the flow of inmates into and out of the facility. By implementing furloughs or good-time options, jail officials would have a release mechanism similar to that which has long been available for prisons (Mattick 1974, 828–830).

The future of American jails clearly lies in our ability to adapt to the changing needs of the criminal-justice system. In the past two decades, we have attempted to build our way out of the current predicament, and as new jails are built, more and more prisoners are held. The inmate-litigation explosion has forced us to take a careful look at what has been a persistent problem. The conditions in today's jails are not new, nor are the solutions something we can implement in the short term. If we are to truly control the jail crisis, we must make a concerted effort to allocate the necessary funds to renovate or replace old and deteriorating facilities. Funds must also be made available for the hiring, retention, and training of correctional personnel. And finally, we must make a concerted effort to clearly define the role and function of jails in the United States.

KEY TERMS

Bridewells—another name for House of correction, named for the location of the first established facility at St. Brigit's Well

Fee system—early jailers charged inmates for their room and board

Functional living units—design where all sleeping, food, hygiene and recreational facilities are in a self-contained unit

Gaol—old English term for places of confinement until the offender was tried

Habeas corpus lawsuits—writ used to challenge legality of confinement

Houses of correction—in 15th and 16th century these facilities held minor debtors, vagrants and beggars

Houses of refuge—private institutions in the mid 19th century for juvenile offenders

Linear design—design where inmate cells are situated along corridors

Lockups—places of short term confinement; found in police stations and courtrooms

Mandamus actions—legal action used to compel government officials to perform their duties

Pretrial detainees—those held in jail while awaiting trail

Rabble—Irwin's term for those individuals who are perceived by society as irksome, offensive and threatening; the population found in jails

Regionalization—two or more governments join forces to build one large regional jail

Remote design—jail design where cells are situated around a central dayroom

ROR—Release on Recognizance

Section 1983 actions—hold governmental officials liable for actions taken under color of law that violate constitutional rights

REVIEW QUESTIONS

1. What are some key differences between jails and prisons (function, population, management)?
2. What are some of the problems associated with housing juveniles in adult jails?
3. What are some problems associated with the fee system?
4. Discuss some of the problems associated with housing both convicted offenders and pretrial detainees in a local jail.
5. What are some important architectural changes that have taken place in the design of modern jails?

6. Discuss some issues related to the operation and administration of jails in the United States.
7. Discuss the role of jail standards in bringing about needed change in American jails.
8. Discuss what has been done to limit overcrowding in American jails.
9. What are some special needs of inmates found in many jails?
10. In what areas do we find our best hope for the future of jails?

FURTHER READING

Chesney-Lind, M. 1997. *The Female Offender: Girls, Women, and Crime.* Thousand Oaks, CA: Sage.

Irwin, J. 1985. *The Jail.* Berkeley: University of California Press.

Kalinich, D., and J. Klofas (Eds.). 1986. *Sneaking Inmates Down the Alley: Problems and Prospects in Jail Management.* Springfield, IL: Charles C Thomas.

Schwartz, I. 1988. *Justice for Juveniles: Rethinking the Best Interest of Children.* Lexington, MA: Lexington Books.

Thompson, J., and G. Mays. 1991. *American Jails: Public Policy Issues.* Chicago, IL: Nelson-Hall.

REFERENCES

Allen-Hagen, B. 1991. *Public Juvenile Facilities: Children in Custody 1989.* Washington, DC: U.S. Government Printing Office.

American Correctional Association. 1981. *Standards for Adult Local Detention Facilities.* Rockville, MD: American Correctional Association.

American Correctional Association. 1985. *Jails in America: An Overview of Issues.* College Park, MD: American Correctional Association.

Bureau of Justice Assistance. 2000. *Juveniles in Adult Prisons and Jails: A National Assessment.* Washington, DC: U.S. Department of Justice.

Bureau of Justice Statistics. 1990. *Census of Local Jails, 1988.* Washington, DC: U.S. Department of Justice.

Bureau of Justice Statistics. 1994. *Pretrial Release of Felony Defendants, 1992.* Washington, DC: U.S. Department of Justice.

Bureau of Justice Statistics. 1995. *Jails and Jail Inmates, 1993–94.* Washington, DC: U.S. Department of Justice.

Bureau of Justice Statistics. 1999. *Women Offenders.* Washington, DC: U.S. Department of Justice.

Bureau of Justice Statistics. 2000. *Correctional Populations in the United States, 1997.* Washington, DC: U.S. Department of Justice.

Bureau of Justice Statistics. 2001. *Census of Jails, 1999.* Washington, DC: U.S. Department of Justice.

Bureau of Justice Statistics. 2002. *Profile of Jail Inmates, 2002.* Washington, DC: U.S. Department of Justice.

Bureau of Justice Statistics. 2003a. *Felony Defendants in Large Urban Counties, 2000*. Washington, DC: U.S. Department of Justice.

Bureau of Justice Statistics. 2003b. *Felony Sentences in State Courts, 2000*. Washington, DC: U.S. Department of Justice.

Bureau of Justice Statistics. 2003c. *Prisoners in 2002*. Washington, DC: U.S. Department of Justice.

Bureau of Justice Statistics. 2004. *Prisoners and Jail Inmates at Midyear 2003*. Washington, DC: U.S. Department of Justice.

Cahalan, M. 1986 . *Historical Corrections Statistics in the United States, 1850–1984*. Washington, DC: U.S. Department of Justice.

Castle, M. 1991. *Alternative Sentencing: Selling It to the Public*. Washington, DC: U.S. Department of Justice.

Champion, D. 1990. *Corrections in the United States: A Contemporary Perspective*. Englewood Cliffs, NJ: Prentice Hall.

Champion, D. 1991. "Jail Inmate Litigation in the 1990s." In J. Thompson and G. L. Mays (Eds.), *American Jails: Public Policy Issues*, pp. 197–215. Chicago: Nelson-Hall.

Chesney-Lind, M. 1988. "Girls in Jail." *Crime and Delinquency* 34: 150–168.

Chesney-Lind, M. 1997. *The Female Offender: Girls, Women, and Crime*. Thousand Oaks, CA: Sage.

Clear, T., and G. Cole. 1994. *American Corrections*, 3d ed. Belmont, CA: Wadsworth.

Cox, N., Jr., and W. Osterhoff. 1991. "Managing the Crisis in Local Corrections: A Publics-Private Partnership Approach." In Joel A. Thompson and G. Larry Mays (Eds.), *American Jails: Public Policy Issues*, pp. 227–239. Chicago: Nelson-Hall.

Flaherty, M. 1980. *An Assessment of the National Incidence of Juvenile Suicide in Adult Jails, Lockups, and Juvenile Detention Centers*. Washington, DC: U.S. Department of Justice.

Flynn, E. E. 1973. "Jails and Criminal Justice." In L. Ohlin (Ed.), *Prisoners in America*, pp. 49–85. Englewood Cliffs, NJ: Prentice-Hall.

Garafalo, J., and R. Clark 1985. "The Inmate Subculture in Jails." *Criminal Justice and Behavior* 12, 4: 415–434.

Gibbs, J. 1986. "When Donkeys Fly: A Zen Perspective on Dealing with the Problem of the Mentally Disturbed Jail Inmate." In D. Kalinich and J. Klofas (Eds.), *Sneaking Inmates Down the Alley: Problems and Prospects in Jail Management*, pp. 149–166. Springfield, IL: Charles C Thomas.

Gibbs, J. 1987. "Symptoms of Psychopathology Among Jail Prisoners: The Effects of Exposure to the Jail Environment." *Criminal Justice and Behavior* 14: 299–310.

Goffman, I. 1959. *Presentation of Self in Everyday Life*. Garden City, NJ: Anchor Books.

Goldfarb, R. 1975. *Jails*. New York: Anchor Press.

Gray, T., G. Mays, and M. Stohr. 1995. "Inmate Needs and Programming in Exclusively Women's Jails." *Prison Journal* 75, 2: 186–202.

Guy, E., J. Platt, I. Zwelling, and S. Bullock. 1985. "Mental Health Status of Prisoners in an Urban Jail." *Criminal Justice and Behavior* 12: 29–33.

Guynes, R. 1988. *Nation's Jail Managers Assess Their Problems*. Rockville, MD: National Institute of Justice.

Hayes, L. 1980. *And Darkness Closes In: A National Study of Jail Suicides*. Alexandria, VA: National Center on Institutions and Alternatives.

Hayes, L. 1983. "And Darkness Closes In . . . A National Study of Jail Suicides." *Criminal Justice and Behavior* 10: 461–484.

Howard, J. 1777. *The State of the Prisons in England and Wales*. London.

Irwin, J. 1985. *The Jail*. Berkeley: University of California Press.

Kalinich, D., P. Embert, and J. Senese. 1991. "Mental Health Services for Jail Inmates: Imprecise Standards, Traditional Philosophies, and the Need for Change." In J. Thompson and G. Mays (Eds.), *American Jails: Public Policy Issues*, pp. 79–99. Chicago: Nelson-Hall.

Kennedy, D., and R. Homant. 1988. "Predicting Custodial Suicides: Problems with the Use of Profiles." *Justice Quarterly* 5, 3: 441–456.

Klofas, J. 1990. "The Jail and the Community." *Justice Quarterly* 7, 1: 69–102.

Leibowitz, M. 1991. "Regionalization in Virginia Jails." *American Jails* 5, 5: 42–43.

Lightfoot, C., L. Zupan, and M. Stohr. 1991. "Jails and the Community: Modeling the Future in Local Detention Facilities." *American Jails* 4, 4: 50–52.

Mattick, H. 1974. "The Contemporary Jails of the United States: An Unknown and Neglected Area of Justice." In D. Glaser (Ed.), *Handbook of Criminology*, pp. 777–848. Chicago: Rand McNally.

Mays, G., and J. Thompson. 1988. "Mayberry Revisited: The Characteristics and Operations of America's Small Jails." *Justice Quarterly* 5, 3: 421–440.

Mays, G., and J. Thompson. 1991. "The Political and Organizational Context of American Jails." In J. Thompson and G. Mays (Eds.), *American Jails: Public Policy Issues*, pp. 3–21. Chicago: Nelson-Hall.

McCoy, C. 1982. "New Federalism, Old Remedies, and Corrections Policy-Making." *Policy Studies Review* 2 (November): 271-278.

McGee, R. 1975. "Our Sick Jails." In P. Cromwell (Ed.), *Jails and Justice*. Springfield, IL: Charles C Thomas.

Miller, R., G. Sexton, and V. Jacobsen. 1991. *Making Jails Productive*. Washington, DC: U.S. Government Printing Office.

Moynahan, J., and E. Stewart. 1980. *The American Jail: Its Development and Growth*. Chicago: Nelson-Hall.

National Advisory Commission on Criminal Justice Standards and Goals. 1973. *Report on Corrections*. Washington, DC: U.S. Department of Justice.

National Institute of Justice. 1984. *Jailing Drunk Drivers: Impact on the Criminal Justice System*. Washington, DC: U.S. Department of Justice.

Nelson, W. 1988. *Cost Savings in New Generation Jails: The Direct Supervision Approach*. Washington, DC: U.S. Government Printing Office.

New York City. 2004. Retrieved January 16, 2005, from www.nyc.gov/html/doc/html/overview.html.

Platt, A. 1969. *The Child Savers: The Invention of Delinquency*. Chicago: University of Chicago Press.

Poole, E., and M. Pogrebin. 1991. "Changing Jail Organization and Management: Toward Improved Employee Utilization." In J. Thompson and G. Mays (Eds.), *American Jails: Public Policy Issues*, pp. 163–179. Chicago: Nelson-Hall.

Schwartz, I. 1988. *Justice for Juveniles: Rethinking the Best Interest of Children*. Lexington, MA: Lexington Books.

Schwartz, I. 1991. "Removing Juveniles from Adult Jails: The Unfinished Agenda." In J. Thompson and G. Mays (Eds.), *American Jails: Public Policy Issues*, pp. 216–226. Chicago: Nelson-Hall.

Schwartz, I., L. Harris, and L. Levi. 1988. "The Jailing of Juveniles in Minnesota: A Case Study." *Crime and Delinquency* 34, 2: 133–149.

Sechrest, D., and W. Collins. 1989. *Jail Management and Liability Issues*. Miami: Coral Gables.

Senese, J., D. Kalinich, and P. Embert. 1989. "Jails in the United States: The Phenomenon of Mental Illness in Local Correctional Facilities." *American Journal of Criminal Justice* 14, 1: 104–121.

Soler, M. 1988. "Litigation on Behalf of Children in Adult Jails." *Crime and Delinquency* 34, 2: 190–208.

Stohr, M., N. Lovrich, Jr., B. Menke, and L. Zupan. 1994. "Staff Management in Correctional Institutions: Comparing DiIulio's 'Control Model' and 'Employee Investment Model' Outcomes in Five Jails." *Justice Quarterly* 11, 3: 471–497.

Stojkovic, S. 1986. "Jails Versus Prisons: Comparisons, Problems, and Prescriptions on Inmate Subcultures." In D. Kalinich and J. Klofas (Eds.), *Sneaking Inmates Down the Alley: Problems and Prospects in Jail Management*, pp. 23–37. Springfield, IL: Charles C Thomas.

Takagi, P. 1975. "The Walnut Street Jail: A Penal Reform to Centralize the Power of the State." *Federal Probation* (December): 18–26.

Thompson, J., and G. Mays. 1988. "State-Local Relations and the American Jail Crisis: An Assessment of State Jail Mandates." *Policy Studies Review* 7, 3: 567–580.

Thompson, J., and G. Mays. 1991a. "Paying the Piper But Changing the Tune: Policy Changes and Initiatives for the American Jail." In J. Thompson and G. Mays (Eds.), *American Jails: Public Policy Issues*, pp. 240–246. Chicago: Nelson-Hall.

Thompson, J., and G. Mays. 1991b. "The Policy Environment of the American Jail." In J. Thompson and G. Mays (Eds.), *American Jails: Public Policy Issues*, pp. 1–2. Chicago: Nelson-Hall.

U.S. Department of Justice. 1980. *Indexed Legislative History of the Juvenile Justice Amendments of 1977*. Washington, DC: U.S. Government Printing Office.

U.S. Department of Justice. 2003. *Arrestee Drug Abuse Monitoring (ADAM)*. Washington, DC: U.S. Government Printing Office.

U.S. Department of Labor. 2004. *Occupational Outlook Handbook*. Washington, DC: U.S. Government Printing Office.

Welch, M. 1991. "The Expansion of Jail Capacity: Makeshift Jails and Public Policy." In J. Thompson and G. Mays (Eds.), *American Jails: Public Policy Issues*, pp. 148–162. Chicago, IL: Nelson-Hall.

Winfree, L. T. 1988. "Rethinking American Jail Death Rates: A Comparison of National Mortality and Jail Mortality, 1978, 1983." *Policy Studies Review* 7: 641–659.

Winfree, L., Jr., and J. Wooldredge. 1991. "Exploring Suicide and Death by Natural Causes in America's Large Jails: A Panel Study of Institutional Change, 1978 and 1983." In J. Thompson and G. Mays (Eds.), *American Jails: Public Policy Issues*, pp. 63–78. Chicago: Nelson-Hall.

Wooden, W., and A. Ballan. 1996. "Jail/Prison Inmate Socialization: One Man's Journey." Paper presented at the annual meeting of the Academy of Criminal Justice Sciences, Las Vegas, Nevada.

Zupan, L. 1991. *Jails: Reform and the New Generation Philosophy.* Cincinnati, OH: Anderson.

Zupan, L., and B. Menke. 1991. "The New Generation Jail: An Overview." In J. Thompson and G. Mays (Eds.), *American Jails: Public Policy Issues*, pp. 180–194. Chicago: Nelson-Hall.

CASES CITED

Bell v. Wolfish, 441 U.S. 520 (1979)

Wyatt v. Stickney, 325 F. Supp. 781 [M.D. Ala. 1971]

10

Looking Toward the Future

Joycelyn M. Pollock
Texas State University–San Marcos

Chapter Objectives

- Be familiar with the impact that imprisonment has had on other state services.
- Be aware of the issues regarding private prisons.
- Understand how prisons might brutalize those who work within them.
- Be aware of the problem of cross-sex supervision.
- Be aware of the emerging initiatives that support reentry and alternatives to prison.

In the first edition of this text, the final chapter opened with a quote from the 1970s wherein Daniel Glaser (1970, 261–266) predicted the prison of the future:

> *[T]he prison of the future will be small. There will be many institutions, of diverse custodial levels, within any metropolitan area. . . . The prisons of the future, because they will be small, can be highly diverse, both in architecture and in program. . . . In the prison of tomorrow there will be much concern with utilizing the personal relationships between staff and inmates for rehabilitative purposes. . . . The prison of the future, clearly, will be part of a society in which rationality is institutionalized, and goodness, truth and beauty are cardinal goals (Glaser 1970, 261–266).*

The quote was written during the height of the **rehabilitative era**, when there was hope and optimism that prisons could be healthy places of individual change. Who would have predicted that shortly after Glaser wrote this description of the future, the Attica riot would occur, Martinson's report would come out, and the tide would turn away from rehabilitation, toward a more punitive approach? In

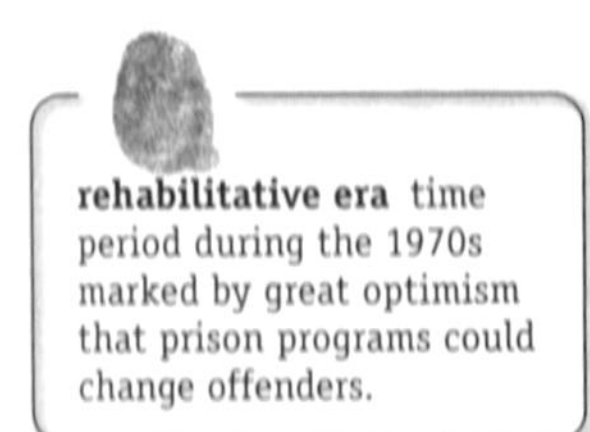

rehabilitative era time period during the 1970s marked by great optimism that prison programs could change offenders.

the next decade (the 1980s), the prison incarceration rate began its dramatic climb upward, and prisons grew larger and ever larger to accommodate the numbers. New facilities did not open in cities; rather, small towns were cajoled and coerced to accept them, and then these towns began in earnest to compete for the prisons in a desperate attempt to jolt depressed rural economies. Today, we have massive prisons, we have more prisons (at last count, more than 1,500), and we have prisons where "goodness, truth, and beauty" are in short supply. In this last chapter, we will explore some important issues that will shape the prisons of the future.

Overcrowded Prisons, Drug Laws, and Race

Today in the United States, there are nearly 7 million adults in prison or under some form of correctional supervision. This number is larger than most state populations. While the nation's population has grown 29 percent between 1980 and 2002, the correctional population has increased 268 percent. The percentage of adults under some form of correctional supervision has tripled (Austin and Fabelo 2004, 5). There are about 2 million people incarcerated in this country, even though we have seen a decade-long pattern of declining crime rates. In fact, we are at 30-year lows. The National Crime Victimization Survey indicates that the rate of violent crime was 22.6 per 1,000 in 2003, down from 49.9 in 1993; for property crimes, the rate (163.2 per 1,000) has declined by about half from 1993 (Wald 2004). The homicide rate is at a 40-year low. It is now about 5.5 per 100,000, compared to more than 10 per 100,000 in the 1980s and early 1990s (Bureau of Justice Statistics 2004). Explanations for the crime decline identify tough drug laws, increased enforcement, the reduction of turf wars between drug-dealing gangs, a reduction in the number of people in the crime-prone age groups, and the fact that so many are incarcerated.

However, there does not seem to be much relationship between crime rates and imprisonment. Those states that had the most dramatic drop in crime were not the ones with the highest rate of increase in imprisonment. In fact, some of the states with lower incarceration rates experienced the most dramatic crime drops (Austin and Fabelo 2004, 11). Part of the increase in the prison population has been because of sentencing. Offenders are now more likely to go to prison for all offenses, and they are more likely to serve a longer sentence. (Note that they may not be receiving longer sentences, but they are serving a greater portion of their sentence.) Another impact on prison populations is the use of probation and parole. The percentage of those who receive parole from those who are eligible goes up and down; when it goes down, more people stack up in prison. Also, the percentage of parole or probation **revocations** that result in a prison commitment (Mauer 2002) affect the level of prison populations.

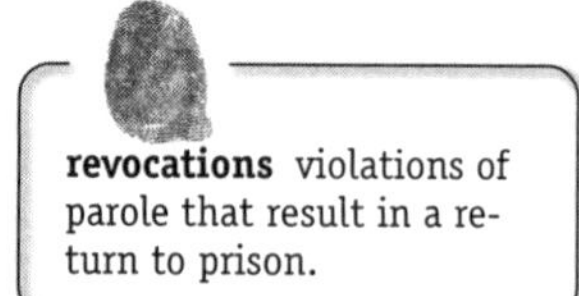

revocations violations of parole that result in a return to prison.

deinstitutionalization a time period during the 1970s when mental hospitals were closed in favor of community mental health treatment centers.

Another factor may be the **deinstitutionalization** of the mentally ill. In the 1970s, a series of court cases created procedural protections for those individuals who were subject to involuntary commitment. Partially as a result of these court cases, large mental hospitals were closed all across the nation. The mentally ill were supposed to be cared for by a network of community mental-health centers. Instead, what has happened is that large numbers of the mentally ill get no help at all—and end up in jails and prisons. As the number of mentally ill in hospitals shrank, their numbers in prison grew through the 1980s and 1990s. There are now three times more mentally ill people in prisons than there are in mental-health hospitals (Human Rights Watch 2003, 1). How many jail and prison inmates today have mental-health issues? Estimates range from 10 percent to more than 20 percent. Two of the largest mental-health providers in the country are Cook County Jail (Chicago) and Los Angeles County Jail (Human Rights Watch 2003, 16). Individuals who have mental-health problems may be unable to conform their behavior to societal expectations, and thus find themselves enmeshed in the criminal-justice system over serious or minor crimes. They may attempt to **self-medicate**, using street drugs to quiet the demons or help them suppress anxiety and depression, and in this way become involved in the justice system as a drug abuser.

self-medicate the practice of individuals with mental health issues using street drugs in lieu of mental health treatment and legally prescribed drugs.

A large factor in the burgeoning prison population has been because of this nation's **War on Drugs**. The largest contributors to the massive numbers sent to prisons in the 1980s and 1990s were state and federal drug laws. As was cited in Chapter 3, the largest single category of prison commitments is for drug crimes. They account for 21 percent of all convictions, and 31 percent of all sentences to state prisons (Durose and Langan 2004, 3). Ironically, our imprisonment pattern for drug users does not seem to be consistent at all with drug use patterns. According to the National Household Survey on Drug Abuse, about 17 percent of individuals 12 and over used any type of illegal drug in 1979. In 1985, the percentage of users was already going down, and continued to decline through the 1990s, to 12.5 percent in 2001. As far as cocaine is concerned, 1979 had the highest number of reported users, declining after that to 2 percent or less of any age group (Office of National Drug Control Policy 2002). The rate of drug users being arrested and sent to prison during this same time period has been steadily increasing. In other words, our "war" against drug users bears little relationship to the actual use patterns, which reflect little change in drug use over 30 years.

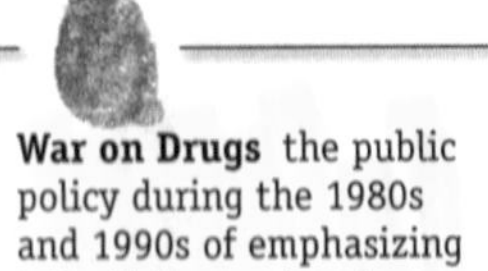

War on Drugs the public policy during the 1980s and 1990s of emphasizing interdiction and enforcement over considering drug abuse a public health problem.

As was discussed in Chapter 3, one effect of the War on Drugs and the incarceration trends fueled by it has been the increasing disproportional percentage of African-Americans in prison. African-Americans make up about 12 percent of our population; however, they represented about 44 percent of all inmates sentenced to one year or more in prison in 2003. Part of this is due to the fact that blacks are more likely to be involved in violent street crimes, which receive long prison sentences, but it is also due to differential sentencing in other crime categories.

Statistics indicate that while 13 percent of drug users are minority members, 60 percent of those convicted of drug crimes are African-American, and 75 percent of those sent to prison for drug convictions are African-American (Schemo 2001; Parenti 1999). What happens to white users? One thing is clear: prison is much more often the solution when drug users and offenders are minorities.

A large percentage of minorities in prison are there, not for violent crimes, but for drug and property offenses. Forty percent of all arrests are for drug and alcohol crimes (Austin and Fabelo 2004, 7). Only 27 percent of all arrests are of blacks (even though they constitute roughly half of the prison population) (Austin and Fabelo 2004, 7). Ironically, the profile of the most likely victim is similar to the profile of those individuals likely to end up in prison—both are likely to be minority and poor. One of every three black men will be sentenced to prison during their lifetime; one in six Hispanic men will be. This compares to one in nine white men (Austin and Fabelo 2004, 6). Whether intentional or not, the sentencing practices in the United States are differentially affecting minorities.

Opportunity Costs

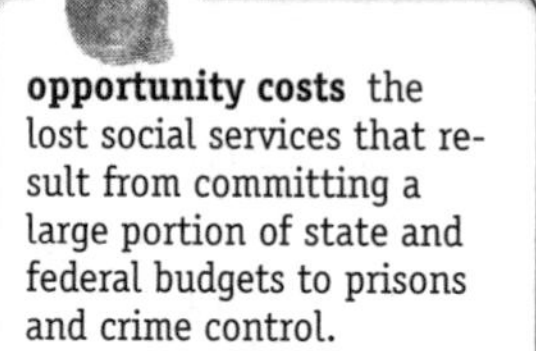

opportunity costs the lost social services that result from committing a large portion of state and federal budgets to prisons and crime control.

One of the arguments that prison critics make is that for every dollar spent on prison construction and maintenance, there is that much less for other social services—in other words, there are **opportunity costs** associated with prison use. States have limited resources, and what they spend on prisons is taking away from what could be spent on education (among other services). From 1977 to 1999, total state and local expenditures for corrections increased by 946 percent, and judicial- and legal-system increases were an astronomical 1,518 percent—compared to the increase in education funding of only 370 percent. In fact, education received less than the average increase for all state-function spending (401 percent) (Gifford 2002). Between 1980 and 2000, corrections spending (both state and local) increased by 104 percent, while education's share of spending dropped by 21 percent (Schiraldi and Ziedenberg 2002).

There is another link between education and prisons. We see that the increased risk of incarceration is differentially concentrated in those who are high school dropouts. It has been reported that 52 percent of African-American male high school dropouts had prison records by their early 30s (Western, Schiraldi, and Ziedenberg 2003), compared to much lower numbers for those who gradated from high school or have some amount of college. It is indeed ironic that as states spend less and less on education, they may be reinforcing the need for greater and greater spending on prisons.

Some argue that, as expensive as prison is, the costs of crime are higher, justifying such expense. This is debatable. The total economic loss to victims has been estimated as $15.6 billion (in 2002), which is substantially less than the total costs of the justice system and the corrections system, which are about 10 times that number (Austin and Fabelo 2004, 9). Obviously, that does not include pain-and-suffering costs, and some estimates place those costs as much higher than the justice/corrections budget. However, it should be noted that the majority of offenders are property or public-order offenders, not violent offenders. Thus, when one considers only those individuals, it does seem financially irresponsible to continue to pay so much more to incarcerate those offenders than the amount that they "cost" society.

Private Prisons, Private Profits

One of the ways that states have met their need for prison beds has been to turn to the private-prison industry. Private companies now hold about 5 percent of all prison beds in this country (Parenti 1999, 218). The companies that build and run prisons under government contracts have experienced a tremendous growth, and the biggest such firms now trade on the New York Stock Exchange. Prisons have become big business. The growth of this industry has outstripped legal and ethical analysis of the wisdom of making a business out of depriving others of their freedom.

A "build-and-manage" contract allows the state to sign up for new-prison construction without going to the voters to approve a bond initiative. The private company uses its money or financial leverage to build the facility, with the agreement that the state will promise the company a certain number of inmates per day (with the attendant **per diem** cost set by contract and paid to the company). Often there is a clause that at the end of a period, the state may (or must) buy the facility from the private contractor at a predetermined amount.

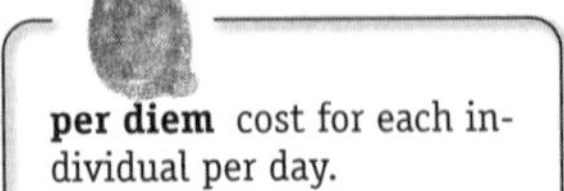

per diem cost for each individual per day.

The private company must find a location for this private prison, and small towns have been the targets of their search. Towns are offered a host of financial enticements, and they often offer incentives of their own, such as tax rebates or even free land. The town's policy makers are encouraged to think that the prison will be a financial boon to the area. Actually, evidence indicates that these prisons do not provide the financial windfall that towns hope for. Several studies found no evidence of economic growth after prisons came to small towns (Lawrence and Travis 2004, 7). On the other hand, there may be funding benefits. Inmates are counted in the small town's census, and therefore a large prison influences all federal monies that are allocated on the basis of population. Counties reap the benefits of large numbers of prisoners that inflate their population rolls without requiring any services. Formula-based monies include such things as Medicaid, foster care, adoption assistance, and social-services block grants (Lawrence and Travis 2004, 7).

The problem with private companies running prisons is that their profit is determined by the difference between the per diem amount they receive from the state for each prisoner and the per diem amount they spend. The pressure to cut corners in order to post larger profits for their shareholders is obvious. Further, there is no incentive to reduce prison populations—in fact, there have been some instances where states have been locked into contracts that promise a certain number of inmates to a private vendor when, at the same time, the state has had empty beds in state facilities. Critics of private prisons do not have to look very hard to find a long list of scandals at private facilities—escapes, assaults, killings, and other problems in these facilities occur with depressing regularity. In fact, in some instances, state officials have determined that the problems are severe enough to remove all state prisons from private facilities. States have canceled contracts and/or not renewed contracts over scandals with private vendors (for review, see: Pollock 2004).

The evaluations of private prisons are highly politicized. Conservative think tanks or think tanks that are funded by private companies find, not surprisingly, that states realize a cost savings when they contract with private-prison vendors. However, other evaluations, and even the Government Accounting Office, have found either very modest cost savings or none at all. Further, these evaluations point to other issues—such as safety, guard training, and the absence of programming—to show that private prisons end up giving the taxpayers less for their money than a state facility would (Pollock 2004).

One of the problems with private prisons is that they sometimes contract to hold prisoners in facilities that are in different states from the one where the inmate was sentenced. Thus, Missouri prisoners may be held in a private prison in Texas, or Hawaii prisoners may find themselves in Colorado. This is troublesome for two reasons. First, officials in the state where the facility is located may not be aware of the type of inmate being held. A state may find that a private facility is housing dangerous offenders (for sex offenses or violent crimes) only after an escape has taken place and the private company has asked for help in recapturing the offender. Second, the prisoners have very little recourse to complain about conditions, since they are so far away from friends, family, or legal assistance. They also have trouble maintaining ties with family and enforcing their right of access to courts for civil matters or constitutional challenges (Pollock 2004).

Private vendors' profits increase when they hold more inmates for longer periods of time. If somehow we could change these incentives, privatization might make sense. If a state could create a contract giving a private vendor a financial incentive for each inmate that does *not* return to prison via a revocation hearing, we might see American ingenuity make a dent in the recidivism rate and come up with effective institutions. Until then, the fact that cannot be ignored is that private-prison companies are financially invested in the growing spiral of incarceration. If one considers our ever-increasing incarceration rate a problem, then one might say that privatization is not the solution, but rather, part of the problem.

The Supermax Prison

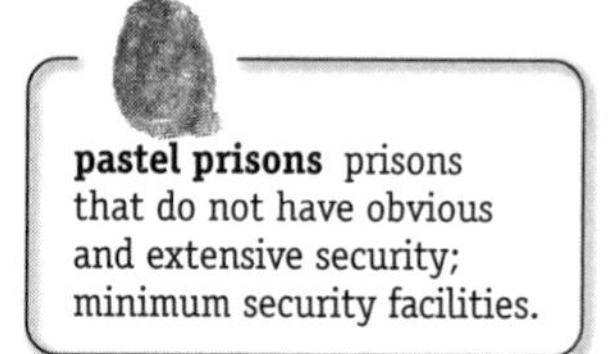

pastel prisons prisons that do not have obvious and extensive security; minimum security facilities.

In all state systems, there are minimum-, medium-, and maximum-security institutions. Minimum facilities are usually work camps or "**pastel prisons**." Security is minimal because inmates are either low risk and/or nearing release. Medium facilities typically have a perimeter fence and restricted movement, but prisoners have quite a bit of freedom of movement within the facility. Maximum-security institutions are what people envision when they think of a prison—the wall is usually imposing and impenetrable, with towers manned by armed guards. Inmates' movement in the facility is controlled—that is, they may move from place to place only at designated times. When inmates commit disciplinary infractions in the lower-security prisons, the punishment is often a transfer to the maximum-security institution in the state. When maximum-security prisoners commit offenses, they are either put into punitive segregation (**the hole**) or, for chronic offenders, transfer to a "supermax" is the solution.

the hole punitive segregation.

The **supermax**, in those state systems that have built such prisons, is for those "worst of the worst" in-prison offenders. A supermax prison limits even the minimal movement that exists inside the facility. Inmates are locked in their cells 23 to 24 hours a day. It is an eerie return to the Pennsylvania system, albeit with a high-tech twist. Video cameras monitor cells and halls 24 hours a day. There are no radios, televisions, or other materials allowed in the cells. Inmates are allowed out of their cells only infrequently because most activities can be conducted in their cells. For example, the newest supermax cells offer mist-spraying nozzles

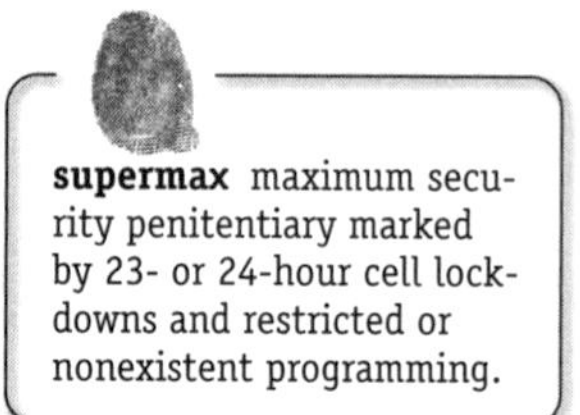

supermax maximum security penitentiary marked by 23- or 24-hour cell lockdowns and restricted or nonexistent programming.

that allow prisoners to shower without leaving. Guards operate supermax prisons from control booths, which are electronic surveillance pods. Literally every aspect of prison operation can be conducted behind bars and 5-inch-thick panes of Plexiglas. By using remote control, correctional officers can open and close doors, speak to inmates, and even control toilets and water flow to cells. If prisoner movement is required, typical policies mandate that no fewer than two guards move the inmate, who will be shackled in every conceivable way (King 1999).

By last count in 1997, there were 57 supermax prisons operated by 36 states and the federal government. Perhaps the best known of these today is a California prison called Pelican Bay, opened in 1989, and its supersecure housing unit called the SHU. Still others—such as the federal supermax in Florence, Colorado, or the supermax in Oak Park Heights, Minnesota—have received much attention. It is unclear whether supermax construction will be slowing in the near future. Several states, including Wisconsin and Virginia, have recently opened supermax facilities, and other states are building new prisons or converting old ones to keep up with demand (Pollock 2004). Still other states, however, have started an opposite trend, mainly because of the costs associated with these special prisons.

Inmates placed in supermax facilities will stay there approximately one to three years before they are released back to the general prison population (King 1999). Some supermax prisoners are released directly to the streets. Critics argue that, after a length of time in such a deprived environment, inmates will be unlikely to return to the general prison environment without incident, and are surely not ready to be released directly to the streets to function in any "normal" way (Pollock 2004). This is especially problematic for young prisoners sent to supermax. Since most young offenders will get out of prison sooner rather than later, this means that the most "sick and impulse-ridden" of supermax releasees may still be in their prime offending years. This spells problems for the rest of society as these prisoners come out of their sentence more frustrated, deprived, and angry than when they went in.

Although the supermax has been promoted as the solution to chronically violent inmates, there is no good evidence to indicate that such facilities have been instrumental in reducing violence (Henningsen, Johnson, and Wells 2003). In fact, one study found that the state that most frequently utilized a supermax had higher rates of violence in their prisons after they began sending prisoners to the supermax (Briggs, Sundt, and Castellano 2003). There are disturbing reports that states send inmates who have mental-health problems to supermax facilities rather than to mental-health facilities. These settings are the worst possible environments for those who have mental-health problems, and the solitary confinement can exacerbate an individual's problems, especially when antipsychotic drugs are also withheld (Human Rights Watch 2003, 3).

Most recent court decisions have reinforced the right of states to run supermax prisons, despite the disturbing reports of the critics. As we learned in Chapter 8, prison officials need no longer provide due-process protections when sending an individual to segregation (*Sandin v. Conner*, 115 S. Ct. 2293 [1995]). Inmates are perceived to have no due-process protections regarding transfer to more-severe prisons (*Meachum v. Fano*, 427 U.S. 215 [1976]), or to prisons that are out of state and far away from family and friends (*Olim v. Wakinekona*, 461 U.S. 238 [1983]). Thus, the legal right of states to send prisoners to supermax prisons is not in question. The conditions that may give rise to prison challenges include 24-hour solitary confinement, double-celling (because crowding is already an issue in some of these prisons), lack of programming, and brutality by officers. In Chapter 8, *Madrid v. Gomez* (889 F. Supp. 1146 [1995]) was discussed. This case dealt with Pelican Bay's medical services, along with other conditions at this supermax facility. In a rare "win" for prisoners, the court issued a sweeping holding that condemned the conditions in which prisoners were held in that facility; however, it remains to be seen whether courts see the conditions of these supermax facilities, absent egregious cases of brutality or lack of services, as unconstitutional.

Abu Ghraib in Our Own Backyard

The year 2004 was a difficult one for the United States. Among the most shocking images of the war in Iraq were grainy pictures of American soldiers torturing prisoners by making them strip naked, posing them in sexually suggestive positions, chaining them to dog collars, and inflicting a range of physical assaults (Horton 2005). While some members of the public were sanguine about the acts, arguing that, after all, "war is hell," others noted that it might not have been a coincidence that some of the soldiers who engaged in these brutal acts were correctional officers in civilian life (Rights Group 2004; Peirce 2004). Lane McCotter, in charge of setting up the Abu Ghraib prison, was forced to resign as head of the Utah prison system after a scandal (Peirce 2004). The connection between prison scandals in the United States and what took place in Abu Ghraib is too obvious to be ignored.

While the military would prefer to portray Corporal Charles Graner, Private Lynndie England, and others as "rotten apples," their actions were not all that different from what has been reported in court cases concerning mistreatment of prisoners in the United States. There are documented cases of beatings, rapes, gladiator fights, and deaths of prisoners at the hands of guards. Why does this occur? The first step toward torture is to perceive the inmate as less than human, and therefore less than deserving of respect and dignity. Once the imprisoned are considered "animals," there is little restraint on one's actions. There is something about the role of "keeper" that exposes the baser instincts of any of us. Those who are put in a position of power over others may abuse that power.

Stanford prison experiment famous college experiment that showed college students could become autocratic and misuse power after only a few days of pretending to be guards (see www.prisonexp.org for more information).

The **Stanford prison experiment**, discussed in Chapter 4, showed how normal college students could turn into autocratic megalomaniacs within a few days while their "prisoners" became depressed and oppositional. This experiment, conducted 34 years ago on the grounds of Stanford University by Phillip Zimbardo, is chilling evidence that good people can behave very badly in a prison environment. The more troubling issue is: Why don't people stop their peers from behaving so badly, especially when their behavior violates the law? One of the ways that individuals can misuse their power is to sexually coerce inmates; unfortunately, these incidents seem to be increasing.

Cross-Sex Supervision

As we saw in Chapter 3, the increase in the number of men in prison in the past 25 years has been dramatic, but the increase in the number of women has been even more dramatic. Even though women still constitute only about 7 percent of the total prison population, their rate of increase has exceeded that of men for every year during the past several decades. What this means is that there are more prisons for women and an ever-increasing guard force to supervise them. In the 1970s, most female prisoners were guarded by women. Today, the majority of correctional officers in prisons for women are men—it is called cross-sex supervision when you have a member of the opposite sex guarding inmates in their living units.

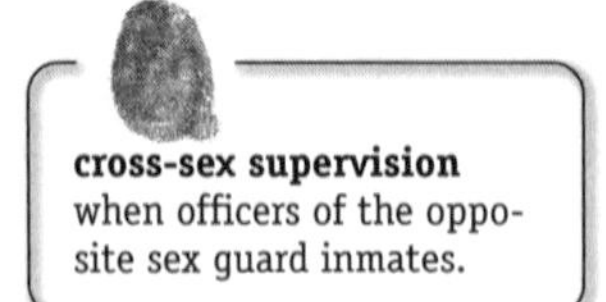

cross-sex supervision when officers of the opposite sex guard inmates.

The reversal of same-sex supervision emerged after equal-protection challenges were made by female officers who were barred from prisons for men. When prisons for men opened to female officers, prisons for women offered available posts for male officers as well (Pollock 2002, 2004). The other reason that there are now more male guards is quite simply that states hire from the pool of applicants to fill guard positions, and the majority of applicants are men. Critics argue that it is a violation of prisoners' right of privacy to have a member of the opposite sex watching them shower or go to the bathroom. In some cases, opposite-sex officers even supervise strip searches. Courts, however, have not been unanimous in their treatment of such challenges. As we saw in Chapter 8, in general, the courts have found that inmates' right of privacy is not sufficient to overcome the state's right to employ guards and place them in any prison. Only a few courts have noted that there might be special issues involved with female prisoners, who are more likely to have been victims of incest or sexual abuse, and therefore suffer more from having male officers conduct pat-downs and strip searches than do male prisoners supervised by female officers (Pollock 2002, 2004).

Unfortunately, with increased opportunities for both male and female correctional officers, there has also been an increase in sexual misconduct. It is true that there are some female officers who become sexually involved with male inmates. The more common problem, however, is that of male officers engaging in sexual misconduct with female inmates. This misconduct has included sexual assault; it is often sexual extortion, and it is always sexual coercion, if one takes the position that the female inmate, by her powerlessness, is not in a position to give consent.

In Georgia, 17 staff members were indicted for having sex with inmates; in Michigan, the state paid litigants $850,000 and changed policies because of a lawsuit over male officers coercing sex from female inmates; in New Jersey, a correctional-officer lieutenant was indicted for sexually assaulting two female inmates. Sex may involve bartering—sexual favors in exchange for cigarettes or drugs or cell assignments—but it can also be rape. Inmates say that they are afraid to report such assaults because officers have powerful friends and no one would believe the inmates. Still, lawsuits have been won, even though inmates make bad witnesses because, as one lawyer observed, they are immature, erratic, prone to exaggeration, and unlikable. However, the lawyer added, these flaws do not mean that the inmates are wrong (New Jersey 2004). Unfortunately, in a New Jersey case, the officer who helped prisoners bring forward incriminating evidence to prove sexual assault was harassed by other officers and punished by his peers and superiors for being a whistle-blower (New Jersey 2004).

The problem of cross-sex supervision is not going to go away. To attempt to go back to a situation where only same-sex guards supervise prisoners is probably not a feasible solution. Equal-employment rights of officers are weightier than privacy rights of inmates, and removing all same-sex guards punishes the majority who do not abuse their position. The solution will be to conduct training and enforce a disciplinary system that punishes mistreatment of prisoners. Whether the culture of the prison, including the tendency to disbelieve inmates and protect errant officers, allows for such a scenario is another question.

A New Era of Prison History?

In 2004, states faced an aggregate $78 billion deficit. Most states faced budget deficits of more than 5 percent of their general funds (Austin and Fabelo 2004, 8). This means that every state function is being scrutinized for budget cuts, including corrections. Crime, criminals, and how they should be punished have been perennially popular issues for political platforms. These issues were largely missing in the 2004 political campaigns, however, since terrorism, the war in Iraq, and the economy eclipsed any lingering fears and concerns about crime. It may be that the budget woes of state policy makers, combined with the decade-long decrease in crime rates, have created the perfect opportunity for a new renaissance of community-correctional alternatives.

There is also evidence that the public has grown weary of being engaged in a War on Drugs with sons and daughters, neighbors and friends as enemies. In a recent poll, 77 percent of respondents agreed with the statement, "Many people in prison today are non-violent drug addicts who need drug treatment, not a prison sentence" (Greene and Schiraldi 2002, 6). Only 31 percent of Americans feel that our current policies are an effective solution to drug problems (Lock, Timberlake, and Rasinki 2002, 384). We see both a recognition that states must do more to provide assistance for reentry into the community from prison, and also a renewed energy toward finding alternatives to prison in the first place.

Reentry

In 2004, lawmakers began to take seriously the fact that more than 650,000 people are released each year back into the community. About 67 percent are returned to prison within three years (Elsner 2005). Congress has allocated money to study the problems of inmates who reenter society without adequate skills or resources to make it on the outside. Legislators have recognized that inmates who are reentering the community need job-placement and mental-health services, substance-abuse treatment and housing assistance (Elsner 2005). It remains to be seen, however, whether these initiatives will be funded anytime soon, considering the national deficit and the cost of the war in Iraq.

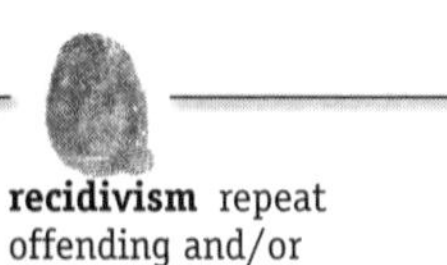

recidivism repeat offending and/or conviction.

technical violations when parolees or probationers violate a rule or condition that is not a new crime.

While **recidivism** numbers are commonly estimated at around 67 percent, released prisoners account for only about 4.7 percent of all arrests for serious crime that occurred in the United States from 1994 to 1997 (Austin and Fabelo 2004, 13). Depending on how recidivism is counted, it might be that prisoners are returned to prison for **technical violations** or for new crimes that are less serious than the initial crime that had resulted in imprisonment. In fact, by some accounts, 30 to 40 percent of prisoners are classified as low security and might be good candidates for early release (Austin and Fabelo 2004, 14).

Some states are realizing that they can save substantial amounts of money by slightly increasing the number of inmates released on parole. Even a modest reduction in the number of months served in prison (for example, bringing parole forward by one or two months) can realize substantial savings for large state systems. Further, if parole-revocation rates were adjusted downward and if returns to prison were reduced (for example, by not sending technical violators back to prison), even greater savings could be achieved, with minor risks to the public. Of course, this would mean political and policy-maker leadership that would have to challenge the timeworn platform of longer terms for more criminals.

Another area of interest is in removing some of the civil disabilities that plague former prisoners. Many jobs requiring state licenses are closed to ex-convicts. Many state laws bar prisoners from voting. This ban may be either for a finite period of time, or, in a few states, for life. There are federal laws that prohibit drug offenders from accessing public housing or any form of welfare. Of course, the biggest hurdle that former inmates face is finding a job that pays a decent wage. Employers are not likely to hire ex-cons. In times of low unemployment, such as the mid- to late 1990s, employers were so desperate for workers that they were more open to hiring ex-convicts—even, in some cases, setting up their business in a prison and hiring inmates. Today, the situation is very different and unemployment rates are much higher, making it harder for inmates to find an employer willing to take a chance, even with some tax incentives that are offered by the federal government (Mauer, Chesney-Lind, and Clear 2002).

The factors that are relevant to successful reentry are pretty much determined by common sense. Those prisoners who have a supportive family, a job, and a place to live are much more likely to succeed. Those inmates who are paroled to a mission or transient-living situation, who do not have a job that pays a decent wage, and who do not see a future for themselves are less likely to withstand the temptation of drugs and/or criminal opportunities. Of course, an offender who never goes to prison in the first place doesn't have the problem of reentry.

Alternatives to Prison

Even conservatives are now realizing that our use of incarceration has exceeded the point where it is reasonable or financially sound. Justice Anthony Kennedy has been quoted as saying, "Our resources are misspent, our punishments too severe, our sentences too long." And James Q. Wilson, a noted conservative, has also stated that our incarceration practices have reached a "tipping point" where there are "diminishing returns" on imprisoning so many offenders (Austin and Fabelo 2004).

In Texas, for instance, it costs roughly $40 per day to hold an inmate in prison, compared to about $2.50 a day for supervision in the community (Ward 2005).

Considering that the prison population of Texas is about 157,000, if 20 percent of these offenders (31,400), could be released into the community—at a savings of, conservatively, $35 per day—the state could realize a savings of more than $1 million a day. Other states' savings are relative to the numbers of citizens they incarcerate. There are good models across the country for how to reduce prison populations. States have instituted drug courts, adjusted sentencing guidelines, repealed mandatory-minimum-sentencing laws, increased the use of probation (sometimes using electronic monitoring, house arrest, or restitution), and have instituted graduated sanctions rather than use prison as the first resort after probation revocation (Pollock 2004).

Conclusions

Despite the opening quote of this chapter, more than 30 years later, American prisons are basically warehouses holding the detritus of society. They are run with the minimum of programs and the minimum of attention—unless they erupt in hostage taking, escapes, or riots. Then, as always, the public's response is to punish—and if that doesn't work, to punish more severely. The worst prisons have gotten appreciably better. Prisoners now enjoy some rights of mail, movement, and due-process protections. However, their legal position is once again in a no-man's-land of being less than fully endowed with rights that most of us hold sacrosanct. Court holdings that might have reinforced the need to protect rights of religion, safety, and privacy have instead swung to the side of "states' rights" that promote order over individual rights. Some observers note that the courts' move back to a type of hands-off approach to prisoner litigation has led inevitably to a worsening of prison conditions.

This chapter opened with a quote from the 1970s that seemed quaint and idealistic in its vision for the future of prisons. Glaser's predictions were painfully inaccurate. In the first edition of this text, the predictions were substantially less optimistic, and, unfortunately, more accurate.

> *. . . prisons will probably continue on pretty much the way they have since their inception. They have always been the repositories for the disenfranchised. They have endured several cycles of overcrowding and corruption, followed by optimism and enthusiasm for their capacity to change the individuals incarcerated within. Their management and objectives continue to be the hostage of politics and public sentiment influenced more by general economics and human caprice than crime rates or the needs of those incarcerated. Change can and does occur in prison. In some cases it is positive change, influenced by a teacher, a sympathetic correctional officer, a work foreman, or a chaplain; in some cases it is negative change, incurred through a prison rape, a mental breakdown, or the constant belittlement that all prisoners endure when they receive the pervasive message that "you are nothing and you are not wanted in our community" (Pollock 1997, 474).*

Since the previous edition, there has been no great groundswell of change in penology, no epiphanies, no new directions. There are more prisons, there are many more prisoners, and there are more states that are finding it difficult to extract themselves from the spiral of increasing costs associated with larger and larger prison populations. Trends we do see are more small towns signing up for the prison business, and more private companies being created to get into the business of incarceration. These economic pressures become influential when a state tries to close a prison or get out of a private-prison contract.

The reentry initiatives set in motion by the federal government are encouraging in that they signal a recognition that some policy makers have realized that incarceration cannot continue to escalate. The social costs are too high. The human costs are even higher. There are initiatives taken at the state level as well that are encouraging. It may be that the climate is right for political leaders to advocate alternatives to prison without fearing political suicide by appearing to be "soft on crime." Certainly, some of the voices being heard at this point advocating funding for treatment are conservatives who have concluded that incarcerating petty drug offenders is counterproductive both fiscally and socially.

Fyodor Dostoyevsky wrote that "the degree of civilization in a society can be judged by entering its prisons." We hope that this book has opened the door to prisons, at least in a small way. The reader is urged to investigate the prison system of his or her own state.

KEY TERMS

cross-sex supervision—when officers of the opposite sex guard inmates.

deinstitutionalization—a time period during the 1970s when mental hospitals were closed in favor of community mental health treatment centers.

the hole—punitive segregation.

opportunity costs—the lost social services that result from committing a large portion of state and federal budgets to prisons and crime control.

pastel prisons—prisons that do not have obvious and extensive security; minimum security facilities.

per diem—cost for each individual per day.

recidivism—repeat offending and/or conviction.

rehabilitative era—time period during the 1970s marked by great optimism that prison programs could change offenders.

revocations—violations of parole that result in a return to prison.

self-medicate—the practice of individuals with mental health issues using street drugs in lieu of mental health treatment and legally prescribed drugs.

Stanford prison experiment—famous college experiment that showed college students could become autocratic and misuse power after only a few days of pretending to be guards (see www.prisonexp.org for more information).

supermax—maximum security penitentiary marked by 23- or 24-hour cell lockdowns and restricted or nonexistent programming.

technical violations—when parolees or probationers violate a rule or condition that is not a new crime.

War on Drugs—the public policy during the 1980s and 1990s of emphasizing interdiction and enforcement over considering drug abuse a public health problem.

REVIEW QUESTIONS

1. Discuss the relationship between crime rates and incarceration rates.
2. Explain how the deinstitutionalization of the mentally ill has affected prison populations.
3. Discuss the relationship between drug-use patterns and enforcement/imprisonment patterns.
4. Define and explain "opportunity costs" as they relate to prisons.
5. Explain why small towns would want prisons to be built there.
6. What is a supermax prison? What do critics say about such a prison?
7. What is the connection between Abu Ghraib and prisons in the United States?
8. What is cross-sex supervision? What is the problem with it?

9. How many people return to the community each year? Why do 67 percent fail to stay out of prison?
10. What are some alternatives to prison? Why do we not use more of them?

FURTHER READING

Austin, J., and T. Fabelo. 2004. *The Diminishing Returns of Increased Incarceration.* Washington, DC: JFA Institute.

Greene, J., and V. Schiraldi. 2002. *Cutting Correctly: New State Policies for Times of Austerity.* Washington, DC: Justice Policy Institute.

Parenti, C. 1999. *Lockdown America: Police and Prisons in the Age of Crisis.* New York: Verso New Left Books.

Pollock, J. 2004. *Prisons and Prison Life: Costs and Consquences.* Los Angeles: Roxbury.

REFERENCES

Austin, J., and T. Fabelo. 2004. *The Diminishing Returns of Increased Incarceration.* Washington, DC: JFA Institute.

Briggs, C., J. Sundt, and T. Castellano. 2003. "The Effect of Supermaximum Security Prisons on Aggregate Levels of Institutional Violence." *Criminology* 41, 4: 1341–1377.

Bureau of Justice Statistics. 2004. "Homicide Trends in the U.S. (B.J.S. Report)." Washington, DC: U.S. Department of Justice. Retrieved September 7, 2004, from www.ojp.usdoj.gov/bjs/homicide/hmrt.htm.

Durose, M., and P. Langan. 2004. *Felony Sentences in State Courts, 2002.* Washington, DC: U.S. Department of Justice.

Elsner, A. 2005. Lawmakers Launch Effort to Help Released Felons. Reuters.com. Retrieved on February 4, 2005 from http://www.reuters.com/printerfriendly popup.jhtm/?type=topNews&storyID=7517188.

Gifford, S. 2002. "Justice Expenditure and Employment in the United States, 1999 (B.J.S. Report)." Washington, DC: U.S. Department of Justice.

Glaser, D. 1970. "The Prison of the Future." In D. Glaser (Ed.), *Crime in the City*, pp. 261–266. New York: Harper and Row.

Greene, J., and V. Schiraldi. 2002. *Cutting Correctly: New State Policies for Times of Austerity.* Washington, DC: Justice Policy Institute.

Henningsen, R., W. Johnson, and T. Wells. 2003. "Supermax Prisons: Panacea of Desperation?" *Corrections Management Quarterly* 3, 2: 57–60.

Horton, S. 2005. "A Scandal That Goes Beyond 'A Few Bad Apples.'" *Austin American Statesman* (January 26, 2005): A13.

Human Rights Watch. 2003. *Ill Equipped: US Prisons and Offenders with Mental Illness.* Retrieved March 3, 2005, from http://www.hrw.org/reports/2003/usa/003/.

King, R. 1999. "The Rise and Fall of Supermax: An American Solution in Search of a Problem." *Punishment & Society* 1: 163–186.

Lawrence, S., and J. Travis. 2004. *The New Landscape of Imprisonment: Mapping America's Prison Expansion.* Washington, DC: Urban Institute, Justice Policy Center.

Lock E., J. Timberlake, and K. Rasinki. 2002. "Battle Fatigue: Is Public Support Waning for War-Centered Drug Control Strategies?" *Crime and Delinquency* 48, 3: 380–398.

Mauer, M. 2002. "Analyzing and Responding to the Driving Forces of Prison Population Growth." *Criminology and Public Policy* 1, 3: 389–393.

Mauer, M., M. Chesney-Lind, and T. Clear. 2002. *Invisible Punishment: The Collateral Consequences of Mass Imprisonment.* New York: Free Press.

"New Jersey's Women Inmates Tell Their Own Stories of Abuse." 2004. *Everything Jersey* (online journal). Retrieved May 25, 2004, from www.nj.com/printer/printer.ssf?/base/news-15/1085294264321300.xml.

Office of National Drug Control Policy. 2002. "Drug Use Trends—October 2002." Retrieved August 31, 2004, www.whitehousedrugpolicy.gov/publications/factsht/druguse/index.html.

Parenti, C. 1999. *Lockdown America: Police and Prisons in the Age of Crisis.* New York: Verso New Left Books.

Peirce, N. 2004. "Abu Ghraib Hits Home." *Seattle Times* (July 5). Retrieved February 20, 2005, from seattletimes.nwsource.com/html/opinion/2001971935_peirce05.html.

Pollock, J. (Ed). 1997. *Prisons: Today and Tomorrow.* Gaithersburg, MD: Aspen Publishing Company.

Pollock, J. 2002. *Women, Prison, and Crime.* Belmont, CA: Wadsworth.

Pollock, J. 2004. *Prisons and Prison Life: Costs and Consequences.* Los Angeles: Roxbury.

"Rights Group Say Abu Ghraib Abuses Extension of U.S. Prison System." 2004. *New Standard.* Retrieved February 20, 2005, from newstandardnews.net/content/index.cfm/items/360.

Schemo, D. 2001. "Students Find Drug Law Has Big Price: College Aid." *New York Times.* Retrieved May 3, 2001, from www.nytimes.com/2001/05/03/politics/03DRUGhtml.

Schiraldi, V., and J. Ziedenberg. 2002. *Cellblocks or Classrooms? The Funding of Higher Education and Its Impact on African American Men.* Washington, DC: Justice Policy Institute.

Wald, M. 2004. "Most Crimes of Violence and Property Hover at 30 Year Lows." *New York Times.* Retrieved September 14, 2004, from www.nytimes.com/2004/09/13/national/13crime.html?pagewanted=print&position.

Ward, M. 2005. "Daily Cost for Felons Down, But Fears Loom." *Austin American Statesman* (January 26): B1, B5.

Western, B., V. Schiraldi, and J. Ziedenberg. 2003. *Education and Incarceration.* Washington, DC: Justice Policy Institute.

CASES CITED

Sandin v. Conner, 115 S. Ct. 2293 (1995)
Madrid v. Gomez, 889 F. Supp. 1146 (1995)
Meachum v. Fano, 427 U.S. 215 (1976)
Olim v. Wakinekona, 461 U.S. 238 (1983)

Index

F

G

H

I

J

K

L

M

N

T

U

V

W